An understanding of the virulence determinants of bacterial pathogens and their regulation, and also of the pathogenesis of bacterial infections, is essential to the development of safe and effective vaccines and therapeutic strategies. This book presents an authoritative and timely account of the contributions being made by molecular genetics and by improved analytical technology in uncovering the molecular basis of bacteria–host interactions in infections. The broad scope of the volume encompasses normal flora and colonization resistance, evolution of pathogenic mechanisms, genetic approaches to understanding bacterial pathogenesis and *in vivo* studies of mutants, antigenic variation, bacterial invasion, toxins, environmental control of gene expression, regulatory networks in gene control, and attenuation of virulence through auxotrophy.

This volume will provide a state-of-the-art reference work for researchers in the areas of bacteriology, medical microbiology, infection, vaccine development, epidemiology, and drug development. It will also be suitable as an advanced text for undergraduate and postgraduate students of these disciplines.

MOLECULAR BIOLOGY OF BACTERIAL INFECTION
Current Status and Future Perspectives

SYMPOSIA OF THE
SOCIETY FOR GENERAL MICROBIOLOGY

Series editor (1991–1996): Dr Martin Collins, Department of Food Microbiology, The Queen's University of Belfast

Volumes currently available:

MOLECULAR BIOLOGY OF BACTERIAL INFECTION
Current Status and Future Perspectives

EDITED BY
C. E. HORMAECHE, C. W. PENN
AND C. J. SMYTH

FORTY-NINTH SYMPOSIUM OF THE
SOCIETY FOR GENERAL MICROBIOLOGY
HELD AT THE UNIVERSITY OF DUBLIN,
TRINITY COLLEGE
SEPTEMBER 1992

Published for the Society of General Microbiology

TRINITY
400

CAMBRIDGE
UNIVERSITY PRESS

Published by the Press Syndicate of the University of Cambridge
The Pitt Building, Trumpington Street, Cambridge CB2 1RP
40 West 20th Street, New York, NY 10011-4211, USA
10 Stamford Road, Oakleigh, Victoria 3166, Australia

First published 1992

Printed in Great Britain by Redwood Press Limited, Melksham, Wiltshire

A catalogue record for this book is available from the British Library

Library of Congress cataloguing in publication data available

ISBN 0 521 43298 7 hardback

CONTENTS

CONTRIBUTORS

ACHTMAN, M., Max-Planck-Institute for Molecular Genetics, Ihnestrasse 73, DW-1000 Berlin 33, Germany

CHATFIELD, S., Vaccine Research Unit, Medeva Group Research, Department of Biochemistry, Imperial College of Science Technology and Medicine, Wolfson Laboratories, London SW7 2AY, UK

CORNELIUS, G., Université Catholique de Louvain, Unite de Microbiologie, UCL 54–90, 54 Avenue Hippocrate, B-1200 Brussels, Belgium

DORMAN, C. J., Department of Biochemistry, Medical Sciences Institute, University of Dundee, Dundee DD1 4HN, UK

DOUCE, G., Department of Biochemistry, Imperial College of Science, Technology and Medicine, Wolfson Laboratories, London SW7 2AY, UK

DOUGAN, G., Department of Biochemistry, Imperial College of Science, Technology and Medicine, Wolfson Laboratories, London SW7 2AY, UK

FINLAY, B. B., The University of British Columbia, Biotechnology Laboratory, Room 237 – Westbrook Building, 6174 University Boulevard, Vancouver BC V6T 1Z3, Canada

FLESCH, I. E. A., Department of Immunology, University of Ulm, Albert-Einstein-Allee 11, D-7900 Ulm, Germany

FOSTER, T. J., Department of Microbiology, Moyne Institute, Trinity College, Dublin 2, Ireland

HAKENBECK, R., Max-Planck Institute for Molecular Genetics, Ihnestrasse 73, DW-1000 Berlin 33, Germany

HAYBELL, J. D., Department of Immunology, St. Mary's Hospital Medical School, Norfolk Place, London W2 1PG, UK

HEWITT, C. R. A., Department of Immunology, St. Mary's Hospital Medical School, Norfolk Place, London W2 1PG, UK

KAUFMANN, S. H. E., Department of Immunology, University of Ulm, Albert-Einstein-Allee 11, D-7900 Ulm, Germany

LAMB, J. R., Department of Immunology, St. Mary's Hospital Medical School, Norfolk Place, London W2 1PG, UK

LI, J. L., Vaccine Research Unit, Medeva Group Research, Department of Biochemistry, Imperial College of Science, Technology and Medicine, Wolfson Laboratories, London SW7 2AY, UK

MASKELL, D., Department of Paediatrics & Institute of Molecular Medicine, John Radcliffe Hospital, Headington, Oxford OX3 9DU, UK

MEYER, T. F., Max-Planck-Institute for Biology, Spemannstrasse 34, D-74 Tübingen, Germany

MOXON, B. R., Department of Paediatrics & Institute of Molecular Medicine, John Radcliffe Hospital, Headington, Oxford OX3 9DU, UK

NÍ BHRIAIN, N., Department of Biochemistry, Medical Sciences Institute, University of Dundee, Dundee D1 4HN, UK

O'HEHIR, R. E., Department of Immunology, St. Mary's Hospital Medical School, Norfolk Place, London W2 1PG, UK

PENN, C. W., School of Biological Sciences, Biology West Building, University of Birmingham, Birmingham B15 2TT, UK

ROBERTSON, B. D., Max-Planck-Institute for Biology, Spemannstrasse 34, D-74 Tübingen, Germany

SANSONETTI, P. J., Unité de Pathogénie Microbienne Moleculaire, U199 INSERM, Institut Pasteur, 25-28 rue du Dr. Roux, F-75724 Paris Cedex 15, France

SMITH, S. G. J., Department of Microbiology, Moyne Institute, Trinity College, Dublin 2, Ireland

SMYTH, C. J., Department of Microbiology, Moyne Institute, Trinity College, Dublin 2, Ireland

SYDENHAM, M., Vaccine Research Unit, Medeva Group Research, Department of Biochemistry, Imperial College of Science, Technology and Medicine, Wolfson Laboratories, London SW7 2AY, UK

VAN DER WAAIJ, D., University of Gröningen, Faculty of Medicine, Department of Medical Microbiology, Oostersingel 59, 9700 RB Gröningen, The Netherlands

WREN, B. W., Department of Medical Microbiology, St. Bartholomew's Hospital Medical College, West Smithfield, London EC1A 7BE, UK

EDITORS' PREFACE

This year, the Society for General Microbiology is marking the Quatercentenary of the foundation of Trinity College, University of Dublin, in 1592 by charter from Queen Elizabeth I of England in holding its autumn meeting in Dublin's fair city. On a microbiological and microbial pathogenicity scale, Trinity College predates Antonie van Leeuwenhoek's demonstration of 'animalcules' in 1680 by almost a century, the publication of Jenner's vaccination studies in 1798 by two centuries, and the discovery of tetanus toxin by Behring and Kitasato in 1880 by three centuries.

This volume represents the fifth in the SGM symposium series concerned with microbial pathogenicity. In 1955, the fifth volume of the series dealt with *Mechanisms of Microbial Pathogenicity*, examining a range of bacteria, protozoa and fungi pathogenic to man. This was followed in 1964 by *Microbial Behaviour*, 'In Vivo' *and* 'In Vitro' which dealt with microbial behaviour in animals and plants compared to examination of microorganisms grown under defined conditions in pure culture *in vitro*. The 22nd Symposium in 1972 was entitled *Microbial Pathogenicity in Man and Animals*. It drew parallels between the pathogenic activities of the various microbial types. To mark the 25th anniversary of the Virus Group, Symposium 40 in 1987 was devoted to the *Molecular Basis of Virus Disease*.

Interest in microbial pathogenesis has boomed since the 1964 and 1972 symposia. The current volume follows this lineage and appropriately deals with bacterial pathogenicity on this occasion, with an emphasis on the contribution of modern molecular biological and molecular genetic techniques to our understanding of pathogenic mechanisms of bacteria. Molecular genetics has been an important key to the enormous strides forward in understanding mechanisms of bacterial pathogenicity. It is particularly fitting to hold the current symposium in Dublin in view of the major role played by W. Hayes, a Trinity graduate and former member of staff of the Microbiology Department, in the unravelling of mechanisms of genetic exchange in bacteria. There is also another strong link with Trinity College. J. P. Arbuthnott, who was professor of Microbiology from 1975–1988 was the first convenor of the Pathogenicity Group. His influence was central to the development of the Pathogenicity Group during its formative years in the 1970s.

This symposium highlights not only areas of bacterial infection where mechanisms underpinning virulence and pathogenesis are well worked out at the molecular genetic level but also pathogenic phenomena such as latency, the carrier state and chronicity which are still poorly understood. The first chapters deal with normal flora and colonization resistance, the evolution of pathogenic clones of bacteria, and genetic approaches to understanding the pathogenesis of complex bacterial pathogens. The next section reviews invasive bacteria, antigenic variation, capsules and LPS, intracellular survival in phagocytes, and latency and the carrier state. This is followed by two contributions on toxins, dealing with toxin–receptor interactions and superantigen action, and a chapter on *in vivo* studies of mutants of pathogenic bacteria. The concluding chapters are devoted to regulation of bacterial virulence with contributions on global regulatory mechanisms of gene expression, regulatory networks in invasive bacteria, and regulation of fimbrial biogenesis. The last chapter shows how advanced genetic manipulation systems can lead to effective attenuated vaccines.

Although these various contributions convincingly demonstrate that molecular genetics provides an elegant tool for dissecting and identifying virulence factors, one must not lose sight of the fact that, no matter how elegant genetic manipulation of bacterial pathogens may be, pathogenesis studies require a sound foundation in experimental pathology. Finally, it is worthwhile emphasizing that some fundamental knowledge of genetic control mechanisms and bacterial genome structure and function has been gained from studies on pathogenic bacteria, e.g. from the mechanisms of rapid switching of gene expression.

We thank the authors for their contributions, and the staff of Cambridge Univerisity Press for their help and patience. We hope that this volume will not only mark the SGM's celebration of Trinity 400 but also provide stimulus to research on bacterial pathogenesis and be a useful teaching text for senior undergraduate and postgraduate students.

C. E. Hormaeche
C. W. Penn
C. J. Smyth

MECHANISMS INVOLVED IN THE DEVELOPMENT OF THE INTESTINAL MICROFLORA IN RELATION TO THE HOST ORGANISM; CONSEQUENCES FOR COLONIZATION RESISTANCE

D. VAN DER WAAIJ

Laboratory for Medical Microbiology, University of Gröningen, Oostersingel 59, 9713 EZ Gröningen, The Netherlands

INTRODUCTION

This chapter deals with some basic information concerning the mechanisms and processes which dictate the composition of an organism's microflora. This also includes the function of this microbial community in animals and in man. As may become clear in the following paragraphs, the microflora may be very beneficial to the host organism. For example, the intestinal flora co-operates with the host organism in maintaining its own composition in a largely apathogenic and stable state. The host organism normally has the capacity and the 'tools' to select and maintain the gastro-intestinal (GI) tract microflora. As will be outlined in this chapter, these 'tools' for cooperation with the microflora involve the capacity of the immune system to respond positively or negatively to foreign antigens and also its regulatory function in mucosal renewal and the production of mucosal secretions. This latter subject, involving the intestinal immune system and its interactions with the microflora, has been reviewed comprehensively by Kagnoff (1987) and more recently by Brandtzaeg and co-authors (1990). Therefore, although the role of the immune system is central in the present review, major emphasis will be given to the development and maintenance of the intestinal flora and its protective functions. The following subjects will be discussed: first, factors involved in the formation of the intestinal ecosystem; secondly, its protective function termed 'colonization resistance'; and thirdly, the role of the immune system in the control of the intestinal ecosystem. Clearly, this role of immune interactions with the intestinal flora may become of medical importance. The state of the immune system (determined genetically and by nutritional status) and the course of host interactions with the flora which starts after birth, may determine whether or not certain autoimmune disorders develop.

THE MICROBIAL ECOSYSTEM OF THE DIGESTIVE TRACT

After birth, an important microbial ecosystem develops in the digestive tract. The major source of this microflora is normally the mother (Schaedler, 1973; Lee, 1985; Van der Waaij, 1986; Kuhn, Tullus & Molby, 1986). The development of a normal flora occurs in several phases. In mice it may take three weeks (Savage, 1977; Lee, 1985) and in man two months (Long & Svenson, 1977; Lundequist, Nord & Winberg, 1985). Once the gut has adopted its numerous largely anaerobic inhabitants, this bacterial community stabilizes (Long & Swenson, 1977; Heidt & Van der Waaij 1979; Costerton *et al.*, 1983; Lee, 1985; Lundequist *et al.*, 1985). Apparently, the bacterial community has meanwhile become a coherent society. These organisms suppress new microbial species that are ingested daily. The composition of this 'adopted bacterial society' in the digestive tract after birth, is largely determined by two factors:
(a) *chance* depending on the bacterial load of the environment, and
(b) active *selection* by various mechanisms.

Chance is largely dependent on hygienic circumstances, while *selection* may be performed either by other bacteria in the gut microflora (competition for nutrients, production of toxic substance(s) or steric hindrance), or by the host organism (immune system, chemical composition of saliva and intestinal mucus).

Chance as a factor which determines the digestive tract microflora

Bacteria which are ingested first in sufficient numbers after birth may generally have better chances of establishment than those which come later, when a stable ecosystem has developed (Long & Swenson, 1977; Costerton *et al.*, 1983, Lundequist *et al.*, 1985). Early in life many ingested bacteria may initially stay for some time. However, once a more complex microflora has developed, many of them may die soon because they lack the capacity to utilize the nutrients available. An important source of nutrients – the intestinal mucus – becomes available only some time after birth. In the gut, only those bacteria which can live in the mucus covering the intestinal epithelial lining, avoid removal mechanically with the intestinal contents by peristaltic movement (Heidt & Van der Waaij, 1979). In mice, it may take as long as three weeks before all parameters indicate that the flora is more or less complete (Savage, 1977; Lee, 1985).

Selection as a factor which determines the digestive tract microflora

To become a member of the bacterial ecosystem of the digestive tract or in other words to become 'indigenous' (resident), persistent colonization at or close to the mucous membrane appears necessary. In the oropharynx it is insufficient to live adjacent to the mucous membrane and adhesion to the mucous membrane is a prerequisite for bacteria to stay for long periods and to become resident. Only a limited number of bacterial species can adhere to the mucosa in the oropharynx or on the teeth, by possession of proper adherence pili or other adhesins (Beachey, 1981; Van Houte, 1982). The composition of the flora in a particular niche, may therefore differ at different locations (Savage, 1980). This has been studied best in mice (Savage, 1977, 1980).

For colonization of the intestines, adhesion to the mucosa is not a strict requirement. In the intestines, another factor may be more dominant: only bacteria which have the proper enzymes to digest the nutrients continuously provided by the host organism can colonize. Mostly in cooperation with other bacteria they live in the intestinal mucus layer for longer periods (if not life-long). Host-provided nutrients are derived from saliva, mucus and desquamated oropharyngeal cells or enterocytes (Allen & Hoskins 1988; Hoskins & Boulding, 1981). In man, the amount of saliva and mucus may be as much as several litres per day.

From the site of primary colonization, respectively on the oropharyngeal mucous membrane as well as in the intestinal mucus gel, indigenous bacteria are continuously released and mix respectively with oral or intestinal contents. Another source of bacteria that may multiply in the intestinal contents, comprises newly ingested bacteria. Numerically, however, the number of bacteria ingested per day is normally far outnumbered by bacteria composing the indigenous flora and which are released from the gut wall. As a consequence, the great majority of newly ingested bacteria is retarded in growth or dies because of lack of adequate nutrients and/or because of toxic products produced by the indigenous flora (Wilhelm & Lee, 1987).

Selection by competition for nutrients in the diet

As explained earlier, newly ingested microorganisms can only survive their transit through the gut to be excreted alive with the faeces if they can compete with the indigenous bacteria for the nutrients ingested by the host with the diet (Drasar *et al.*, 1973; Finegold *et al.*, 1974; Drasar, Jenkins & Cummings, 1976). Conversely, the composition of the diet may obviously also have its limitations for some members of the indigenous bacteria shed from the gut wall. Only indigenous bacteria which have the proper set of enzymes to utilize dietary components may survive in the intestinal contents. This could explain why the flora released from the rather stable intestinal mucosal ecosystem in the gut lumen can be *modulated* to some extent by the composition of the diet. Bacteria which flourish at the mucosal site with the nutrients provided by the host organism, may not necessarily grow equally well in the intestinal contents. Quantitative changes in the composition of the faecal flora therefore can be expected when the diet of the host organism is changed in composition. However, the mucosal flora appears not to change qualitatively after exposure to different diets (Gorbach et al., 1967, Van der Waaij & Van der Waaij, 1990). Yet, there is an obvious stable interindividual difference in the composition of the faecal flora (Meijer-Severs & Van Santen, 1986). In general, the composition of the faecal flora consists predominantly of Gram-positive anaerobic bacteria (Gossling & Slack, 1974).

While living in the intestines, indigenous bacteria produce volatile fatty acids and other substances which may be toxic to a number of pathogenic and potentially pathogenic (opportunistic) bacteria.

Mechanisms which interfere with mucosal colonization by newly ingested bacteria

An important (first) step to infectious disease is colonization by pathogens at the mucosal site (Reed & Williams, 1978). Colonization of the epithelial layer is a prerequisite for invasion of tissues underlying the mucosa. As discussed before, colonization by foreign (possibly pathogenic) microorganisms fortunately often appears to be interfered with by the mucosal bacterial ecosystem. By their presence in the intestinal mucus, the indigenous bacteria may form a kind of conveyer belt which covers the mucosa. In this way, they do hinder colonization by newly ingested bacteria not only by competition for nutrients but also sterically (Savage, 1977). The mechanisms by which the indigenous microorganisms in concert with the host keep the intestinal ecosystem quite stable in composition, is called colonization resistance (Van der Waaij, Berghuis-de Vries & Lekkerkerk-Van der Wees, 1971). This subject will be discussed later in greater detail.

Involvement of the immune system in the control of composition of the indigenous flora (Kagnoff, 1987, Brandtzaeg et al., 1990)

The immune system is in continuous contact with intestinal contents through its gateways: the Peyer's patches (small intestines) and the more numerous lymphoid follicles in the colon. The epithelial cells overlying these small lymphoid organs, called the M cells in case of Peyer's patches, are specialized for absorption of larger molecules, and phagocytosis of larger particles or even bacteria from the intestinal lumen. After uptake from the intestinal contents, this antigen-rich material is transported through the cells to the underlying tissues packed with macrophages, B cells and T cells. It requires some knowledge of immunology to understand why there is not a continuous aggressive interaction between bacteria which live in close contact with the host in the mucus layer and the immune system of the host. The following is an attempt to provide sufficient background information for understanding this rather complex matter. Much of the intestinal antigenic load transferred by the M cells, appears to be digested and destroyed by the macrophages to become nonimmunogenic in an aspecific way. The higher the degree to which this occurs the less correspondingly is the amount of aspecific (polyclonal) stimulating activity by endotoxin or other bacterial cell wall substance such as peptidoglycan. Further, there is strong evidence that the majority (varying between individuals) of the bacterial representatives of the indigenous flora are essentially nonimmunogenic to the host organism (Foo & Lee, 1974). This could partially be ascribed to cross-antigenicity between host tissue antigens and major antigenic determinants of the bacterial cell wall. In addition, much antigenic material in the intestines becomes 'immunologically neutralized' in the Peyer's patches by the induction of *hyporesponsiveness*. This may occur by the production of enhancing (neutralizing) antibodies (polyspecific IgM?) or even by induction of *immunologic tolerance* by specific T suppressor cells. These different ways to render an individual immunologically hyporesponsive to foreign antigen are generally indicated in the immunological literature by the term 'oral tolerance' (Kagnoff 1987; Brandtzaeg *et al.*, 1990; André *et al.*, 1975; Challacombe & Tomasi 1980).

To a number of bacteria and nutritional antigens which do contact the intestinal immune system, largely through Peyer's patches (and colonic lymphatic follicles?), a *humoral response* (antibody production) and/or cellular immunity does nevertheless ensue. Antibodies of the IgA class may develop in this way and are, before secretion with saliva and intestinal mucus, linked to 'secretory component'; a molecule which is believed to protect secretory IgA from inactivation by proteolytic enzymes. This complex is usually indicated by the abbreviation sIgA. Secretory IgA may interfere with adhesion of bacteria to the mucous membrane or with absorption of larger molecules that otherwise may cause allergy (Kagnoff, 1987, Brandtzaeg *et al.*, 1990). In addition, IgA is an agglutinating antibody (McCleland *et al.*, 1972). By agglutinating bacteria when they 'enter' the mucus layer and/or by preventing adherence, these antibodies may promote clearance by intestinal flow and thus interfere with colonization (Kagnoff, 1987; Brandtzaeg *et al.*, 1990; Lundequist *et al.*, 1985).

The *cellular immunity* mediated by T cells, also appears to play an important role in control of the colonization pattern of the mucous layer in the gut (Mowat & Ferguson, 1981, Mowat & Felstein, 1987). The epithelial cells (enterocytes) covering the villi of the intestines, are formed in the crypts of Lieberkuhn. Normally, there exists a balance between cell reproduction in the crypts and cell extrusion at the tip of the villi. Increased cell extrusion, such as takes place during intestinal infection, may be an efficient means to get rid of unwanted (pathogenic and immunogenic) bacteria, yeasts or parasites which adhere to the tip of the villi. If an infectious stimulus is present at the tip of the villi, T cells in the submucosa appear to respond with the production of certain lymphokines which enhance crypt cell proliferation. In this way enterocytes are moved faster along the villi to the tip. If production of enterocytes does not follow the desquamation rate at the villus tip sufficiently fast, the villi become shorter. This efficient 'early response' mechanism of the intestines to pathogenic microorganisms is normal in intestinal infections. In experimental animals which have congenitally no thymus and thus practically no mature T cells, this protective response mechanism is deficient. As a result they must be maintained isolated since otherwise, they will suffer from chronic intestinal infections.

COLONIZATION RESISTANCE OF THE DIGESTIVE TRACT

Introduction

The term 'colonization resistance' (CR) was introduced in 1971 (Van der Waaij et al., 1971). The original definition of CR regarding facultatively anaerobic bacteria was: 'the logarithm of the number of Gram-negative bacteria which, when given as an oral dose, results in colonization of 50% of a sufficiently large group of experimentally contaminated mice for minimally two weeks'. Later, other ways to determine the CR were developed (Van der Waaij & Berghuis, 1974; Van der Waaij, 1982a; Apperloo-Renkema, Van der Waaij & Van der Waaij, 1990; Meijer-Severs & Van Santen, 1990; Welling et al., 1991; Meijer et al., 1991). Each CR-test has provided insight into the degree of stability of the indigenous ecosystem. It also appeared, that other bacteria such as enterococci and staphylococci encounter a CR to their colonization of the gut (Van der Waaij et al., 1972a). This also applies to yeasts (Van der Waaij, 1982b).

The influence which antimicrobial treatment may have

In conventional mice with a stable microflora, this flora exerts usually a strong suppressive effect on newly ingested Gram-negative and Gram-positive organisms. This means that they have a high CR; much higher than is required for living in an environment with an artificially maintained low contamination rate, i.e. under so-called specific pathogen free (SPF) conditions. For example 10^8–10^9 of most potentially pathogenic bacteria are required in such animals to establish long lasting colonization of the colon in 50% of the group of host individuals (Van der Waaij et al., 1971). These high contamination numbers will rarely – if ever – occur in animals maintained under classical SPF-conditions (Van der Waaij & Van der Waaij, 1990). Among conventionally maintained animals and man, however, this may occur and may have implications. Some individuals in each animal or human population may have an indigenous flora of relatively low protective quality, e.g. may have a relatively low CR (Van der Waaij & Van der Waaij, 1990). In such individuals an oral inoculation with 10^5–10^6 potentially pathogenic bacteria may 'take' for more than two weeks (Van der Waaij et al., 1971). In individuals with an extremely low CR-value, small oral contamination doses with $(10–10^2)$ bacteria are required for long lasting colonization. A low CR can be seen during suppression of active flora by antimicrobial treatment (Van der Waaij et al., 1971; Van der Waaij et al., 1972a; Van der Waaij 1982b; Van der Waaij, 1987). This has been studied best in experimental animals, which were orally or parenterally treated with broad-spectrum antibiotics to suppress the indigenous flora (Van der Waaij et al., 1972; Van der Waaij, 1987). However, there is much evidence that inter-individual differences in the quality of the CR as well as the negative effect of some antibiotics, also apply to man (Meijer-Severs & Van Santen 1990; Welling et al., 1991; Meijer et al., 1991; Van der Waaij, 1984; Van der Waaij, 1987). Obviously, the extremely low level of colonization resistance of the digestive tract during treatment with flora-suppressive antibiotics, is in all cases restricted to bacteria and/or yeasts, which are resistant to the antibiotic(s) used for treatment (Van der Waaij, 1984).

As outlined above, there is an inverse relationship between the quality of the CR and the number of newly ingested bacteria required for digestive tract colonization by these organisms. Furthermore, during periods with decreased CR due to suppression of the oropharyngeal/intestinal microflora, the concentration at which resistant microorganisms may subsequently grow out in the oropharynx or in the gut, is high. The lower the CR of the digestive tract of an individual, for example, because of some antibiotic therapy, the higher the number of newly ingested resistant strains may become in the intestinal contents. This is reflected in the faecal cultures (Van der Waaij et al., 1971, 1972, 1984, 1987).

The influence of the (intestinal) immune system

Other reasons for substantial changes in the microflora and therewith of the CR involve diminished contribution of the host individual to the CR. When the (cellular) immune system functions suboptimally, the amount of the secreted nutrients available to bacteria in the saliva and mucus may decrease. Immune-suppressive factors such as stress, aging, severe illness and most obviously chemotherapy or radiation are involved (Van der Waaij et al., 1978). Recovery of the host organism mostly involves recovery of the indigenous flora and thus of the CR (Van der Waaij & Heidt, 1977).

TRANSLOCATION OF (POTENTIALLY) PATHOGENIC BACTERIA

The phenomenon of translocation

Another event that is of practical importance in relation to CR, is that potentially pathogenic microorganisms present in high numbers at the mucosal site may 'translocate' from the intestinal mucosa to the submucosa and, from there, they may reach the mesenteric lymph nodes, the liver and the spleen (Van der Waaij *et al.*, 1972*b*; Berg & Carlington, 1979; Wells *et al.*, 1988*a,b*). Translocation of potentially pathogenic bacteria occurs in higher numbers, the higher is their number that colonizes at the mucosal membrane. This applies also to the much more numerous indigenous anaerobic bacteria, although they appear to translocate to a much lower degree (Wells *et al.*, 1988*a,b*). High numbers of a certain (potentially) pathogenic bacterium in the mucus are generally reflected in the intestinal contents. Translocation of potentially pathogenic organisms in a subject with a normal defence capacity is, in general, clinically unnoticed. Particularly after a first encounter with the immune system in Peyer's patches, the fate of translocating bacteria is that they become phagocytized when they enter the submucosal tissues. A number of these translocating bacteria however, may get to more remote locations such as the lymph nodes draining the mucosa or even more distant organs involved in bacterial clearance from the blood stream (Van der Waaij *et al.*, 1972*b*; Berg & Carlington, 1979, Wells *et al.*, 1988*a,b*). Evidence of translocation has also been found in man (Tancrède & Andremont, 1985).

The response of the immune system to translocating bacteria

As was described in the previous paragraph, dependent on their numbers at the mucosa, some bacteria – pathogens as well as potentially pathogenic bacteria – may translocate. This implies that some bacteria can penetrate the mucosal barrier either more or less passively through the M cells covering the Peyer's patches or, perhaps in a more active way, through the mucous membrane. Enhanced by opsonising antibodies (IgA,IgM), induced in Peyer's patches, the great majority of these bacteria will be phagocytized in the submucosal area. A fraction of the translocating bacteria however, for which these antibodies have not or not sufficiently been formed, may reach the mesenteric lymph nodes and the spleen. In these 'central lymphoid organs', they may induce humoral immunity with antibody of another isotype than IgA or polyspecific IgM. If they subsequently translocate at any time in sufficient numbers to reach the 'central immune system', in most cases this may evoke a memory response by IgA-producing B cells which meanwhile have migrated to the spleen (Kagnoff, 1987). In the spleen, potentially damaging or 'reactive' complement-fixing antibodies of the IgG-isotype are generally formed. These antibodies initiate an inflammatory reaction upon specific binding. However, because translocation by potentially pathogenic bacteria involves mostly a relatively low number of bacteria, the immune response is moderate and the antibody titre is low. This may explain why inflammatory reactions to translocating bacteria are rare if they ever occur regardless of translocation to central organs. If they did occur regularly, however, they might cause a *nonspecific intestinal infection* or an *auto-immune reaction*.

Invasive essentially entero-pathogenic bacteria such as *Salmonella* or *Shigella* species, on the other hand, may invade massively and rapidly. Because of the speed of this process, they circumvent the mucosa associated lymphoid tissues to induce central immunity. Thus they may stimulate high titres of circulating 'reactive' (complement fixing) IgG antibodies to various bacterial surface antigens. These bacterial invasions are normally associated with serious inflammatory responses.

CONCLUSIONS

The foregoing paragraphs indicate the potential involvement of a conventional intestinal microflora (including transient bacteria) in the development and the activity of the gut associated lymphoid tissues. In these paragraphs, it has been indicated, that the intestinal microflora could be divided into the following two parts:

- a predominant indigenous (resident or syngenous) part
- a much smaller nonindigenous (transient) part. Bacteria belonging to this fraction may originate from other individuals of the same species (allogenous) or from other animal species, plants, soil or water (xenogenous).

Both parts could immunologically be subdivided into:

1. bacteria which are *essentially non-immunogenic* to the host organism (predominantly indigenous bacteria, non-immunogenic through antigenic mimicry?). Which bacterial species this concerns does not only differ between animal species, but also between individuals of the same species,
2. bacteria which *induce hyporesponsivesness* by B cell or T cell dependent mechanisms,
3. bacteria which are *immunogenic* (induce specific specific T- and B cells responses). Whether this involves 'reactive' antibodies or non-complement fixing IgA (IgM) antibodies may depend on the massiveness of invasion and the site of the primary response.

It is still uncertain whether circulating IgM isotype antibodies to intestinal indigenous bacteria are involved in the mechanism of hyporesponsiveness or just represent a sign of first encounter by the immune system. Following further stimulation, a switch to other isotypes may occur. The majority of the anti-bacterial IgM, however, may belong to the category of polyspecific antibodies (Scott Rodkey, 1980; Fougereau and Schiff, 1988; Coutinho, 1989). This would imply that these antibodies bind to bacteria with antigens common to the host organism. These IgM-type antibodies may enhance phagocytosis and thus clearance of indigenous intestinal bacteria once they have penetrated actively the gut wall or, which seems more likely, passively through lesions, without elicting an inflammatory reaction.

REFERENCES

Allen, A. & Hoskins, L. C. (1988). Colonic mucus in health and disease. In *Diseases of the Colon, Rectum and Anal Canal*. Kirsner, J. B., Shorter, R. E.; Williams and Wilkins, eds, Baltimore, pp. 65–94.

André, C., Heremans, J. F., Vaerman, J. P. & Cambiaso, C. I. (1975). A mechanism for the induction of immunological tolerance by antigen feeding: antigen–antibody complexes. *Journal of Experimental Medicine*, **142**, 1509–19.

Apperloo-Renkema, H. Z., Van der Waaij, B. D. & Van der Waaij, D. (1990). Determination of colonization resistance of the digestive tract by biotyping of Enterobacteriaceae. *Epidemiology and Infection*, **105**, 355–61.

Beachey, E. H. (1981). Bacterial adherence: adhesion–receptor interactions mediating the attachment of bacteria to mucosal surfaces. *Journal of Infectious Diseases*, **143**, 325–45.

Berg, R. D. & Carlington, A. W. (1979). Bacterial translocation of certain indigenous bacteria from the gastrointestinal tract to mesenteric lymph nodes and other organs in a gnotobiotic mouse model. *Infection and Immunity*, **23**, 403–11.

Brandtzaeg, P., Bjerke, K., Halstensen, T. S. *et al.* (1990). Local immunity: The human mucosa in health and disease. In *Advances in Mucosal Immunology*. MacDonald, T. T., Challacombe, S. J., Bland, P. W., Stokes, Ch. R., Heatley, R. V., Mowat, A. McI., eds, Kluwer Academic Publishers, Dordrecht, Boston, London. pp. 1–12.

Challacombe, S. J. & Tomasi, T. B. (1980). Systemic tolerance and secretory immunity after oral immunization. *Journal of Experimental Medicine*, **152**, 1459–72.

Costerton, J. W., Rozee, K. R. & Cheng, K. J. (1983). Colonization of particulates, mucus and intestinal tissue. *Progress in Food and Nutritional Sciences*, **7**, 191–5.

Coutinho, A. (1989). Beyond clonal selection and network. *Immunological Reviews*, **110**, 63–87.

Drasar, B. S., Crowther, J. S., Goddard, P., Hawksworth, G., Hill, M. J., Peach, S. & Williams, R. E. O. (1973). The relation between diet and the gut microflora in man. *Proceedings of the Nutritional Society*, **32**, 49–52.

Drasar, B. S., Jenkins, D. J. A. & Cummings, J. H. (1976). The influence of a diet rich in wheat fibre on the human faecal flora. *Journal of Medical Microbiology*, **9**, 423–31.

Finegold, S. M., Attebery, H. R. & Sutter, V. L. (1974). Effect of diet on human fecal flora: comparison of Japanese and American diets. *American Journal of Clinical Nutrition*, **27**, 1446–69.

Foo, M. C. & Lee, A. (1974). Antigenic cross-reaction between mouse intestine and a member of the autochthonous microflora. *Infection and Immunity*, **9**, 1066–9.

Fougereau, M. & Schiff, C. (1988). Breaking the first circle. *Immunological Reviews*, **105**, 69–84.

Gossling, J. & Slack, J. M. (1974). Predominant Gram-positive bacteria in human feces: numbers, variety and persistence. *Infection and Immunity*, **9**, 719–29.

Guiot, H. F. L. (1982). Role of competition for substrate in bacterial antagonism in the gut. *Infection and Immunity*, **38**, 887–92.

Heidt, P. J. & Van der Waaij, D. (1979). Induction and maintenance of gnotobiotic states in experimental animals. *Zentralblatt für Bakteriologie Mikrobiologie und Hygiene, Abteilung I., Supplement*, **7**, 67–72.

Hoskins, L. C. & Boulding, E. T. (1981). Mucin degradation in the colon ecosystem. Evidence for the existence and role of bacterial subpopulations producing glycosidases as extracellular enzymes. *Journal of Clinical Investigation*, **67**, 163–72.

Kagnoff, M. F. (1987). Immunology of the digestive system. In *Physiology of the Gastrointestinal Tract*. 2nd edn. Johnson, L. R., ed. Raven Press, New York, pp. 1699–1728.

Kuhn, I., Tullus, K. & Molby, R. (1986). Colonization and persistence of *Escherichia coli* phenotypes in the intestines of children aged 0 to 8 months. *Infection*, **14**, 7–12.

Lee, A. (1985). Neglected niches. The microbial ecology of the gastrointestinal tract. In *Advances in Microbial Ecology*. vol. 8. Marshall, K. C., ed., Plenum Publishing Company, New York, pp. 115–161.

Long, S. S. & Swenson, R. M. (1977). Development of anaerobic faecal flora in healthy newborn infants. *Journal of Pediatriatics*, **72**, 298–301.

Lundequist, B., Nord, C. E. & Winberg, J. (1985). The composition of the faecal microflora in breastfed and bottle fed infants from birth to eight weeks. *Acta Paediatriaca Scandinavica*, **74**, 45–51.

McCleland, D. B. L., Samson, R. R., Parkin, D. M. & Shearman, D. J. C. (1972). Bacterial agglutination studies with secretory IgA prepared from human gastrointestinal secretions and colostrum. *Gut*, **13**, 450–8.

Meijer-Severs, G. J. & Van Santen, E. (1986). Variations in the anaerobic faecal flora of ten healthy human volunteers with special reference to the *Bacteroides fragilis* group and *Clostridium difficile. Zentralblatt für Bakteriologie und Hygiene, A* **261**, 43–52.

Meijer-Severs, G. J. & Van Santen, E. (1990). Short-chain fatty acid and organic concentrations in feces of healthy human volunteers and their correlation with anaerobe cultural counts during systemic ceftriaxone administration. *Scandinavian Journal of Gastroenterology*, **25**, 698–704.

Meijer, B. C. Kootstra, G. J., Geertsma, D. G. & Wilkinson, M. H. F. (1991). Effects of ceftriaxone on faecal flora; analysis by micromorphometry. *Epidemiology and Infection*. In press.

Mowat, A. McI. & Ferguson, A. (1981). Intraepithelial lymphocyte count and crypt hyperplasia measure the mujcosal component of graft-versus-host reaction in mouse small intestine. *Gastroenterology*, **83**, 417–23.

Mowat, A., McI & Felstein, M. V. (1987). Experimental studies of immunologically mediated enteropathy. II. Role of natural killer cells in the intestinal phase of murine graft-versus-host reaction. *Immunology*, **61**, 179–83.

Reed, W. P. & Williams, R. C. (1978). Bacterial adherence: a first step in pathogenesis of certain infections. *Journal of Chronic Diseases*, **31**, 67–72.

Savage, D. C. (1977). Interactions between host and its microbes. In *Microbial Ecology of the Gut*. Clarke, R. T. J. and Bauchop, T., eds, Academic Press, London, pp. 277–310.

Savage, D. C. (1980). Adherence of normal flora to mucosal surfaces. In *Bacterial Adherence*. Beachey, E. H., ed., Chapman and Hill, London, pp. 23–59.

Schaedler, R. W. (1973). Symposium on gut microflora and nutrition in the nonruminant. *Proceedings of the Nutritional Society*, **32**, 41–7.

Scott Rodkey, L. (1980). Autoregulation of immune response via idiotype network interactions. *Microbiological Reviews*, **44**, 631–59.

Tancrède, C. H. & Andremont, A. O. (1985). Bacterial translocation and Gram-negative bacteremia in patients with hematological malignancies. *Journal of Infectious Diseases*, **152**, 99–103.

Van der Waaij, D., Berghuis-de Vries, J. M. & Lekkerkerk-van der Wees, J. E. C.

(1971). Colonization resistance of the digestive tract in conventional and antibiotic-treated mice. *Journal of Hygiene*, **69**, 405–11.

Van der Waaij, D., Berghuis, J. M. & Lekkerkerk, J. E. C. (1972*a*). Colonization resistance of the digestive tract of mice during systemic antibiotic treatment. *Journal of Hygiene*, **70**, 605–10.

Van der Waaij, D., Berghuis-De Vries, J. M. & Lekkerkerk-Van der Wees, J. E. C. (1972*b*). Colonization resistance of the digestive tract and spread to lymphatic organs in mice. *Journal of Hygiene*, **70**, 335–42.

Van der Waaij, D. & Berghuis, J. M. (1974). Determination of the colonization resistance of the digestive tract in individual mice. *Journal of Hygiene*, **72**, 379–87.

Van der Waaij, D. & Heidt, P. J. (1977). Intestinal bacterial ecology in relation to immunological factors and other defense mechanisms. In: *Food and Immunology*. Hambraeus, L., Hanson, L. A., McFarlane, H., eds., Almquvist and Wiksell International, Stockholm, pp. 133–141.

Van der Waaij, D., Tieleman-Speltie, T. M. & De Roeck-Houben, A. M. J. (1978). Relation between the faecal concentration of various potentially pathogenic microorganisms and infections in individuals (mice) with severely decreased resistance to infection. *Antonie van Leeuwenhoek*, **44**, 395–405.

Van der Waaij, D. (1982*a*). The digestive tract in immunocompromised patients: importance of maintaining its resistance to colonization, especially in hospital inpatients and those taking antibiotics. In *Action of Antibiotics in Patients*. Sabath, L. D., ed., Hans Huber Publishers, Bern, Stuttgart, Vienna, pp. 104–118.

Van der Waaij, D. (1982*b*). Gut resistance to colonization: clinical usefulness of selective use of orally administered antimicrobial and antifungal drugs. In *Infection in Cancer Patients*. Klastersky, J., ed., Raven Press, New York, pp. 73–85.

Van der Waaij, D. (1984). The digestive tract in immunocompromized patients: importance of maintaining its resistance to colonization, especially in hospital in-patients. *Antonie van Leeuwenhoek*, **50**, 745–61.

Van der Waaij, D. (1986). The apparent role of the mucous membrane and the gut-associated lymphoid tissue in the selection of the normal resident flora of the digestive tract. *Clinical Immunology Newsletter*, **7**, 4–7.

Van der Waaij, D. (1987). Antibiotics, anaerobes and colonization resistance. In *Recent Advances in Anaerobic Bacteriology*. Borriello, S. P. and Hardie, J. M., eds., Martinus Nijhoff Publishers, Dordrecht/Boston/Lancaster, pp. 100–107.

Van der Waaij, D. & Van der Waaij, B. D. (1990). The colonization resistance of the digestive tract in different animal species and in man: a comparative study. *Epidemiology and Infection*, **105**, 237–43.

Van Houte, J. (1982). Bacterial adherence and plaque formation. *Infection*, **4**, 252–60.

Welling, G. W., Meijer-Severs, G. J., Helmus, G., Van Santen, E., Tonk, R. H. J., De Vries-Hospers, H. G., Van der Waaij, D. (1991). The effect of ceftriaxone on the anaerobic bacterial flora and the enzymatic activity in the intestinal tract. *Infection*, **19**, 313–16.

Wells, C. L., Maddaus, M. A., Simmons, R. L. (1988*a*). Proposed mechanisms for the translocation of bacteria. *Reviews of Infectious Diseases*, **10**, 958–79.

Wells, C. L., Jechorek, R. P. & Maddaus, M. A. (1988*b*). The translocation of intestinal facultative and anaerobic bacteria in defined flora mice. *Microbial Ecology in Health and Disease*, **1**, 227–35.

Wilhelm, M. P. & Lee, D. T. (1987). Bacterial interference by anaerobic species isolated from human faeces. *European Journal of Microbiology*, **6**, 266–70.

RECENT DEVELOPMENTS REGARDING THE EVOLUTION OF PATHOGENIC BACTERIA

MARK ACHTMAN AND REGINE HAKENBECK

Max-Planck Institut für molekulare Genetik, Ihnestr. 73, DW-1000 Berlin 33, Germany

INTRODUCTION

This review attempts to summarize concisely selected recent observations on evolutionary mechanisms for bacterial pathogens and makes no claim to be comprehensive. After an introduction to recent developments regarding clonal analyses of pathogenic bacteria, evolutionary principles in prokaryotes and inter-species genetic transfer, unusually incisive analyses of DNA sequence alterations which can be correlated with the clonal structure of natural populations are focused upon. It may seem unusual that pathogenesis, virulence factors and plasmids are largely ignored but the current state of knowledge about their evolution does not seem to have advanced very far yet (and the current body of literature on other aspects of both topics is too extensive for a short resumé such as this to do it justice). Thinking about bacterial pathogens as a limited series of clones has changed our viewpoints regarding the evolution of bacterial pathogens.

CLONES

The fact that, in some cases bacterial pathogens represent widespread members of a single clone was first observed by Duguid and collaborators in the 1960s (Morgenroth & Duguid, 1968). They were struck by the uniform properties of so-called FIRN *Salmonella typhimurium* bacteria, which are non-fimbriate and cannot ferment rhamnose or inositol. Three independent mutations leading to inability to ferment rhamnose and to lack of piliation are conserved in independent bacteria isolated world-wide and the original strain has diversified sufficiently for 27 phage types and 22 biotypes to have been recognized among FIRN strains (Morgenroth & Duguid, 1968; Old & Duguid, 1979).

In the mid-1970s the Ørskov's concluded that enterotoxigenic *Escherichia coli* strains from diarrhoeal disease represented a few distinct clones because they belonged to only few O:H serotypes (O for lipopolysaccharide [LPS] antigens and H for flagellar antigens) which were otherwise exceedingly rare among the 20 000 *E. coli* isolates from other sources that they had examined (Ørskov *et al.*, 1976). In 1980–81, Selander and Levin introduced the techniques used for population genetics of eukaryotic organisms to bacteria.

They used the method of starch gel electrophoresis to examine numerous *E. coli* bacteria for variation of cytoplasmic isoenzymes (called MLEE for *M*ulti-*L*ocus *E*nzyme *E*lectrophoresis; Selander & Levin, 1980; Caugant, Levin & Selander, 1981) and found that some bacteria were indistinguishable from one another even when 20 enzymes and/or the plasmid profile were examined. Such a group of bacteria whose electrophoretic profile is indistinguishable is referred to as an electrophoretic type or ET and corresponds to a clone whose members are believed to have all descended from a single ancestral cell. Within a set of bacteria containing many dissimilar ETs, a clone may encompass several closely related ETs due to minor evolution (Caugant, Levin & Selander, 1981). Numerous fairly closely related clones have been designated as a clone complex (Caugant *et al.*, 1986*b*) or a subgroup (Olyhoek, Crowe & Achtman, 1987).

It seems that many disease syndromes are caused by pathogenic bacteria which belong to only few clones/complexes, even if the species to which those clones belong is/are very diverse. Most *E. coli* bacteria isolated from newborn septicemia or meningitis express the K1 capsular polysaccharide and these usually belong to one of only six widespread clones (Achtman *et al.*, 1983; Ochman & Selander, 1984*a*; Achtman & Pluschke, 1986) which are associated with particular O:K:H serotypes, outer membrane protein electrophoretic variants, nutritional requirements, and often biotype (Achtman *et al.*, 1983). Most epidemics of meningococcal meningitis since 1960 have been caused by one of five clones of serogroup A (Olyhoek, Crowe & Achtman, 1987; Achtman *et al.*, 1992) or one clone of serogroup B or C (Caugant *et al.*, 1986*b*). Other examples are a single clone of *Staphylococcus aureus* associated with toxic shock syndrome (Musser *et al.*, 1990), a single clone of *Haemophilus influenzae* biogroup aegyptius associated with Brazilian purpuric fever, one clone of *E. coli* O157:H7 associated with hemorrhagic colitis and hemolytic uremic syndrome (Whittam & Wilson, 1988), and all isolates of *S. typhi* consisting of a single clone (Reeves *et al.*, 1989).

SPECIES SUBDIVISIONS

Numerous analyses have been performed with various methods to subdivide bacterial species. Some recent analyses have depended on restriction fragment length polymorphism (RFLP) of total chromosomal DNA with conventional or pulsed-field gel electrophoresis whereas others have analysed RFLP after hybridization with selected DNA probes to reveal the RFLP patterns of single or multiple genes (Olyhoek *et al.*, 1988; Loos *et al.*, 1989; Kuijper *et al.*, 1989; Kapperud *et al.*, 1989; Beger, Heuzenroeder & Manning, 1989; Arthur *et al.*, 1990; Arbeit *et al.*, 1990; Blumberg, Kiehlbauch & Wachsmuth, 1991; Van Soolingen *et al.*, 1991). Although these methods may be extremely useful for analysing a limited number of strains, they are inadequate for analysing numerous isolates from a species with extensive genetic variation. Traditional studies have used serological typing based on capsular polysaccharide, LPS and/or protein antigens; biochemical properties (biotyping), bacteriophage or colicin typing; or the protein polymorphism revealed by analysis of total cell extracts or membrane proteins by polyacrylamide gel electrophoresis. These methods may provide markers highly useful for microepidemiology (Achtman *et al.*, 1983; Brindle, Bryant & Draper, 1989; Kapperud *et al.*, 1989; Kuijper *et al.*, 1989) but are not based on characters which can be correlated directly with the genetic diversity of the bacteria. Subdividing a species should be based on methods which can be used for the analysis of 1000s of isolates, which index diverse and numerous chromosomal genes and which are susceptible to mathematical analysis. None of the above methods fulfils all these criteria although they will continue to be used because of tradition, scientific modes and because most analyses involve only relatively few strains. MLEE involving 15 or more polymorphic cytoplasmic isoenzymes does fulfil all the requirements and can be recommended for analysis of the genetic variation within a species (Selander *et al.*, 1986). Very extensive analyses using MLEE have been reported for *E. coli* (summarized in Selander, Caugant & Whittam, 1987*a*), *H. influenzae* (Musser *et al.*, 1985, 1988*a,b*), *N. meningitidis* (Caugant *et al.*, 1986*a,b*, 1987*a,b*; Olyhoek *et al.*, 1987) and *Salmonella* (Beltran *et al.*, 1988; Reeves *et al.*, 1989) and somewhat more limited studies have been reported for *Yersinia enterocolitica* (Caugant *et al.*, 1989), *Streptococcus agalactiae* (group B *Streptococcus*; Musser *et al.*, 1989), *Listeria monocytogenes* (Piffaretti *et al.*, 1989), *Haemophilus pleuropneumoniae* (Musser, Rapp & Selander, 1987*b*), *Bordetella* spp. (Musser *et al.*, 1986, 1987*a*), and *Legionella pneumophilia* (Selander *et al.*, 1985; McKinney *et al.*, 1989). All these analyses demonstrated that these species consisted of a limited number of ETs and clones or clone complexes, rather than an pseudo-infinite collection of heterogeneous ETs, although the genetic variability can differ markedly from species to species. Many of these analyses have been reviewed elsewhere (Selander *et al.*, 1987*a,b*).

EVOLUTIONARY MECHANISMS

Numerous analyses have attempted to test whether the neutral theory of molecular evolution (Kimura, 1983) can account for the differences between species. According to this theory, mutations with neutral selective effects will accumulate with time and provide the substrate for selection to yield genetic variation and speciation. Then differences between the genetic sequences of any one gene within a group of bacteria should correlate with the genetic distance between each bacterial pair as defined by MLEE. The last common ancestor of *E. coli* and *S. typhimurium* probably existed about 140 million years (MYrs) ago (Wilson, Ochman & Prager, 1987). During that period of time, silent substitutions (mutations primarily affecting the third codon position which have no effect on amino acid composition) have accumulated in numerous genes in these two species, and the average silent substitution rate for 22 genes was 95% (with a few exceptions resulting in the wide range of 4–158%; Ochman & Wilson, 1987). The differences between any two *E. coli* isolates should be much less.

Indeed, when a 1300 bp portion of the *trp* region was sequenced from 12 natural isolates of *E. coli*, 8 were essentially indistinguishable from *E. coli* strain K-12, 3 were indistinguishable from each other but differed from K-12 by 10 bp and 1 isolate differed by 44 bp (Milkman & Crawford, 1983). K-12 and *S. typhimurium* differed by 40–50% of third codon positions in that region. In contrast to *trp* which is fairly strongly conserved within *E. coli*, the *gnd* locus, encoding 6-phosphogluconate dehydrogenase, was one of the most polymorphic detected by MLEE (Ochman & Selander, 1984*b*; Selander *et al.*, 1987*a*). The sequences of the *gnd* gene from nine natural isolates and from K-12 and *S. typhimurium* have been determined (Bisercic, Feutrier & Reeves, 1991). Variation was detected at 184 of 1404 bp within *E. coli*, mostly at third codon positions and up to 6% of the nucleotides varied between pairs of bacteria. This variation represents about one-third of the variation between *E. coli* and *S. typhimurium*. The results showed that there was only a poor correlation between sequence differences and differences in electrophoretic migration. Furthermore, the genetic dendrogram relating genetic distance between these bacteria according to electrophoretic profiles for 35 enzymes did not correlate with the dendrogram relating sequence divergence of the *gnd* gene (Bisercic *et al.*, 1991). And phylogenetic trees constructed for descent of the *gnd* gene did not correspond to those constructed for the *trp* genes in the same organisms (Dykhuizen & Green, 1986). Based on these, and results summarized below, it is currently believed that dendrograms based on MLEE do not reflect genetic descent but only genetic relatedness, that evolution of bacteria reflects a combination of mutational variation and recombinational events due to horizontal genetic exchange, and that the sequence of any one gene may represent a different evolutionary path than that of the chromosome as a whole.

Possibly the best set of data for exemplifying these conclusions is represented by two recent and superb analyses of Milkman's (Stoltzfus, Leslie & Milkman, 1988; Milkman & Bridges, 1990). A 3500 bp region of *E. coli* K-12 lying between *trp* and *tonB* was sequenced and six open reading frames (ORFs) which extend over 95% of that region were recognized (Stoltzfus *et al.*, 1988). The sequences for numerous natural isolates, whose relatedness had been determined by MLEE were also obtained and compared (Stoltzfus *et al.*, 1988; Milkman & Bridges, 1990). Differences corresponding to a few per cent were observed between the sequences. These differences mostly corresponded to synonymous substitutions within the ORFs but were scattered over all three codon positions in non-coding regions. Dramatic large-scale DNA rearrangements were also observed, including replacement by an insertion sequence (IS1), deletion of parts of ORFI, II and/or III, (Stoltzfus *et al.*, 1988) and insertions of 'Atlas'. Atlas is a generic name for huge insertions corresponding to one of a series of lambdoid prophages which are homologous at the right end, vary at the left end and which share a unique attachment site within this DNA region (Milkman & Bridges, 1990). In the second of these publications (Milkman & Bridges, 1990), the analysis was extended to 15 chromosomal regions of 1500 bp each and to bacteria chosen to be relatively similar or very different according to MLEE analysis. The data showed that the sequence relatedness between strains varies from DNA region to region. Some of the sequence differences can best be explained by recombinational events because the nucleotide replacements are clustered and occur at frequencies well above those observed in other DNA regions of the same bacterial pairs. The data are interpreted as showing that clonality only extends to specific DNA segments and bacteria may be clonal for one segment while differing at others (Milkman & Bridges, 1990): Level I clones contain bacteria whose sequences are almost indistinguishable over the 16.5 kb analysed. Level II clones include closely related bacteria which may have one or a few different regions although most regions are identical or very similar. Level III clones are 'generally similar and occasionally identical' (Milkman & Bridges, 1990). With few exceptions, most of the *E. coli* studied fall into 'one grand clone'. Recombination events were concluded to result in an average replacement size of 1000 bp, and mathematical models indicated that recombination involving so little DNA would derange the similarity or identity of the remainder of the chromosome only very slowly. Thus MLEE patterns would reflect the clonal relationships of the entire chromosome even after a number of small recombinational replacements had occurred.

These analyses ignored selection mechanisms specific for bacterial pathogens. But they do supply a theoretical framework which needs to be expanded by examination of events in pathogenic populations. Furthermore, the results do indicate that recombinational events should be considered as an important factor in evolution of bacterial pathogens. However, *E. coli* cannot be taken as typical of all pathogenic bacterial species, and it is likely that different species will show different evolutionary patterns when compared to those of *E. coli*.

Recent observations on conjugation

Although conjugation has formerly been thought to occur only between generally similar bacteria, new observations have shown that DNA can be transferred between very distinct organisms. Conjugative transposition in Gram-positive bacteria (for historical review see Clewell & Gawron-Burke, 1986) seems to resemble lambdoid phage integration (Poyart-Salmeron *et al.*, 1990). It involves two transposon proteins, Xis-Tn and Int-Tn, which are

responsible for excision and integration of a non-replicative circular intermediate that can then transpose elsewhere in the same cell or in a different
cell after conjugation. Although conjugative transposition seemed to be
restricted to Gram-positive bacteria, it has now been shown that it can occur
between highly diverse organisms. Conjugative transposons have been
linked synthetically to plasmids capable of replicating in the recipient
species and then used to show that DNA transfer can occur from the Gram-
positive *Enterococcus faecalis* to *E. coli* (Trieu-Cuot, Carlier & Courvalin,
1988) or from *E. coli* to *Enterococcus faecalis*, *Streptococcus lactis*, *Streptococcus agalactiae*, *Bacillus thuringiensis*, *Listeria monocytogenes* and
Staphylococcus aureus (Trieu-Cuot *et al.*, 1987). Transfer between diverse
bacteria seems also to have occurred in nature. *E. coli* have been isolated
carrying plasmids with an *ermBC* gene encoding constitutive resistance to
macrolide-lincosamide-streptogramin (MLS) antibiotics (Brisson-Noël,
Arthur & Courvalin, 1988). When introduced in the laboratory to *Bacillus
subtilis*, *ermBC* renders those bacteria constitutively resistant to these
antibiotics. *ermBC* is highly homologous to *ermB* in the conjugative transposon Tn917 from *Enterococcus faecalis* and only 5 bp substitutions leading to
three amino acid changes were detected over 1139 bp when *ermBC* and
ermB were compared. The kanamycin resistance gene *aphA-3* from *Campylobacter coli* is identical to that from streptococci and is also expressed in *B.
subtilis* and *E. coli* (Trieu-Cuot *et al.*, 1985).

Even more unexpectedly, evidence has been presented that conjugation is
responsible for the DNA transfer to, and malignant transformation of,
plants by *Agrobacterium tumefaciens* (Buchanan-Wollaston, Passiatore &
Cannon, 1987) and that conjugation can be observed between *E. coli* and
yeast (Heinemann & Sprague, 1989).

DNA transformation

Some bacterial species including *Neisseria meningitidis*, *Neisseria gonorrhoeae*, *Haemophilus influenzae*, and *Streptococcus pneumoniae* are naturally transformable (for review see Stewart & Carlson, 1986). Whereas
Neisseria are continuously competent for DNA uptake (Scocca, 1990), both
Haemophilus and *Streptococcus pneumoniae* are competent only at certain
phases of their growth cycle (Goodgal, 1982; Morrison, Mannarelli &
Vijayakumar, 1982). Encapsulated pathogenic strains of *S. pneumoniae* are
apparently inactive in competence development but can be readily transformed upon addition of competence factor, a soluble, excreted protein,
albeit at a slightly lower efficiency than found with non-encapsulated
derivatives (Yother, McDaniel & Briles, 1986). *Neisseria* and *Haemophilus*
have DNA uptake systems that interact specifically with short defined DNA
sequences preferentially found in these species, thus restricting the accessible gene pool (Danner *et al.*, 1980*a,b*; Goodman & Scocca, 1991). In

contrast, pneumococci do not discriminate between different DNAs during uptake nor is transforming DNA affected by restriction (Lacks & Greenberg, 1977). Rather, sufficient homology with the chromosome for recombination to occur is the limiting factor for DNA integration. For all three genera, conjugation seems to be rare and transformation appears to be the primary mechanism for transfer of chromosomal DNA.

Pilin variation of *N. gonorrhoeae* is at least partially due to transformation with defective *pilS* genes during growth (Seifert *et al.*, 1988; Gibbs *et al.*, 1989) and such variation occurs rapidly and frequently during infection (Swanson *et al.*, 1987). In fact, it has been proposed (Scocca, 1990; in the absence of experimental evidence) that DNA in the environment informs *N. gonorrhoeae* during uptake that sister cells are dying and stimulates recombination enzymes. These enzymes supposedly then reshuffle partially homologous DNA sequences within the chromosome and result in genetic variation independently of the DNA taken up.

Examples will be presented below where transformation has been directly invoked to account for the formation of mosaic genes.

EXAMPLES OF DNA ALTERATIONS

Mosaic genes

The existence of gene mosaics whose formation can be best attributed to recombination has been recently reported. The *iga* gene encoding an IgA1 protease has been partially sequenced from four diverse strains of *N. gonorrhoeae* (Halter, Pohlner & Meyer, 1989). The sequence differences were such that no single phylogenetic tree could have given rise to the different genes and instead recombinational events in which the genes represented insertion of stretches from different genes ('mosaic') must have occurred.

The penicillin-binding proteins (PBPs) in penicillin-resistant *Neisseria* species and in *S. pneumoniae* represent some of the most extensively studied examples of mosaic genes due to work by B. Spratt and his colleagues. PBPs are the target proteins of the penicillin family of antibiotics and intrinsic resistance (as opposed to that caused by β-lactamases) is a result of the reduction of the affinity of PBP's for penicillin. Clinical isolates of *Neisseria* and pneumococci contain PBP genes with one or more blocks of sequences ('resistant sequences') which differ by 8–>20% from the comparable sequences in sensitive strains. The resistant sequences are interspersed with 'sensitive sequences' that vary by less than 1% within any one species (Smith, Dowson & Spratt, 1991).

Intrinsic penicillin resistance in *N. gonorrhoeae* can be caused by changes in any one of four genes (Dougherty, Koller & Tomasz, 1980; Cannon & Sparling, 1984; Faruki & Sparling, 1986), two of which encode PBP1 and

PBP2 (*penA*). *penA* from three different gonococci contained stretches diverging by up to 14%, with the resistant blocks of two strains being identical. The source of these divergent blocks was attributed to recombination after transformation with DNA from a related species (Spratt, 1988). Subsequent analyses have shown that *penA* from both penicillin-resistant gonococci and meningococci contain fragments which probably came from *Neisseria flavescens*, a commensal species which is naturally more resistant to penicillin (MIC = 0.4 mg/ml versus 0.004 mg/ml). Apparent transfer of resistant blocks from *N. flavescens* to the non-pathogenic *Neisseria lactamica* has also been documented (Lujan *et al.*, 1991). Recently, a second resistant donor species, *Neisseria cinerea*, has also been identified (Spratt *et al.*, 1992). Sixteen penicillin-resistant and 12 susceptible meningococci from Spain were assigned to two distinct lineages by MLEE (Mendelman *et al.*, 1989) and resistant as well as sensitive strains were represented in both lineages. Thus, the emergence of penicillin-resistant strains is not due to spread of a single clone.

The PBPs involved in penicillin-resistance in pneumococci are called 1a, 2a, 2b and 2x (Laible, Spratt & Hakenbeck, 1991). Two different patterns of resistance blocks were detected in PBP 2b and therefore two distinct DNA donors have been invoked (Dowson *et al.*, 1989). The DNA sequences of PBP's 1a and 2x from resistant bacteria differed even more strongly from the sequences of sensitive genes than was the case for PBP 2b (Laible *et al.*, 1991, C. Martin, C. Sibold & R. Hakenbeck, manuscript in preparation). Five resistant PBP 2x genes differed from each other as much as from the sensitive prototype, indicating strong heterogeneity of the gene pool, and the sequences of PBP 1a genes from three South African isolates suggested that at least four different DNA donors had been involved. The putative donors for the above sequences have not been identified. However, the PBP 2b gene has already spread to viridans streptococci (Dowson *et al.*, 1990). In the laboratory, it has been possible to select penicillin-resistant pneumococci after transformation with DNA from resistant *Streptococcus sanguis II* or *Streptococcus mitis* (Chalkley *et al.*, 1991) and PBP 1a and 2x resistant sequences found in resistant pneumococci also were found in low-level resistance *S. mitis* strains (Sibold, C., and R. Hakenbeck, unpublished observations).

rfb loci

Recent analyses have contributed to an explanation of the tremendous variability of the lipopolysaccharide (LPS) O antigens in *E. coli* and *Salmonella enterica* (the currently suggested name for all *Salmonellae* according to Le Minor & Popoff, 1987). In 1985, 60 and 160 antigenically distinguishable O antigens had been described in these two species, respectively (Ewing, 1986). Some of these O antigens are chemically similar and differ by one monosaccharide in the polysaccharide repeating unit while others are quite distinct. The DNA sequence of the 23 kb encompassing the *rfb* genes, which encode the enzymes for LPS-related polysaccharide synthesis, has been determined for a strain of serovar *typhimurium* (Jiang *et al.*, 1991) and differed from the comparable regions from strains of serovars *typhi, paratyphi* (Verma & Reeves, 1989; Liu *et al.*, 1991) and *anatum* (Wang, Romana & Reeves, 1992). In addition, the *rfb* regions from O2 and O111 *E. coli* bacteria (Bastin, Romana & Reeves, 1991; Neal *et al.*, 1991) and from *Salmonella* serovars *montevideo* (Lee, Romana & Reeves, 1992) and *muenchen* (Brown, Romana & Reeves, 1991) have been cloned, mapped and compared to serovar *typhimurium* by southern hybridization. The results were interpreted as showing that the central portion of *rfb* genes in *typhimurium* could be assigned to three to five different regions of low GC content. Because of their low GC content, it was concluded that these regions probably evolved in an (unspecified) low GC species other than *Salmonella*. Analyses of the differences between these various bacterial sequences led to the suggestion that the sequences originally all evolved in the (unspecified) low GC species, followed by recombinational events approximately 1 MYrs ago (Reeves, 1991).

Capsular polysaccharide

Haemophilus influenzae expresses six distinct capsular polysaccharides termed types a through f; most invasive disease is caused by type b organisms. 2209 strains of encapsulated bacteria fell into 280 ETs which were assigned to 12 lineages, called A through L. Lineages A–G were separated from lineages H–L at a genetic distance (0.66) often associated with distinct species (Musser *et al.*, 1988*a,b*). Serotype b bacteria were present in lineages A, B, C and in J. Serotype a bacteria were present in lineages B and H and I. Lineage B included bacteria of serotypes a, b and d. Thus, unrelated bacteria could express the same capsular type and unrelated capsular types could be expressed by related bacteria. The so-called *cap* DNA region is a 20 to 40 kb segment of the chromosome containing essential loci for biosynthesis of capsular polysaccharide. For serotypes a and b, the RFLP pattern of *cap* DCNA differed strongly between the two divisions (Musser *et al.*, 1988*a,b*). Analysis of serotype b representatives from both divisions showed that the central 4.8 kb of the *cap* genes are identical and that the flanking *cap* DNA is homologous but different (Kroll & Moxon, 1990). However, the chromosomal location of the *cap* genes differed in the two divisions. Within the homologous flanking regions, a 1 kb insertion was found in one of the divisions, there was only 88% identity within the 800 bp which were sequenced and some restriction sites differed in the unsequenced DNA. It was concluded that the *cap* flanking regions within the lineages in the two divisions have diverged by evolution and that a serotype b specific fragment from the central region was transferred between the two lineages, probably by transformation (Kroll & Moxon, 1990).

Indirect evidence suggests that horizontal exchange of capsular genes may have occurred in other bacteria as well: *N. meningitidis* bacteria of serogroups A, B and C were isolated in East Germany in the late 1980s which were indistinguishable by MLEE (Wedege *et al.*, 1991, Caugant, D. A., pers. comm.). Similarly, 11 serogroup B and 34 serogroup C *N. meningitidis* strains which were responsible for outbreaks in US army recruits in the late 1960's (Mandrell & Zollinger, 1989) belonged to 1 ET within the ET37 complex according to MLEE (Wang *et al.*, 1991, Wang, J., Caugant, D. A., and Achtman, M., man in prep.).

Protein antigens

One of the major outer membrane proteins of *N. meningitidis*, called the Class 1 protein, has been used as the basis of an antigenic typing scheme called serosubtyping (Frasch, Zollinger & Poolman, 1985; Abdillahi & Poolman, 1988). The *porA* genes encoding these proteins contain two variable DNA sequences, each of which can be recognized independently by specific monoclonal antibodies (MAbs; McGuinness *et al.*, 1990; Maiden *et al.*, 1991) and presumably a (future) complete serosubtyping scheme will be based on identification of each strain by two distinct MAbs. Experiments where the serosubtype was still only defined by one MAb specificity per strain showed that serosubtype can vary independently of genetic relatedness (Caugant *et al.*, 1986*b*, 1987*a,b*). Moreover, comparison of bacteria which reacted with certain MAbs indicated that the *porA* genes can only have arisen by independent genetic exchanges followed by recombination (Feavers *et al.*, 1992).

Unstable DNA segments

Some *E. coli* strains from urinary tract infection contain two distinct regions encoding hemolysins called *hlyI* and *hlyII*. In some bacterial strains these regions are also each linked to operons for pilus synthesis. *hlyI* and *hlyII* can be lost concurrently or independently due to deletion mutations, both after growth *in vitro* and *in vivo* (Hacker *et al.*, 1990). The deleted DNA comprises a 75 kb region flanked by two 16 bp direct repeats for *hlyI* (Knapp *et al.*, 1986) and a 190 kb region for *hlyII* (Hacker *et al.*, 1992). Representatives of two relatively closely related clonal groups differed in the *hlyII* region as if one of the groups had suffered such a deletion early in its history (Hacker *et al.*, 1990).

N. gonorrhoeae contain six DNA regions called *pilS1* to *pilS6* which contain 17 so-called silent pilin genes (Haas, Velt & Meyer, 1992). The chromosome also contains one to two active pilin genes, called *pilE* (Meyer, Mlawer & So, 1982) which vary at an extremely high rate through chromosomal rearrangements involving recombinational events with the silent copies (Meyer, Mlawer & So, 1982; Haas & Meyer, 1986; Swanson *et al.*, 1987; Meyer & van Putten, 1989). The rearrangements occur at least partially due to transformation with DNA released from other bacteria rather than by recombination between two copies from the same molecule (Seifert *et al.*, 1988; Gibbs *et al.*, 1989). These bacteria also contain 11 to 12 *opa* genes at distinct locations on the bacterial chromosome (Dempsey *et al.*, 1991; Bihlmaier *et al.*, 1991; Bhat *et al.*, 1991) which are very similar except for two hyper-variable regions. The number of *opa* genes can vary to a limited extent even in the laboratory (Bihlmaier *et al.*, 1991) and the sequences of the genes can be best interpreted as their having arisen by recombinational events between a lesser number of primeval alleles (Bhat *et al.*, 1991; Dempsey *et al.*, 1991).

BACTERIAL SPREAD

For most of the bacterial species just discussed, little is known about their patterns of spread. Indistinguishable bacteria have been isolated from different continents and decades apart (Achtman *et al.*, 1983, 1992; Olyhoek *et al.*, 1987; Beltran *et al.*, 1988) but few of the analyses available can make the claim to have sampled even a significant proportion of the global diversity inferred to exist. The speed with which a particularly fit bacterial strain will spread from its original source is unpredictable. It may therefore be illustrative to examine the data available for epidemic meningitis where spread of the bacteria can be documented by examining spread of reportable epidemics. Epidemics caused by subgroup I of serogroup A *N. meningitidis* broke out in 1967 in North Africa and had spread by 1968–1972 throughout most of West Africa and the Mediterranean (Olyhoek *et al.*, 1987). A variant of these bacteria caused disease in North America in the early and mid-1970s (Olyhoek *et al.*, 1987). Clone III-1 spread quickly from China (1966) to Scandinavia and Russia (1969) and from Saudi Arabia (1987) to all East Africa (1988–1990; Achtman *et al.*, 1992). Clearly, spread can be very rapid and very extensive. Yet not all serogroup A bacteria have spread as dramatically: although clone IV–1 has caused repeated disease and epidemics in West Africa between the early 1960s and the late 1980's (Olyhoek *et al.*, 1987), it has not yet been isolated from any other site than India (Wang *et al.*, man. in prep.) and still other serogroup A bacteria have only been isolated rarely and from unique locations or seem to have disappeared after World War II (Olyhoek *et al.*, 1987).

A multi-resistant serotype 23F clone of pneumococci has become common in recent years in the United Kingdom, Spain, South Africa and the USA (Munoz *et al.*, 1991). PBPs 1a, 2b and 2x which were seemingly identical to those in the 23F clone were found in a serogroup 9 clone in Spain, indicating that once mosaic chromosomal genes are formed, they may rapidly spread across clonal barriers within a species (Coffey *et al.*, 1991). In contrast, *H. influenzae* seems usually to be geographically fixed (Musser *et al.*, 1988*a*) and geographical specificities were found for certain invasive K1 *E. coli* as well (Achtman *et al.*, 1983).

CONCLUSIONS

The data presented here have been selected to illustrate the very striking DNA changes which seem to have occurred in diverse bacterial species. Genetic mechanisms exist which can account for these changes and the importance of horizontal genetic exchange has been illustrated. Yet the authors wish to reiterate that bacterial populations are clonal in nature and that usually numerous isolates from a common source are either indistinguishable at the genetic level or nearly so. Most pathogenic bacteria within certain species belong to a limited number of ETs and some of the genetic diversity described here may be a property of otherwise unsuccessful and rarely isolated bacteria. The factors affecting spread and selection of certain widespread clones remain obscure.

Essentially all of the examples cited which invoke horizontal genetic exchange are impossible to date, i.e. the exchange might have happened between last year or over 1 MYr ago although those authors who have been willing to speculate have tended toward the latter estimate except for antibiotic resistance. The geographical location of the genetic exchange is unknown: guesses at the location have only been made for antibiotic-resistant PBPs and even there the data are circumstantial. The supposed DNA donor is unknown except for neisserial PBPs where homologous sequences were found in related species. Again with the exception of PBPs, the importance of these exchanges for selection pressures is uncertain. We believe that such answers may perhaps first be available when molecular genetics are applied to bacterial isolates whose epidemiological sources are well documented and anticipate that such results may soon be available. The availability of extensive strain collections which have been subjected to MLEE should aid in these efforts.

REFERENCES

Abdillahi, H. & Poolman, J. T. (1988). *Neisseria meningitidis* group B serosubtyping using monoclonal antibodies in whole-cell ELISA. *Microbial Pathogenesis*, **4**, 27–32.

Achtman, M., Kusecek, B., Morelli, G. *et al.* (1992). A comparison of the variable antigens expressed by clone IV–1 and subgroup III of *Neisseria meningitidis* serogroup A. *Journal of Infectious Diseases*, **165**, 53–68.

Achtman, M., Mercer, A., Kusecek, B. *et al.* (1983). Six widespread bacterial clones among *Escherichia coli* K1 isolates. *Infection and Immunity*, **39**, 315–35.

Achtman, M. & Pluschke, G. (1986). Clonal analysis of descent and virulence among selected *Escherichia coli*. *Annual Reviews of Microbiology*, **40**, 185–210.

Arbeit, R. D., Arthur, M., Dunn, R., Kim, C., Selander, R. K. & Goldstein, R. (1990). Resolution of recent evolutionary divergence among *Escherichia coli* from related lineages: The application of pulsed field electrophoresis to molecular epidemiology. *Journal of Infectious Diseases*, **161**, 230–5.

Arthur, M., Arbeit, R. D., Kim, C. *et al.* (1990). Restriction fragment length polymorphisms among uropathogenic *Escherichia coli* isolates: *pap*-related sequences compared with *rrn* operons. *Infection and Immunity*, **58**, 471–9.

Bastin, D. A., Romana, L. K. & Reeves, P. R. (1991). Molecular cloning and expression in *Escherichia coli* K-12 of the *rfb* gene cluster determining the O antigen of an *E. coli* O111 strain. *Molecular Microbiology*, **5**, 2223–31.

Beger, D. W., Heuzenroeder, M. W. & Manning, P. A. (1989). Demonstration of clonal variation amongst O-antigen serotype variants of *Escherichia coli* O2 and O18 using DNA probes to the *rfb* region of the *E. coli* strain B41 (O101:K99/F41). *FEMS Microbiology Letters*, **57**, 317–22.

Beltran, P., Musser, J. M., Helmuth, R. *et al.* (1988). Toward a population genetic analysis of *Salmonella*: Genetic diversity and relationships among strain of serotypes *S. choleraesuis*, *S. derby*, *S. dublin*, *S. enteritidis*, *S. heidelberg*, *S. infantis*, *S. newport*, *and S. typhimurium*. *Proceedings of the National Academy of Sciences, USA*, **85**, 7753–7.

Bhat, K. S., Gibbs, C. P., Barrera, O. *et al.* (1991). The opacity proteins of *Neisseria gonorrhoeae* strain MS11 are encoded by a family of 11 complete genes. *Molecular Microbiology*, **5**, 1889–1901.

Bihlmaier, A., Römling, U., Meyer, T. F., Tümmler, B. & Gibbs, C. P. (1991). Physical and genetic map of the *Neisseria gonorrhoeae* strain MS11–N198 chromosome. *Molecular Microbiology*, **5**, 2529–39.

Bisercic, M., Feutrier, J. Y. & Reeves, P. R. (1991). Nucleotide sequences of the *gnd* genes from nine natural isolates of *Escherichia coli*: evidence of intragenic recombination as a contributing factor in the evolution of the polymorphic *gnd* locus. *Journal of Bacteriology*, **173**, 3894–900.

Blumberg, H. M., Kiehlbauch, J. A. & Wachsmuth, I. K. (1991). Molecular epidemiology of *Yersinia enterocolitica* O:3 infections: use of chromosomal DNA restriction fragment length polymorphisms of rRNA genes. *Journal of Clinical Microbiology*, **29**, 2368–74.

Brindle, R. J., Bryant, T. N. & Draper, P. W. (1989). Taxonomic investigation of *Legionella pneumophila* using monoclonal antibodies. *Journal of Clinical Microbiology*, **27**, 536–9.

Brisson-Noël, A., Arthur, M. & Courvalin, P. (1988). Evidence for natural gene transfer from Gram-positive cocci to *Escherichia coli*. *Journal of Bacteriology*, **170**, 1739–45.

Brown, P. K., Romana, L. K. & Reeves, P. R. (1991). Cloning of the *rfb* gene cluster of a group C2 *Salmonella* strain: comparison with the *rfb* regions of groups B and D. *Molecular Microbiology*, **5**, 1873–81.

Buchanan-Wollaston, V., Passiatore, J. E. & Cannon, F. (1987). The *mob* and *oriT* mobilization functions of a bacterial plasmid promote its transfer to plants. *Nature (London)*, **328**, 172–5.

Cannon, J. G. & Sparling, P. F. (1984). The genetics of the gonococcus. *Annual Reviews of Genetics*, **38**, 111–33.

Caugant, D. A., Aleksic, S., Mollaret, H. H., Selander, R. K. & Kapperud, G.

(1989). Clonal diversity and relationships among strains of *Yersinia enterocolitica*. *Journal of Clinical Microbiology*, **27**, 2678–83.

Caugant, D. A., Bøvre, K., Gaustad, P. *et al.* (1986*a*). Multilocus genotypes determined by enzyme electrophoresis of *Neisseria meningitidis* isolated from patients with systemic disease and from healthy carriers. *Journal of General Microbiology*, **132**, 641–52.

Caugant, D. A., Frøholm, L. O., Bøvre, K. *et al.* (1986*b*). Intercontinental spread of a genetically distinctive complex of clones of *Neisseria meningitidis* causing epidemic disease. *Proceedings of the National Academy Sciences USA*, **83**, 4927–31.

Caugant, D. A., Levin, B. R. & Selander, R. K. (1981). Genetic diversity and temporal variation in the *E. coli* population of a human host. *Genetics*, **98**, 467–90.

Caugant, D. A., Mocca, L. F., Frasch, C. E., Frøholm, L. O., Zollinger, W. D. & Selander, R. K. (1987*a*). Genetic structure of *Neisseria meningitidis* populations in relation to serogroup, serotype, and outer membrane protein pattern. *Journal of Bacteriology*, **169**, 2781–92.

Caugant, D. A., Zollinger, W. D., Mocca, L. F. *et al.* (1987*b*). Genetic relationships and clonal population structure of serotype 2 strains of *Neisseria meningitidis*. *Infection and Immunity*, **55**, 1503–13.

Chalkley, L., Schuster, C., Potgieter, E. & Hakenbeck, R. (1991). Relatedness between *Streptococcus pneumoniae* and viridans streptococci: transfer of penicillin resistance determinants and immunological similarities of penicillin-binding proteins. *FEMS Microbiology Letters*, **90**, 35–42.

Clewell, D. B. & Gawron-Burke, C. (1986). Conjugative transposons and the dissemination of antibiotic resistance in streptococci. *Annual Reviews of Microbiology*, **40**, 635–59.

Coffey, T. J., Dowson, C. G., Daniels, M. *et al.* (1991). Horizontal transfer of multiple penicillin-binding protein genes, and capsular biosynthetic genes, in natural populations of *Streptococcus pneumoniae*. *Molecular Microbiology*, **5**, 2255–60.

Danner, D. B., Deich, R. A., Siseo, K. L. & Smith, H. O. (1980*a*). An eleven-basepair sequence determines the specificity of DNA uptake in *Haemophilus* transformation. *Gene*, **11**, 311–18.

Danner, D. B., Deich, R. A., Siseo, K. L. & Smith, H. O. (1980*b*). An eleven-basepair sequence determines the specificity of DNA uptake in *Haemophilus* transformation. *Gene*, **11**, 311–18.

Dempsey, J. A. F., Litaker, W., Madhure, A., Snodgrass, T. L. & Cannon, J. G. (1991). Physical map of the chromosome of *Neisseria gonorrhoeae* FA1090 with locations of genetic markers, including *opa* and *pil* genes. *Journal of Bacteriology*, **173**, 5476–86.

Dougherty, T. J., Koller, A. E. & Tomasz, A. (1980). Penicillin-binding proteins of penicillin-susceptible and intrinsically resistant *Neisseria gonorrhoeae*. *Antimicrobial Agents and Chemotherapy*, **18**, 730–7.

Dowson, C. G., Hutchison, A., Brannigan, J. A. *et al.* (1989). Horizontal transfer of penicillin-binding protein genes in penicillin-resistant clinical isolates of *Streptococcus pneumoniae*. *Proceedings of the National Academy Sciences USA*, **86**, 8842–6.

Dowson, C. G., Hutchison, A., Woodford, N., Johnson, A. P., George, R. C. & Spratt, B. G. (1990). Penicillin-resistant viridans streptococci have obtained altered penicillin-binding protein genes from penicillin-resistant strains of *Streptococcus pneumoniae*. *Proceedings of the National Academy Sciences USA*, **87**, 5858–62.

Dykhuizen, D. E. & Green, L. (1986). DNA sequence variation, DNA phylogeny, and recombination in *E. coli*. *Genetics*, **113**, s71 (Abstract).

Ewing, W. H. (1986). In *Edwards and Ewing's Identification of Enterobacteriaceae*, 4th edn, p. 181. New York: Elsevier Science Publishing, Inc.

Faruki, H. & Sparling, P. F. (1986). Genetics of resistance in a non-β-lactamase-producing gonococcus with relatively high-level penicillin resistance. *Antimicrobial Agents and Chemotherapy*, **30**, 856–60.

Feavers, I. M., Heath, A. B., Bygraves, J. A. & Maiden, M. C. J. (1992). Role of horizontal genetical exchange in the antigenic variation of the Class 1 outer membrane protein of *Neisseria meningitidis*. *Molecular Microbiology*, **6**, 489–95.

Frasch, C. E., Zollinger, W. D. & Poolman, J. T. (1985). Serotype antigens of *Neisseria meningitidis* and a proposed scheme for designation of serotypes. *Reviews of Infectious Diseases*, **7**, 504–10.

Gibbs, C. P., Reimann, B.-Y., Schultz, E., Kaufmann, A., Haas, R. & Meyer, T. F. (1989). Reassortment of pilin genes in *Neisseria gonorrhoeae* occurs by two distinct mechanisms. *Nature (London)*, **338**, 651–2.

Goodgal, S. H. (1982). DNA uptake in Haemophilus transformation. *Annual Reviews of Genetics*, **16**, 169–92.

Goodman, S. D. & Scocca, J. J. (1991). Factors influencing the specific interaction of Neisseria gonorrhoeae with transforming DNA. *Journal of Bacteriology*, **173**, 5921–3.

Haas, R. & Meyer, T. F. (1986). The repertoire of silent pilus genes in *Neisseria gonorrhoeae*: evidence for gene conversion. *Cell*, **44**, 107–15.

Haas, R., Velt, S. & Meyer, T. F. (1992). Silent pilin genes of *Neisseria gonorrhoeae* MS11 and occurrence of related hypervariant sequences among other gonococcal isolates. *Molecular Microbiology*, **6**, 197–208.

Hacker, J., Bender, L., Ott, M. *et al.* (1990). Deletions of chromosomal regions coding for fimbriae and hemolysins occur *in vitro* and *in vivo* in various extraintestinal *Escherichia coli* isolates. *Microbial Pathogenesis*, **8**, 213–25.

Hacker, J., Ott, M., Blum, G. *et al.* (1992). Genetics of *Escherichia coli* uropathogenicity: analysis of the O6:K15:H31 isolate 536. *Zentralblatt für Bakteriologie*, **276**, 165–75.

Halter, R., Pohlner, J. & Meyer, T. F. (1989). Mosaic-like organization of IgA protease genes in *Neisseria gonorrhoeae* generated by horizontal genetic exchange *in vivo*. *EMBO Journal*, **8**, 2737–44.

Heinemann, J. A. & Sprague, G. F., Jr. (1989). Bacterial conjugative plasmids mobilize DNA transfer between bacteria and yeast. *Nature (London)*, **340**, 205–9.

Jiang, X.-M., Neal, B., Santiago, F., Lee, S. J., Romana, L. K. & Reeves, P. R. (1991). Structure and sequence of the *rfb* (O antigen) gene cluster of *Salmonella* serovar *typhimurium* (strain LT2). *Molecular Microbiology*, **5**, 695–713.

Kapperud, G., Lassen, J., Dommarsnes, K. *et al.* (1989). Comparison of epidemiological marker methods for identification of *Salmonella typhimurium* isolates from an outbreak caused by contaminated chocolate. *Journal of Clinical Microbiology*, **27**, 2019–24.

Kimura, M. (1983). *The Neutral Theory of Molecular Evolution*. Cambridge: Cambridge University Press.

Knapp, S., Hacker, J., Jarchau, T. & Goebel, W. (1986). Large, unstable inserts in the chromosome affect virulence properties of uropathogenic *Escherichia coli* O6 strain 536. *Journal of Bacteriology*, **168**, 22–30.

Kroll, J. S. & Moxon, E. R. (1990). Capsulation in distantly related strains of *Haemophilus influenzae* type b: Genetic drift and gene transfer at the capsulation locus. *Journal of Bacteriology*, **172**, 1374–9.

Kuijper, E. J., van Alphen, L., Leenders, E. & Zanen, H. C. (1989). Typing of *Aeromonas* strains by DNA restriction endonuclease analysis and polyacrylamide gel electrophoresis of cell envelopes. *Journal of Clinical Microbiology*, **27**, 1280–5.

Lacks. S. & Greenberg, B. (1977). Complementary specificity of restriction endonucleases of *Diplococcus pneumoniae* with respect to DNA methylation. *Journal of Molecular Biology*, **114**, 153–68.

Laible, G., Spratt, B. G. & Hakenbeck, R. (1991). Inter-species recombinational events during the evolution of altered PBP 2x genes in penicillin-resistant clinical isolates of *Streptococus pneumoniae*. *Molecular Microbiology*, **5**, 1993–2002.

Le Minor, L. & Popoff, M. Y. (1987). Designation of *Salmonella enterica* sp. nov., nom. rev. as the type and only species of the genus *Salmonella*. *International Journal of Systematic Bacteriology*, **37**, 465–8.

Lee, S. J., Romana, L. K. & Reeves, P. R. (1992). Cloning and structure of group C1 O antigen (*rfb* gene cluster) from *Salmonella enterica* serovar montevideo. *Journal of General Microbiology*, **138**, 305–12.

Liu, D., Verma, N. K., Romana, L. K. & Reeves, P. R. (1991). Relationships among the *rfb* regions of *Salmonella* serovars A, B, and D. *Journal of Bacteriology*, **173**, 4814–19.

Loos, B. G., Bernstein, J. M., Dryja, D. M., Murphy, T. F. & Dickinson, D. P. (1989). Determination of the epidemiology and transmission of nontypeable *Haemophilus influenzae* in children with otitis media by comparison of total genomic DNA restriction fingerprints. *Infection and Immunity*, **57**, 2751–7.

Lujan, R., Zhang, Q., Sáez Nieto, J. A., Jones, D. M. & Spratt, B. G. (1991). Penicillin-resistant isolates of *Neisseria lactamica* produce altered forms of penicillin-binding protein 2 that arose by interspecies horizontal gene transfer. *Antimicrobial Agents and Chemotherapy*, **35**, 300–4.

Maiden, M. C. J., Suker, J., McKenna, A. J., Bygraves, J. A. & Feavers, I. M. (1991). Comparison of the class 1 outer membrane proteins of eight serological reference strains of *Neisseria meningitidis*. *Molecular Microbiology*, **5**, 727–36.

Mandrell, R. E. & Zollinger, W. D. (1989). Human immune response to meningococcal outer membrane protein epitopes after natural infection or vaccination. *Infection and Immunity*, **57**, 1590–8.

McGuinness, B., Barlow, A. K., Clarke, I. N., et al. (1990). Deduced amino acid sequences of class 1 protein (PorA) from three strains of *Neisseria meningitidis*. Synthetic peptides define the epitopes responsible for serosubtype specificity. *Journal of Experimental Medicine*, **171**, 1871–82.

McKinney, R. M., Kuffner, T. A., Bibb, W. F., et al. (1989). Antigenic and genetic variation in *Legionella pneumophila* serogroup 6. *Journal of Clinical Microbiology*, **27**, 738–42.

Mendelman, P. M., Caugant, D. A., Kalaitzoglou, G. et al. (1989). Genetic diversity of penicillin G-resistant *Neisseria meningitidis* from Spain. *Infection and Immunity*, **57**, 1025–9.

Meyer, T. F., Mlawer, N. & So, M. (1982). Pilus expression in *Neisseria gonorrhoea* involves chromosomal rearrangement. *Cell*, **30**, 45–52.

Meyer, T. F. & van Putten, J. P. M. (1989). Genetic mechanisms and biological implications of phase variation in pathogenic Neisseriae. *Clinical Microbiology Reviews*, **2** Suppl., S139–45.

Milkman, R. & Bridges, M. M. (1990). Molecular evolution of the *Escherichia coli* chromosome. III. Clonal frames. *Genetics*, **126**, 505–17.

Milkman, R. & Crawford, I. P. (1983). Clustered third-base substitutions among wild strains of *Escherichia coli*. *Science*, **221**, 378–80.

Morgenroth, A. & Duguid, J. P. (1968). Demonstration of different mutational sites controlling rhamnose fermentation in FIRN and non-FIRN rha⁻ strains of *Salmonella typhimurium*: an essay in bacterial archaeology. *Genetical Research (Cambridge)*, **11**, 151–69.

Morrison, D. A., Mannarelli, B. & Vijayakumar, M. N. (1982). Competence for transformation in *Streptococcus pneumoniae*: an inducible high-capacity system for genetic exchange. In *Microbiology–1982*, ed. D. Schlessinger, pp. 136–138. Washington, DC: American Society for Microbiology.

Munoz, R., Coffey, T. J., Daniels, M. *et al.* (1991). Intercontinental spread of a multiresistant clone of serotype 23F *Streptococcus pneumoniae*. *Journal of Infectious Diseases*, **164**, 302–6.

Musser, J. M., Bemis, D. A., Ishikawa, H. & Selander, R. K. (1987a). Clonal diversity and host distribution in *Bordetella bronchiseptica*. *Journal of Bacteriology*, **169**, 2793–803.

Musser, J. M., Granoff, D. M., Pattison, P. E. & Selander, R. K. (1985). A population genetic framework for the study of invasive diseases caused by serotype b strains of *Haemophilus influenzae*. *Proceedings of the National Academy of Sciences USA*, **82**, 5078–82.

Musser, J. M., Hewlett, E. L., Peppler, M. S. & Selander, R. K. (1986). Genetic diversity and relationships in populations of *Bordetella* spp. *Journal of Bacteriology*, **166**, 230–7.

Musser, J. M., Kroll, J. S., Moxon, E. R. & Selander, R. K. (1988a). Clonal population structure of encapsulated *Haemophilus influenzae*. *Infection and Immunity*, **56**, 1837–45.

Musser, J. M., Kroll, J. S., Moxon, E. R. & Selander, R. K. (1988b). Evolutionary genetics of the encapsulated strains of *Haemophilus influenzae*. *Proceedings of the National Academy of Sciences USA*, **85**, 7758–7762.

Musser, J. M., Mattingly, S. J., Quentin, R., Goudeau, A. & Selander, R. K. (1989). Identification of a high-virulence clone of type III *Streptococcus agalactiae* (group B *Streptococcus*) causing invasive neonatal disease. *Proceedings of the National Academy of Sciences USA*, **86**, 4731–5.

Musser, J. M., Rapp, V. J. & Selander, R. K. (1987b). Clonal diversity in *Haemophilus pleuropneumoniae*. *Infection and Immunity*, **55**, 1207–15.

Musser, J. M., Schlievert, P. M., Chow, A. W. *et al.* (1990). A single clone of *Staphylococcus aureus* causes the majority of cases of toxic shock syndrome. *Proceedings of the National Academy of Sciences, USA*, **87**, 225–9.

Neal, B. L., Tsiolis, G. C., Heuzenroeder, M. W., Manning, P. A. & Reeves, P. R. (1991). Molecular cloning and expression in *Escherichia coli* K-12 of chromosomal genes determining the O antigen of an *E. coli* O2:K1 strain. *FEMS Microbiology Letters*, **82**, 345–52.

Ochman, H. & Selander, R. K. (1984a). Evidence for clonal population structure in *Escherichia coli*. *Proceedings of the National Academy of Sciences, USA*, **81**, 198–201.

Ochman, H. & Selander, R. K. (1984b). Standard reference strains of *Escherichia coli* from natural populations. *Journal of Bacteriology*, **157**, 690–93.

Ochman, H. & Wilson, A. C. (1987). Evolution in bacteria: evidence for a universal substitution rate in cellular genomes. *Journal of Molecular Evolution*, **26**, 74–86.

Old, D. C. & Duguid, J. P. (1979). Transduction of fimbriation demonstrating common ancestry in FIRN strains of *Salmonella typhimurium*. *Journal of General Microbiology*, **112**, 251–9.

Olyhoek, T., Crowe, B. A. & Achtman, M. (1987). Clonal population structure of

Neisseria meningitidis serogroup A isolated from epidemics and pandemics between 1915 and 1983. *Reviews of Infectious Diseases*, **9**, 665–92.

Olyhoek, T., Crowe, B. A., Wall, R. A. & Achtman, M. (1988). Comparison of clonal analysis and DNA restriction analysis for typing of *Neisseria meningitidis*. *Microbial Pathogenesis*, **4**, 45–51.

Ørskov, F., Ørskov, I., Evans, D. J., Jr., Sack, R. B., Sack, D. A. & Wadström, T. (1976). Special *Escherichia coli* serotypes among enterotoxigenic strains from diarrhoea in adults and children. *Medical Microbiology and Immunology*, **162**, 73–80.

Piffaretti, J.-C., Kressebuch, H., Aeschbacher, M. *et al.*, (1989). Genetic characterization of clones of the bacterium *Listeria monocytogenes* causing epidemic disease. *Proceedings of the National Academy Sciences, USA*, **86**, 3818–22.

Poyart-Salmeron, C., Trieu-Cuot, P., Carlier, C. & Courvalin, P. (1990). The integration-excision system of the conjugative transposon Tn*1545* is structurally and functionally related to those of lambdoid phages. *Molecular Microbiology*, **4**, 1513–21.

Reeves, M. W., Evins, G. M., Heiba, A. A., Plikaytis, B. D. & Farmer, J. J., III (1989). Clonal nature of *Salmonella typhi* and its genetic relatedness to other salmonellae as shown by multilocus enzyme electrophoresis, and proposal of *Salmonella bongori* comb. nov. *Journal of Clinical Microbiology*, **27**, 313–20.

Reeves, P. (1991). The O antigen of Salmonella. *Today's Life Science*, **3**, 30–40.

Scocca, J. J. (1990). The role of transformation in the variability of the *Neisseria gonorrhoeae* cell surface. *Molecular Microbiology*, **4**, 321–7.

Seifert, H. S., Ajioka, R. S., Marchal, C., Sparling, P. F. & So, M. (1988). DNA transformation leads to pilin antigenic variation in *Neisseria gonorrhoeae*. *Nature (London)*, **336**, 392–5.

Selander, R. K., Caugant, D. A., Ochman, H., Musser, J. M., Gilmour, M. N. & Whittam, T. S. (1986). Methods of multilocus enzyme electrophoresis for bacterial population genetics and systematics. *Applied and Environmental Microbiology*, **51**, 873–84.

Selander, R. K., Caugant, D. A. & Whittam, T. S. (1987*a*). Genetic structure and variation in natural populations of *Escherichia coli*. In *Escherichia coli and Salmonella typhimurium cellular and molecular biology*, ed. F. C. Neidhardt, J. L. Ingraham, K. B. Low, B. Magasanik, M. Schaechter & H. E. Umbarger, pp. 1625–1648. Washington, D.C.: American Society for Microbiology.

Selander, R. K. & Levin, B. R. (1980). Genetic diversity and structure in *Escherichia coli* populations. *Science*, **210**, 545–7.

Selander, R. K., McKinney, R. M., Whittam, T. S. *et al.* (1985). Genetic structure of populations of *Legionella pneumophilia*. *Journal of Bacteriology*, **163**, 1021–37.

Selander, R. K., Musser, J. M., Caugant, D. A., Gilmour, M. N. & Whittam, T. S. (1987*b*). Population genetics of pathogenic bacteria. *Microbial Pathogenesis*, **3**, 1–7.

Smith, J. M., Dowson, C. G. & Spratt, B. G. (1991). Localized sex in bacteria. *Nature (London)*, **349**, 29–31.

Spratt, B. G. (1988). Hybrid penicillin-binding proteins in penicillin-resistant *Neisseria gonorrhoeae*. *Nature (London)*, **332**, 173–6.

Spratt, B. G., Bowler, L. D., Zhang, Q., Zhou, J. & Smith, J. M. (1991). Role of inter-species horizontal transfer of chromosomal genes in the evolution of penicillin resistance in pathogenic and commensal *Neisseria* species. *Journal of Molecular Evolution*, **34**, 115–25.

Stewart, G. J. & Carlson, C. A. (1986). The biology of natural transformation. *Annual Reviews of Microbiology*, **40**, 211–35.

Stoltzfus, A., Leslie, J. F. & Milkman, R. (1988). Molecular evolution of the *Escherichia coli* chromosome. I. Analysis of structure and natural variation in a previously uncharacterized region between *trp* and *tonB*. *Genetics*, **120**, 345–58.

Swanson, J., Robbins, K., Barrera, O. *et al.* (1987). Gonococcal pilin variants in experimental gonorrhoea. *Journal of Experimental Medicine*, **165**, 1344–57.

Trieu-Cuot, P., Carlier, C. & Courvalin, P. (1988). Conjugative plasmid transfer from *Enterococcus faecalis* to *Escherichia coli*. *Journal of Bacteriology*, **170**, 4388–91.

Trieu-Cuot, P., Carlier, C., Martin, P. & Courvalin, P. (1987). Plasmid transfer by conjugation from *Escherichia coli* to Gram-positive bacteria. *FEMS Microbiology Letters*, **48**, 289–94.

Trieu-Cuot, P., Gerbaud, G., Lambert, T. & Courvalin, P. (1985). *In vivo* transfer of genetic information between Gram-positive and Gram-negative bacteria. *EMBO Journal*, **4**, 3583–7.

Van Soolingen, D., Hermans, P. W. M., De Haas, P. E. W., Soll, D. R. & Van Embden, J. D. A. (1991). Occurrence and stability of insertion sequences in *Mycobacterium tuberculosis* complex strains: evaluation of an insertion sequence-dependent DNA polymorphism as a tool in the epidemiology of tuberculosis. *Journal of Clinical Microbiology*, **29**, 2578–86.

Verma, N. & Reeves, P. (1989). Identification and sequence of *rfbS* and *rfbE*, which determine antigenic specificity of group A and group D *Salmonellae*. *Journal of Bacteriology*, **171**, 5694–701.

Wang, J.-F., Morelli, G., Bopp, M. *et al.* (1991). Clonal and antigenic analyses of *Neisseria meningitidis* bacteria belonging to the ET37 complex isolated from Mali and elsewhere. In *Neisseriae – 1990*, ed. M. Achtman, P. Kohl, C. Marchal, G. Morelli, A. Seiler & B. Thiesen, pp. 141–146. Berlin: Walter de Gruyter & Co.

Wang, L., Romana, L. K. & Reeves, P. R. (1992). Molecular analysis of a *Salmonella enterica* group E1 *rfb* gene cluster: O antigen and the genetic basis of the major polymorphism. *Genetics*, **130**, 429–43.

Wedege, E., Caugant, D. A., Frøholm, L. O. & Zollinger, W. D. (1991). Character-ization of serogroup A and B strains of *Neisseria meningitidis* with serotype 4 and 21 monoclonal antibodies and by multilocus enzyme electrophoresis. *Journal of Clinical Microbiology*, **29**, 1486–92.

Whittam, T. S. & Wilson, R. A. (1988). Genetic relationships among pathogenic *Escherichia coli* of serogroup O157. *Infection and Immunity*, **56**, 2467–73.

Wilson, A. C., Ochman, H. & Prager, E. M. (1987). Molecular time scale for evolution. *Trends in Genetics*, **3**, 241–7.

Yother, J., McDaniel, L. S. & Briles, D. E. (1986). Transformation of encapsulated *Streptococcus pneumoniae*. *Journal of Bacteriology*, **168**, 1463–5.

MOLECULAR GENETIC APPROACHES TO UNDERSTANDING BACTERIAL PATHOGENESIS

B. BRETT FINLAY

Biotechnology Laboratory and the Departments of Biochemistry and Microbiology, University of British Columbia, Vancouver, BC, Canada V6T 1Z3

INTRODUCTION

The ability of bacteria to colonize a host, and under some conditions cause disease, usually depends on a complex series of functions requiring several bacterial factors (for reviews, see Finlay & Falkow, 1989; Mims, 1987). Even in the case of what appears to be disease caused by a single virulence factor such as a toxin, many bacterial loci in addition to that encoding the toxin are necessary to potentiate the disease. These loci encode products involved in adherence, regulation, bacterial growth, and other functions. The task of microbiologists studying bacterial pathogenesis is to identify and characterize these factors and to attempt to determine the contribution which each factor makes to virulence. Prior to the application of molecular genetic approaches, pathogenicity studies usually involved the biochemical purification of putative virulence factors and assay of the effects of these factors on animals or in *in vitro* model systems. These studies were difficult to interpret, since purified factors rarely potentiated the disease when removed from the pathogenic organism. The advent of molecular genetic tools has made dissection of virulence factors experimentally feasible *in situ* and has given scientists the unprecedented opportunity to create isogenic strains with which to study virulence factors.

There are traditionally two approaches to defining virulence factors (Fig. 1). One approach is to construct a gene bank of the pathogen of interest and to use this to clone a gene encoding a putative virulence factor, assuming a suitable screening method is available. The cloned locus and encoded product can be studied in molecular detail in a vector host. Mutations in the gene encoding the putative virulence factor are then constructed and the mutated gene moved into the virulent parental strain to replace its wild-type allele. The virulence of the mutants is assayed and compared with that of the parental strain. Alternatively, banks of mutants (usually constructed with a transposon) are screened for loss of a virulence property and appropriate mutants identified. The mutated gene can be cloned (using the transposon as a tag) by recombinant DNA techniques, and the parental gene subsequently

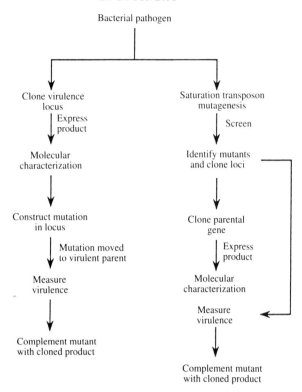

Fig. 1. Common approaches used to study bacterial virulence factors.

isolated and studied. The original transposon mutants can also be studied for virulence. In both cases, attempts should also be made to complement the defect in the mutant. If virulence is restored, the molecular biologist's 'Koch's postulates', as defined by Falkow (1988), have been fulfilled, and the cloned locus indeed encodes a virulence factor. The remainder of this review outlines some of the molecular genetic strategies and techniques used to study bacterial pathogenesis and provides examples of their applications. These techniques and new ones which will be described herein hold much in prospect for the investigation of bacterial pathogenesis.

The field of bacterial pathogenesis is large, and to try to encompass within this review all of the applications of molecular genetics to bacterial pathogenesis would be a task outside its intended scope. Instead a few examples, in several areas of bacterial pathogenesis with which the author is familiar, have been chosen to illustrate various aspects of the use of molecular genetics in the study of virulence factors. Most of the examples are drawn from the field of Gram-negative pathogens (especially enteric bacteria) for two reasons. Firstly, molecular genetic techniques are better defined for these pathogens and secondly, the author's knowledge is largely in the Gram-negative arena. Molecular genetic techniques in Gram-positive pathogens are discussed by Foster (this symposium). Similar principles and techniques with special adaptations apply to this group of pathogens.

MOLECULAR GENETIC STRATEGIES

Recombinant DNA technology

The ability to cleave DNA with restriction enzymes, to ligate the cleaved foreign DNA into a vector, and to propagate the DNA in a suitable host (collectively known as recombinant DNA technology) has affected virtually every aspect of the biological sciences. The field of bacterial pathogenesis has been no exception. This powerful technique has been extremely useful in studying bacterial virulence factors. The ability to clone out the gene encoding a virulence factor, and to study the encoded factor in a "clean" background has greatly facilitated work on bacterial pathogenesis. Since other virulence factors are usually lacking in the vector host which receives the cloned gene encoding the virulence factor under investigation, interpretation of data is usually less complex. Continual improvements in plasmid and other cloning vectors, restriction enzymes, ligations, and transformation frequencies have made this approach feasible for many bacterial pathogens. Additionally the use of electroporation as a technique to introduce DNA into non-transformable species (Fiedler & Wirth, 1988; Miller, Dower & Tompkins, 1988) has broadened the scope of recombinant DNA technology, and has greatly increased the number of feasible experiments.

The literature abounds with examples of the cloning of genes encoding virulence factors and, to a lesser extent, of expression of these in vector hosts such as *Escherichia coli*. One such example is the cloning of the gene encoding invasin from *Yersinia pseudotuberculosis* into *E. coli* (Isberg & Falkow, 1985). *Yersinia* species are capable of entering host eucaryotic cells. This entry is thought to contribute to virulence. In contrast, laboratory strains of *E. coli* do not enter host cells. Isberg & Falkow (1985) constructed a *Y. pseudotuberculosis* gene bank in *E. coli* and then screened this bank for recombinant *E. coli* which now had the capacity to enter cultured epithelial cells. This research relied upon the assumption that the *Yersinia* gene would be expressed and functional in *E. coli*, and the fact that a powerful enrichment screening procedure was available to select the recombinants. In this example, extracellular bacteria were killed with gentamicin, and the viable intracellular organisms isolated, successfully enriching for recombinants which were able to express invasin. This study emphasizes the need for a suitable *in vitro* screening method (hopefully a positive enrichment) and the use of an appropriate vector and host.

Often whole operons that are needed to express a functional organelle, such as a pilus, can be cloned into *E. coli*, providing an ideal setting to study the structure/function of the organelle, including its expression, assembly, and role in pathogenesis. Work by Normark and coworkers on uropathogenic *E. coli* provide an excellent example of this approach (Normark *et al.*, 1983; Uhlin *et al.*, 1985). Using this system they have systematically dissected the P-pilus structure, defining the role of the various genetic loci and their gene products in pilus assembly and adherence to epithelial cell surfaces (see for further discussion Smyth & Smith, this symposium).

Although molecular genetic cloning of a gene encoding a virulence factor greatly facilitates study of the cloned gene and its gene product, it also creates problems, since it removes the virulence factor from its normal bacterial host setting. This often has consequences if regulation, expression, or biosynthesis are being studied. It can create problems in studying virulence, since other factors may also be required for expression of full virulence which are not usually present in the recombinant organism. Only rarely does the cloning of the gene for a virulence trait cause the recipient recombinant strain to become pathogenic. If two genes, which are not located in close proximity, are needed for expression of the virulence factor, other recombinant methods need to be used. One such example can be found with *Salmonella typhimurium* and *Salmonella choleraesuis* invasion. These pathogens, like *Yersinia* species, can enter eucaryotic cells and are considered invasive. However, when an approach similar to that employed for *Y. pseudotuberculosis* was used for these *Salmonella* species, invasive recombinants could not be identified, presumably because of the complexity of the *Salmonella* invasion loci. Instead, other methods such as transposon mutagenesis and molecular genetic cloning into non-invasive *Salmonella* have been used to identify and characterize invasion loci from these species (Finlay *et al.*, 1988*a*; Galen & Curtiss, 1989).

Another problem associated with the cloning of virulence factor genes is that they are often not expressed in the bacterial host strain used as the recipient of the cloned genes. Many possible manipulations can be done to overcome this problem, including cloning of the gene into expression vectors, placing the gene under the control of an inducible promoter, altering the gene to utilize preferential codons, or possibly choosing a different host which may be more related to the pathogen, or a non-virulent derivative of the pathogen under investigation (see below and Foster, this symposium).

Transposon mutagenesis

The identification of mobile genetic elements that can insert non-specifically into foreign DNA has greatly facilitated the study of bacterial pathogenesis. Transposons allow genetic lesions to be constructed that eliminate ('knock out') expression of the gene into which they are inserted. The added advantage that transposon mutagenesis has over classical mutagenesis methods is that the transposons contain an identifiable marker, often an antibiotic resistance gene, that allows identification and cloning of the interrupted gene and the corresponding inserted transposon. A specific assay or screening protocol must exist to test the transposon mutants in order to use this method of mutagenesis successfully. This assay can be as complex as measuring loss of virulence in an animal model of infection or in an *in vitro* model system. However, the assay is usually more specific, such as assaying for loss of enzymic activity, cytotoxicity, adherence or invasion.

One excellent example of the use of transposons to identify virulence factors utilized the transposon Tn5 and *Bordetella pertussis*, the causative agent of whooping cough. Little was known about the genetic factors contributing to the pathogenesis of this organism until Weiss *et al.* (1983) saturated its chromosome with Tn5 and screened for transposon mutants which exhibited loss of various virulence-associated traits. By using several screening methods, they identified transposon insertion mutants lacking the haemolysin, pertussis toxin, and filamentous haemagglutinin. They also identified a regulatory gene for virulence which coordinately regulates expression of these and several other virulence factors. This ground-breaking research provided the foundation for a wealth of studies that have gone on to characterize, at the molecular level, these virulence factors and the genes that encode them, elevating *B. pertussis* from a pathogen whose virulence factors were not well studied to one which is now extensively characterized.

A similar method has been used successfully to identify factors in *S. typhimurium* which are needed for survival inside macropages (Fields *et al.*, 1986). This method involved screening 9516 Tn10 mutants (individually!) for survival inside macrophages. Of these, 83 had decreased survival rates and were also avirulent in mice. Characterization of these mutants has yielded much information on how this organism survives inside the macrophage, including resistance mechanisms to the bactericidal activity of defensins, and led to the identification of a global regulatory locus (*phoP/ phoQ*) which controls expression of several genes needed for intracellular survival. This study underscores the volume of work necessary when using transposon mutagenesis, since each mutant must be screened individually, in contrast to the positive selection that is often available using recombinant DNA techniques.

Since transposons encode a transposase which is necessary for the trans-position mechanisms, they also possess the capacity to insert additional copies after the initial mutational insertion, or, in some cases to excise and reinsert elsewhere. To prevent this additional genetic mutagenesis and associated rearrangements, mini-transposons have been constructed that contain an antibiotic resistance marker flanked by the ends of the transposon which are needed for successful transposition, but no transposase gene. Transposase activity can be supplied in *trans* by a gene carried on another plasmid. Thus, when the mini-transposon inserts, it becomes 'locked in' and cannot transpose (Elliott & Roth, 1988; Hughes & Roth, 1988).

Many other specialized transposons have been developed. One such transposon, Tn*phoA*, has been used extensively in the study of virulence factors. This transposon encodes a modified alkaline phosphatase gene which lacks its signal sequence (Manoil & Beckwith, 1985). Insertion of the Tn*phoA* transposon downstream of and in-frame with a signal sequence results in export of alkaline phosphatase outside of the bacterial cytoplasm, giving rise to a colour change in colonies when bacteria are grown on agar plates containing an appropriate substrate. This method allows the investigator to identify insertions in genes which encode a signal sequence, thereby identifying genes encoding membrane, periplasmic, and secreted products. Since most virulence factors are expressed on the bacterial surface where they can interact with host factors, Tn*phoA* has been invaluable for identifying products involved in bacterial pathogenesis.

For example, this procedure has been used to identify loci in *S. cholerae-suis* that are necessary for this organism to penetrate through polarized host epithelial cells (Finlay *et al.*, 1988*a*). Of the 626 Pho$^+$ insertion mutants screened (representing about 2% of the transposon mutants, most of which were Pho$^-$), 42 were unable to penetrate through polarized epithelial monolayers cultured on permeable supports. Several of these non-penetrating mutants showed reduced virulence for mice or were avirulent on the basis of oral LD$_{50}$ values. One disadvantage of Tn*phoA* (and most other transposons) is that they do not insert randomly. In the example given, of the 42 Tn*phoA* gene fusions identified in *S. choleraesuis*, 20 were within the same gene, while in the remainder Tn*phoA* had inserted at unique sites. Another problem with Tn*phoA* mutants is that a foreign protein (alkaline phosphatase) is now expressed at the surface of the pathogen and can affect virulence and other pathogenic traits through alterations in the bacterial surface composition and surface charge. However, the powerful selection afforded by Tn*phoA* far outweighs these disadvantages in identifying virulence factors, as illustrated by the numerous investigations that have used this technique.

Suicide plasmids

Several techniques exist to deliver transposons to virulent organisms. One of the more common methods utilizes vectors called suicide plasmids. These plasmids have special requirements for replication that are only present in the host bacterium (examples can be found in Miller & Mekalanos, 1988, and Weiss *et al.*, 1983). Once a suicide plasmid containing a transposon is moved (usually through conjugation) into a recipient organism, the trans-poson inserts a copy into the recipient DNA. The suicide plasmid cannot replicate in the recipient organism, and it is soon diluted out. Appropriate antibiotic selection allows identification of transposon mutants, without the presence of the delivery vehicle.

Suicide plasmids can also be used to deliver genes that undergo homolo-gous recombination in the recipient, replacing the resident gene. To ulti-mately demonstrate that a particular gene encodes a virulence factor, that gene needs to be deleted or mutated in an otherwise virulent strain, and loss of virulence demonstrated. This is usually achieved by constructing a mutant allele and moving it onto a suicide plasmid, followed by delivery into a virulent parental strain. Recombination via a *recA*-mediated event occurs, thereby replacing the parental locus with an altered one which can be selected for. Such recombinants can then be tested for virulence. As mentioned previously, when this mutation is complemented in *trans* by cloned parental DNA, virulence should be restored in order to fulfil the molecular Koch's postulates (Falkow, 1988).

Conjugation and mapping

A more classical approach used to identify virulence factors utilizes conju-gation to mobilize chromosomally located genes encoding virulence factors into avirulent hosts. This technique has been used mainly with *E. coli* and *Salmonella* species. Obtaining a set of strains posessing well distributed HFR (high frequency of recombination) sites to use as donors is usually difficult. Although a set of these strains exists for *E. coli*, only a poor collection exists for *S. typhimurium* (Sanderson & Roth, 1988). Recently an approach has been developed which uses a transposon that also encodes an origin of transfer. Insertion of this transposon randomly into the chromo-some creates several sites which can then be mobilized if the transfer functions are supplied in *trans* (Simon, 1984).

Once a virulence factor is identified, it is often useful to map its position on the chromosome. This information can lead to clues about the gene product's function, and may reveal whether the gene in question is clustered near those for other mapped traits, as is often the case with virulence factor genes. Although classical mapping using HFR strains and conjugation has been performed (Sanderson & Roth, 1988), this procedure is tedious. New molecular techniques are now being developed which facilitate gene map-ping. For example, a series of transposon insertions distributed equally around the *S. typhimurium* chromosome, each of which contains a locked-in Mu*d*-P22 prophage, can be used to amplify about 3 minutes of the chromo-some adjacent to the prophage (Youderian *et al.*, 1988). DNA is isolated from each of these transposon insertion mutants, immobilized on membrane supports and probed with DNA from the virulence gene. This technique allows the investigator to map a gene to within 3 min on the chromosomal map. However, this technique is currently limited to *S. typhimurium*.

At present the above mapping techniques are only applicable to a few pathogens. A more general mapping technique is to construct a physical map of the chromosome using pulse-field gel electrophoresis, a technique that allows resolution of large chromosomal DNA fragments (Smith *et al.*, 1987). Restrictions enzymes which cut infrequently are used, and the resulting large DNA fragments resolved using this gel system. Construction of the map is done utilising methods similar to those used to generate conventional restriction maps, including double digests, and probing with known DNA fragments. Once a map of the chromosome of the pathogen is constructed, Southern hybridization analysis with labelled probe DNA from the virulence factor gene can be used to identify the restriction fragment bearing the virulence-associated gene and thus the map position of that gene.

Other recombinant techniques

DNA sequencing has generated much information about many virulence factors. Besides allowing determination of the predicted primary structure of the virulence factor, it can often yield clues about the function of the gene product if amino acid sequence homology is found with better characterized proteins. Amino acid sequence homology searches often help define families of related proteins. Discrete areas of amino-acid sequence homology may provide clues about the functions of various domains, including prediction of the cellular location of the final product, and the extent of conservation between organisms. Examples of this can be found in the two-component regulatory systems (Miller, Mekalanos & Miller, 1989; see also Dorman & Ní Bhriain, this symposium) and in the haemolysin family (Welch, 1991). However, often DNA homology searches do not provide clues about the function of the predicted gene product. Nevertheless, once a DNA sequence is stored in a data bank, other investigators often find, at a later date, related sequences which then provide significant clues about the structure and function of the original locus.

Southern hybridization analysis can provide information about the distribution of virulence factors, and the extent of DNA sequence conservation. This distribution can be between bacterial species, or within a particular species. An example of intraspecies distribution was found with hybridization studies done on various *Yersinia* species probed with invasion genes (*inv* and *ail*) from *Y. enterocolitica* (Miller *et al.*, 1989). The results of this study indicated that only virulent, invasive strains of *Yersinia* contain the *ail* invasion gene, implicating it in virulence. In contrast, the other invasion gene, *inv*, was not well correlated with virulence, although there were some

DNA restriction fragment patterns associated with *inv* which correlated with tissue culture adhesion.

The recent advent of polymerase chain reaction (PCR) is revolutionizing molecular biology. Many new applications are being discovered, but it is difficult to predict the potential future uses of this technique as a tool in molecular genetic investigation of bacterial pathogenesis. Its greatest power is in amplifying trace amounts of DNA. Thus, it has many basic uses in recombinant DNA work. Because of the amplification, it will almost certainly become essential in advancing the field of diagnostics for pathogenic organisms, especially for those that are difficult to culture. An excellent example of the use of PCR to identify an uncultured pathogen was recently published by Relman *et al.* (1990). These workers employed universal 16S rRNA probes to amplify DNA sequences in tissue infected with the causative agent of bacillary angiomatosis. Using this method, they were able to establish without culturing that the pathogen was unique, but related to other intracellular pathogens such as *Rickettsia* species and *Brucella abortus*.

FUTURE PROSPECTS

The prospects for studying bacterial pathogenesis using molecular genetics are indeed promising. As improvements in cloning vectors, transposons, and vector hosts continue, the principles discussed above can be applied to other previously uncharacterized pathogens. Additionally, the use of avirulent strains as cloning hosts provides new avenues for research. For example, the molecular identification of a *S. typhimurium* invasion locus was achieved by using a non-invasive *S. typhimurium* isolate as a recipient for a DNA bank made from an invasive strain (Galan & Curtiss, 1989). Isolation of the most invasive recombinants led to the identification of this *S. typhimurium* invasion locus. Related non-virulent strains (commensals) can also be used as recipients to express virulence factors. For example, internalin is a product which is required for invasion of *Listeria monocytogenes* (a Gram-positive organism) into eucaryotic cells (Gaillard *et al.*, 1991). The gene encoding internalin, *inlA*, was identified by transposon mutagenesis, and cloned. The cloned gene was introduced into *Listeria innocua*, a non-invasive, avirulent organism, and found to confer the ability to enter cultured epithelial cells, thereby confirming the role of internalin in invasion (see also Foster, this symposium).

Perhaps one of the most exciting areas of study in bacterial pathogenesis is examination of the interactions that occur between the host cell and pathogen, including the signals that are triggered in both the bacterium and host cell in response to these interactions ('cross-talk'), and the regulation of expression of bacterial virulence factors that are induced or repressed by host cells or the micro-environment surrounding host cells (for a review see Miller *et al.*, 1989; see also Dorman & Ní Bhriain, this volume). Several new genetic tools have been developed that allow the investigator to insert 'reporter genes' into those encoding virulence factors and thus study regulation of these loci. For example, a Tn5 derivative, Tn5-VB32 (Bellofatto, Shapiro & Hodgson, 1984), has been used to identify *S. typhimurium* genes which are induced when the bacterium binds to the host epithelial cell. This transposon encodes a promoterless kanamycin resistance gene at one end of the transposon. If the transposon inserts downstream of an active promoter, kanamycin resistance is expressed. If the kanamycin-sensitive transposon mutants are collected and exposed to epithelial cells, Tn5-VB32 insertions downstream of the induced genes should confer kanamycin resistance. By adding kanamycin, a bactericidal antibiotic, the only survivors of this treatment are mutants which contain insertions downstream of the induced promoters. This study identified three inducible loci, two of which are induced by bacterial binding to any solid support, and one which is induced only when bacteria adhere to eucaryotic cell surfaces (B. B. Finlay and coworkers, unpublished data).

TnphoA has been used extensively for identification of virulence factors. Alkaline phosphatase activity can be assayed, giving a measurement of the level of induction of the promoter of the fused gene, although this assay is not as convenient as others. A new vector has recently been constructed that utilizes homologous recombination to insert a chloramphenicol acetyltransferase (CAT) gene into the TnphoA DNA, disrupting the *phoA* gene, but fusing *cat* into the locus originally disrupted by TnphoA (Knapp & Mekalanos, 1988). CAT assays, which are more convenient than alkaline phosphatase assays, can then be used to accurately measure regulation of the virulence factors under various conditions.

lacZ gene fusions have traditionally been used to study regulation of bacterial transcription and translation. Pollack, Straley & Klempner (1986) used a strain with a calcium-regulated *lacZ* gene fusion as a probe to measure the calcium levels in the phagolysosome environment containing intracellular *Yersinia pestis*. Recently the β-galactosidase assay has been modified such that enzymic activity can be measured from bacteria that are inside host cells by using a fluorescent substrate (B. B. Finlay and coworkers, unpublished data). This substrate significantly improves the sensitivity of the β-galactosidase assay. The technique has been applied to *S. typhimurium* with known *lacZ* gene fusions for use as bioprobes of the intracellular vacuolar environment to measure such parameters as pH and oxygen, iron, magnesium, glucose and mannose levels. This information has allowed definition of several aspects of the intracellular environment.

Other promising molecular genetic techniques have recently been developed which will probably have several applications in studying bacterial pathogenesis. The production of monoclonal antibodies in *E. coli* will greatly enhance the utility of these powerful immunological reagents (Huse *et al.*, 1989). Similarly, the development of an expression system for expression of random peptide sequences on the surface of a bacteriophage has many potential applications in the field of bacterial pathogenesis (Scott & Smith, 1990). These include identifying host receptors, and uses in the development of vaccines, as well as many uses that are unforeseen at present.

Although molecular genetic techniques have greatly advanced the investigation of bacterial pathogenesis, the parallel development of animal models and of *in vitro* assay systems has been essential to this work. The use of cultured cells to study bacterial adherence and invasion has been very rewarding. Polarized epithelial cell monolayers are useful for related studies (Finlay *et al.*, 1988*b*). Various transgenic animals, serving as host model systems, also offer promise in the future for research on bacterial pathogenesis.

The field of bacterial pathogenesis has blossomed by applying molecular genetics to various problems. The tools are now available for dissecting, at a molecular level, virulence factors and the genes encoding them. The future holds additional promise, as imagination and creativity are used to design and execute experiments which define more precisely how bacteria cause disease. Undoubtedly many new therapeutic agents will also arise from these studies.

ACKNOWLEDGEMENTS

Research in the author's laboratory is supported by the Medical Research Council of Canada, the British Columbia Health Care Research Foundation, the Canadian Bacterial Diseases Network Centre of Excellence, and the Howard Hughes International Research Scholars Programme.

REFERENCES

Bellofatto, V., Shapiro, L. & Hodgson, D. A. (1984). Generation of a Tn5 promoter probe and its use in the study of gene expression in *Campylobacter crescentus*. *Proceedings of the National Academy of Sciences, USA*, **81**, 1035–9.

Elliott, T. & Roth, J. R. (1988). Characterization of Tn*10d*-Cam: a transposition-defective Tn*10* specifying chloramphenicol resistance. *Molecular and General Genetics*, **213**, 332–8.

Falkow, S. (1988). Molecular Koch's postulates applied to microbial pathogenicity. *Reviews of Infectious Diseases*, **10**, Suppl. 2, S274–8.

Fiedler, S. & Wirth, R. (1988). Transformation of bacteria with plasmid DNA by electroporation. *Analytical Biochemistry*, **170**, 38–44.

Fields, P. I., Swanson, R. V., Haidaris, C. G. & Heffron, F. (1986). Mutants of *Salmonella typhimurium* that cannot survive within the macrophage are virulent. *Proceedings of the National Academy of Sciences, USA*, **83**, 5189–93.

Finlay, B. B. & Falkow, S. (1989). Common themes in microbial pathogenicity. *Microbiological Reviews*, **53**, 210–30.

Finlay, B. B., Starnbach, M. N., Francis, C. L., Stocker, B. A., Chatfield, S., Dougan, G. & Falkow, S. (1988*a*). Identification and characterization of Tn*phoA* mutants of *Salmonella* that are unable to pass through a polarized MDCK epithelial cell monolayer. *Molecular Microbiology*, 2, 757–66.

Finlay, B. B., Gumbiner, B. & Falkow, S. (1988*b*). Penetration of *Salmonella* through a polarized Madin–Darby canine kidney epithelial cell monolayer. *Journal of Cell Biology*, 107, 221–30.

Gaillard, J.-L., Berche, P., Frehel, C., Gouin, E. & Cossart, P. (1991). Entry of *L. monocytogenes* into cells is mediated by internalin, a repeat protein reminiscent of surface antigens from Gram-positive cocci. *Cell*, 65, 1127–41.

Galan, J. E. & Curtiss, R., III. (1989). Cloning and molecular characterization of genes whose products allow *Salmonella typhimurium* to penetrate tissue culture cells. *Proceedings of the National Academy of Sciences, USA*, 86, 6383–7.

Hughes, K. T. & Roth, J. R. (1988). Transitory *cis* complementation: a method for providing transposition functions to defective transposons. *Genetics*, 119, 9–12.

Huse, W. D., Sastry, L., Iverson, S. A., Kang, A. S., Alting, M. M., Burton, D. R., Benkovic, S. J. & Lerner, R. A. (1989). Generation of a large combinational library of the immunoglobulin repertoire in phage lambda. *Science*, 246, 1275–81.

Isberg, R. R. & Falkow, S. (1985). A single genetic locus encoded by *Yersinia pseudotuberculosis* permits invasion of cultured animal cells by *Escherichia coli* K-12. *Nature*, 317, 262–4.

Knapp, S. & Mekalanos, J. J. (1988). Two *trans*-acting regulatory genes (*vir* and *mod*) control antigenic modulation in *Bordetella pertussis*. *Journal of Bacteriology*, 170, 5059–66.

Manoil, C. & Beckwirth, J. (1985). Tn*phoA*: a transposon probe for protein export signals. *Proceedings of the National Academy of Sciences, USA*, 82, 8129–33.

Miller, J. F., Dower, W. J. & Tompkins, L. S. (1988). High voltage electroporation of bacteria: genetic transformation of *Campylobacter jejuni* with plasmid DNA. *Proceedings of the National Academy of Sciences, USA*, 85, 856–60.

Miller, J. F., Mekalonos, J. J. & Falkow, S. (1989). Coordinate regulation and sensory transduction in the control of bacterial virulence. *Science*, 243, 916–22.

Miller, V. L. & Mekalanos, J. J. (1988). A novel suicide vector and its use in construction of insertion mutations: osmoregulation of outer membrane proteins and virulence determinants in *Vibrio cholerae* requires *toxR*. *Journal of Bacteriology*, 170, 2575–83.

Miller, V. L., Farmer, J. J., Hill, W. E. & Falkow, S. (1989). The *ail* locus is found uniquely in *Yersinia enterocolitica* serotypes commonly associated with disease. *Infection and Immunity*, 57, 121–31.

Mims, C. A. (1987). *The Pathogenesis of Infectious Disease*. Academic Press: London.

Normark, S., Lark, D., Hull, R., Norgren, M., Båga, M., O'Hanley, P., Schoolnik, G. & Falkow, S. (1983). Genetics of digalactoside-binding adhesin from a uropathogenic *Escherichia coli* strain. *Infection and Immunity*, 41, 942–9.

Pollack, C., Straley, S. C. & Klempner, M. S. (1986). Probing the phagolysosomal environment of human macrophages with a Ca^{2+} responsive operon fusion in *Yersinia pestis*. *Nature*, 322, 834–6.

Relman, D. A., Loutit, J. S., Schmidt, T. M., Falkow, S. & Tompkins, L. S. (1990). The agent of bacillary angiomatosis. An approach to the identification of uncultured pathogens. *New England Journal of Medicine*, 323, 1573–80.

Sanderson, K. E. & Roth, J. R. (1988). Linkage map of *Salmonella typhimurium*, edition VII. *Microbiological Reviews*, 52, 485–532.

Scott, J. K. & Smith, G. P. (1990). Searching for peptide ligands with an epitope library. *Science*, **249**, 386–90.

Simon, R. (1984). High frequency mobilization of Gram-negative bacterial replicons by the *in vitro* constructed Tn5-Mob transposon. *Molecular and General Genetics*, **196**, 413–20.

Smith, C. L., Econome, J. G., Schutt, A., Klco, S. & Cantor, C. R. (1987). A physical map of the *Escherichia coli* K12 genome. *Science*, **236**, 1448–53.

Uhlin, B.-E., Båga, M., Göransson, M., Lindberg, F. P., Lund, B., Norgren, M. & Normark, S. (1985). Genes determining adhesin formation in uropathogenic *Escherichia coli*. *Current Topics in Microbiology and Immunology*, **118**, 163–78.

Weiss, A. A., Hewlett, E. L., Myers, G. A. & Falkow, S. (1983). Tn5-induced mutations affecting virulence factors of *Bordetella pertussis*. *Infection and Immunity*, **42**, 33–41.

Welch, R. A. (1991). Pore-forming cytolysins of Gram-negative bacteria. *Molecular Microbiology*, **5**, 521–8.

Youderian, P., Sugiono, P., Brewer, K. L., Higgins, N. P. & Elliott, T. (1988). Packaging specific segments of the *Salmonella* chromosome with locked-in Mud-P22 prophages. *Genetics*, **118**, 581–92.

MOLECULAR AND CELLULAR BIOLOGY OF EPITHELIAL INVASION BY *SHIGELLA FLEXNERI* AND OTHER ENTEROINVASIVE PATHOGENS

P. J. SANSONETTI

Unité de Pathogénie Microbienne Moléculaire, U199 INSERM, Institut Pasteur, 25–28 rue du Dr Roux, F–75724 PARIS Cedex 15

INTRODUCTION

The successful colonization of host mucosal surfaces by pathogenic bacteria is the consequence of a complex series of events involving both bacterial and host-cell products. Three types of interactions can be considered (Finlay, 1990). Association involves interaction of the pathogen with host products such as mucus or matrix proteins which are secreted by mucosal cells. Adhesion involves highly specific interactions between bacterial adhesins and receptors which are located on the apical surface of the epithelium. Depending on the ligand and receptor that interact, adhesion may proceed to a further step, invasion of epithelial cells. It is still unclear, at the molecular level, how bacterial adherence either leads to extracellular colonization or to epithelial invasion (Isberg, 1991). Analysis of model systems, such as invasion of epithelial cells by *Yersinia* (Isberg, 1989) or entry of *Bordetella pertussis* into macrophages (Relman *et al.*, 1990), in comparison with various models of extracellular colonization, indicate that binding of the bacteria to polysaccharidic domains of glycolipids or glyco-proteins mediate extracellular colonization, whereas binding to cell ad-hesion proteins such as integrins mediate entry. Bacterial enteric infections offer remarkable models for studying these processes. It must be stressed, however, that most of the data available on invasion have been obtained in *in vitro* assay systems. These data have to be validated in more definitive assays including experimental infections in animals whenever possible.

Among the numerous bacterial enteric pathogens, only a limited number can be considered as enteroinvasive, i.e. bacteria that have the ability to penetrate into intestinal epithelial cells in order to develop their pathogenic potential. However, this concept of invasion is too simplistic. It must be extended to the whole mucosa and not simply restricted to invasion of the epithelial layer. Therefore, invasion not only applies to interactions be-tween bacteria and enterocytes, but also to interactions between bacteria and the phagocytic cells present in the *lamina propria*. Enteroinvasive organisms can be broadly grouped into three overlapping categories or pathotypes based on clinical, histopathological and experimental findings. Certain microorganisms cause a dysenteric syndrome, due to their capacity to efficiently invade the epithelial layer of the intestinal mucosa. Others simply cross the epithelial layer in order to reach deeper tissues. They may cause diarrhoea and/or slight dysentery as the invasive process goes on. Finally, certain invasive organisms do not cause symptoms while crossing the epithelium. Shigellosis represents the paradigm of an infection which primarily affects the intestinal epithelial layer. Salmonellosis and yersiniosis, in most of the cases, correspond to diseases in which the microorganisms simply cross the epithelial barrier in order to reach deeper tissues. The *Shigella* model has been a major research topic of our group and will be developed with reference to the other models whenever physiopathological differences may account for significant differences between the disease processes.

SHIGELLOSIS, A PARADIGM OF INTESTINAL INVASION

Bacillary dysentery is an invasive disease of the human colon. It is mostly prevalent in tropical zones, especially in the developing world where *Shigella flexneri* causes the endemic form of the· disease and *Shigella dysenteriae* 1 the epidemic form. Infants and children are the major victims of this disease which is mostly transmitted via the faeco-oral route in areas where hygiene standards and sanitation are insufficient. It is characterized in its typical form by fever, painful abdominal cramps, tenesmus and bloody, mucopurulent stools (LaBrec *et al.*, 1964). Invasion encompasses two major steps: invasion of the cells which form the epithelial lining of the colon (Takeuchi *et al.*, 1965), and invasion of the resident phagocytes within the connective tissues that constitute the *lamina propria* of intestinal villi. The overall process leads to confluent foci of strong inflammation which cause abscesses and ulcerations (Takeuchi, Formal & Spring, 1968). In addition, systemic complications such as the haemolytic uraemic syndrome (HUS) can be observed (Koster *et al.*, 1978). The role of Shiga toxin, a potent cytotoxin produced in high quantity by *S. dysenteriae* 1, in the pathogenesis of HUS is still debated (O'Brien & Holmes, 1987). Its role in enhancing the severity of the invasion process by causing lesions of the capillaries in intestinal villi has also been suggested (Fontaine, Arondel & Sansonetti, 1988).

In vitro assays provide a simplified view of the pathogenesis of shigellosis, allowing a detailed study of the molecular and cellular basis of cell invasion. Several aspects can be explored such as entry of the bacteria into cells, intracellular multiplication, intracellular and cell to cell spread of the bacteria. These are key properties which have to be assessed by testing mutants in more definitive virulence assays such as the guinea pig keratoconjunctivitis assay or Sereny test (Sereny, 1957), the rabbit ligated ileal

loop assay and the simian model in which rhesus monkeys infected intragastrically with *Shigella* develop a dysenteric syndrome (Takeuchi *et al.*,
1968).

ENTRY INTO EPITHELIAL CELLS

Nature of the entry process

Transmission electron microscopy (TEM) performed on continuous lines of
epithelial cells, such as HeLa cells, after infection by *S. flexneri*, has shown
that the bacteria are internalized via an active endocytic process (Hale &
Bonventre, 1979; Hale, Morris & Bonventre, 1979) which is inhibited by
cytochalasins (Tanenbaum, 1978). This indicates that *S. flexneri* is internalized via phagocytosis. In professional phagocytes, microfilaments of polymerized actin, and actin-binding proteins such as myosin form a molecular
complex which is the motor of the phagocytic process (Sheterline, Rickard
& Richards, 1984; Stendahl *et al.*, 1980).

The involvement of actin microfilaments of the invaded cell during the
phagocytic process triggered by *S. flexneri* has been demonstrated using
fluorescent probes such as NBD phallacidin (Barak *et al.*, 1980) which binds
specifically to filamentous actin (F-actin) and monoclonal antibodies
directed against actin-binding proteins, especially myosin. F-actin and
myosin accumulate as aggregates subjacent to the cytoplasmic membrane in
areas of the cell surface that interact with invasive shigellae (Clerc &
Sansonetti, 1987). It can be concluded that *S. flexneri* has the capacity to
induce epithelial cells to perform a phagocytic process similar to that
observed in professional phagocytes. Evidence based on similar experiments indicate that *Yersinia* (Finlay & Falkow, 1988) and *Salmonella* also
invade cultures of epithelial cells by a process of directed phagocytosis
(Finlay, Ruschkowski & Dedhar, 1991).

Identification of the genes and products which trigger the entry process

Presence of a large plasmid of 220 kb is necessary for expression of the
invasive phenotype of *S. flexneri* (Sansonetti, Kopecko & Formal, 1982;
Sansonetti *et al.*, 1983). Cloning of the genes necessary for the induction of
phagocytosis by epithelial cells was first achieved in a strain of *S. flexneri*
serotype 5a. It identified a plasmid sequence of 30 kb (Maurelli *et al.*, 1985).
Subsequent analysis of this sequence by Tn5 mutagenesis (Baudry *et al.*,
1987; Sasakawa *et al.*, 1988) identified five independent contiguous loci. This
inv region is still under study to characterize the numerous genes that are
involved. Locus 1 is now known to contain a positive regulatory gene termed
virB in *S. flexneri* serotype 2a (Adler *et al.*, 1989), and invE in *S. sonnei*
(Watanabe *et al.*, 1990). Under the control of virF, another positive
regulatory gene (Sakai *et al.*, 1986), *virB* positively regulates sets of invasion

genes contained in the other loci, especially Locus 2 which has been extensively studied. Four proteins which are immunogenic both in humans and monkeys convalescing from shigellosis have been identified and called Ipa for 'invasion plasmid antigens' (Hale, Oaks & Formal, 1985; Oaks, Hale & Formal, 1986). The genes encoding IpaA (70 kD), IpaB (62 kD), IpaC (42 kD), and IpaD (38 kD) are located in this locus. Eight genes have been identified which are, in their order of transcription, *icsB*, *ipgA*, and *ipgB* which encode polypeptides of respective molecular weights 57, 15, and 21 kD. Then *ipgC*, a gene encoding a 17 kD polypeptide, and the four *ipa* genes (*ipaB*, *C*, *D*, and *A*), are organized in an operon with a weak additional promoter before ipaD (Baudry, Kaczorek & Sansonetti, 1988, Baudry *et al.*, 1987; Buysse *et al.*, 1987; Sasakawa *et al.*, 1989; Allaoui *et al.*, manuscript submitted).

Investigations on the role of these genes in the invasion process have begun (High *et al.*, manuscript submitted, Allaoui *et al.*, manuscript submitted, Menard *et al.*, manuscript in preparation). Mutations have been generated via double allelic exchange of the wild type plasmid gene with its *in vitro*-mutagenized counterpart. Deletion mutants engineered in *ipaB* are non-invasive in the HeLa cell assay system. They appear adhesive to the surface of the cells but do not trigger significant polymerization of actin. This is evidence that the phagocytic process is no longer elicited, and that an entry step can be differenciated from an adhesion step. The adhesin of *S. flexneri* has yet to be characterized. Based on this genetic evidence, and on the observation that IpaB is expressed on the surface of invasive bacteria (Andrews *et al.*, 1991; Mills, Buysse & Oaks, 1988), this protein is currently considered as a major invasin of *S. flexneri*. In addition to eliciting phagocytosis, this protein is also the contact hemolysin that accounts for lysis of the phagocytic vacuole (High *et al.*, manuscript submitted). This other aspect will be considered in another paragraph.

Gene organization and functions of 'loci 3' to '5' are currently being studied. Genes called *mxi* which belong to 'locus 4' and 'locus 5' appear to be involved in the proper location of Ipa proteins, particularly IpaB (Hromockyj & Maurelli, 1989).

Nature of the host cell components responsible for the phagocytic process

Molecular aspects of the entry process of invasive bacteria into eucaryotic cells have recently been reviewed (Falkow, 1991). The precise nature of the molecules involved in bacterial entry has been identified in a few instances. Evidence is accumulating that certain members of the integrin superfamily are major components that mediate binding and entry of some invasive pathogens. Integrins are dimeric integral transmembrane proteins which bind cell matrix proteins by their extracellular domain and actin filaments via proteins such as talin and vinculin by their cytoplasmic domain (Ruoslahti & Pierschbacher, 1987). Inv, the 103 kD invasin of *Yersinia pseudotuberculosis*, binds to the beta-1 subunit of several integrins (Isberg & Leong, 1990), and the fibrillar haemagglutinin (FHA) of *Bordetella pertussis* uses the CR3 receptor of complement to invade monocytes (Relman *et al.*, 1990). *Legionella pneumophila* and *Mycobacterium tuberculosis* either directly or indirectly use the same CR3 receptor to penetrate macrophages.

A major point that remains to be understood is how integrins mediate entry and more specifically, what are the factors that transform a binding process into an entry process. Data suggest that the *Y. pseudotuberculosis* Inv protein binds to integrins with much higher affinity than fibronectin. Increased affinity may be the crucial factor that will decide between binding and entry (Isberg, 1991).

In HeLa cells (Sansonetti, unpublished observations) as well as in chick embryo fibroblasts (Vasselon *et al.*, 1991), shigellae penetrate essentially at the level of the cells adhesion plaques which are structures by which cells adhere to their matrix. These regions of the cell are particularly rich in actin filaments and in integrins. It is therefore possible that *S. flexneri* also uses integrins at a stage of its entry process.

INTRACELLULAR MULTIPLICATION OF *S. FLEXNERI*

The ability of *S. flexneri* to grow intracellularly can be evaluated in infected HeLa cells. The generation time is about 40 minutes (Sansonetti *et al.*, 1986). This capacity to grow early and efficiently within the intracellular compartment seems to be unique to shigellae. Salmonellae and yersiniae hardly grow at all intracellularly within the first 6 hours of infection (Sansonetti *et al.*, 1986; Small, Isberg & Falkow, 1987). Lysis of the phagocytic vacuole surrounding shigellae, shortly after phagocytosis has occurred, is probably an essential prerequisite for efficient intracellular multiplication. A virulent strain of *S. flexneri* that multiplies efficiently within HeLa cells lyses its phagocytic vacuole within 30 minutes after entry (Sansonetti *et al.*, 1986) and reaches the cytoplasm which offers optimal conditions for rapid bacterial growth. This mechanism is also expected to prevent phagolysosomal fusion and to protect bacteria against killing in phagocytic cells. A correlation has been shown between lysis of the phagocytic vacuole and expression of a contact haemolytic activity (Clerk, Baudry & Sansonetti, 1986; Sansonetti *et al.*, 1986). As mentioned earlier, it has recently been shown that *IpaB* accounted for the contact haemolytic activity. In addition, when the non-invasive, non-haemolytic *ipaB* mutant was used to infect J774 mouse macrophages in order to bypass the entry process, intracellular microorganisms failed to lyse their phagocytic vacuole, thus indicating that IpaB is a unique protein that accounts both for entry and escape from the phagocytic vacuole (High *et al.*, manuscript submitted).

Listeria monocytogenes is quite similar to *S. flexneri*, since it also lyses its phagocytic vacuole shortly after invasion of eucaryotic cells. Listeriolysin O, a potent cytolysin, accounts for rapid lysis as demonstrated by using *hly* mutants obtained by transposition (Gaillard *et al.*, 1987; Katariou *et al.*, 1987; Portnoy, Jacks & Hinrichs, 1988). Recently, this step of vacuole destruction has been confirmed as a prerequisite for rapid intracellular growth by showing that a recombinant strain of *Bacillus subtilis* expressing Listeriolysin O acquired the capacity to lyse its phagocytic vacuole and subsequently grow within macrophages, as compared to the wild-type strain which hardly survived (Bielecki *et al.*, 1990). On the other hand, other enteroinvasive microorganisms such as salmonellae and yersiniae do not lyse their phagocytic vacuole (Sansonetti *et al.*, 1986; Small, Isberg & Falkow, 1987).

These observations provide new insights on the actual role of haemolysins in the pathogenesis of invasive diseases. Bacteria such as shigellae, which lyse their phagocytic vacuole early after the entry process has occurred, can grow rapidly within the intracellular compartment and establish themselves in the cells they encounter. Inversely, bacteria that do not lyse their phagocytic vacuole had to evolve a more sophisticated means of adaptation to the harsh environment of phagosomes or phagolysosomes in order to survive and eventually grow within infected cells.

INTRACELLULAR MOVEMENT OF *S. FLEXNERI* AND CELL TO CELL SPREAD

Shigellae are non-motile microorganisms lacking flagellae. In the intracellular compartment however, they acquire the capacity to move and spread from one cell to another, another essential feature that determines the outcome of epithelial cell infection. Early after entry into cells, once bacteria have lysed the phagocytic vacuole, they effect a sliding movement along the actin stress cables. This phenotype is particularly striking when chick embryo fibroblasts are infected by *S. flexneri* (Vasselon *et al.*, 1991). This Olm phenotype (Organelle like Movement) has not yet been characterized at a molecular level. From the entry site level located in the focal contact plaques, bacteria move toward the nucleus and start forming a microcolony. Unlike the Ics phenotype which will be described in the next paragraph, bacteria are not covered with F-actin in those cells.

Giemsa stains performed on infected HeLa cells after 2 hours of incubation show numerous bacteria spread all over the cell cytoplasm and not simply localized near their site of entry or near the nucleus. Bacteria also appear to form long protrusions initiating from the focal contacts. This Ics (Intra–Inter Cellular Spread) phenotype was demonstrated more than 20 years ago (Ogawa, Natamura & Nakaya, 1968). Using phase-contrast microcinematography, it was shown that intracellular shigellae affected rapid, random movements in the cytoplasm which sometimes led to the formation of protrusions. Later on (Makino *et al.*, 1986) *virG*, a plasmid locus independent of the entry sequences, was shown to be necessary for permanent reinfection of adjacent cells in the Sereny test. More recently, this movement was shown to be based on the interaction between free intracellular shigellae and the host cell actin. Treatment of infected cells by cytochalasin D which prevents actin polymerization blocks intracellular spread of shigellae as well as their passage from one cell to another (Pal *et al.*, 1989). Direct demonstration of the interaction of intracellular shigellae with actin was provided by labelling infected HeLa cells with NBD-phalloidin (Bernardini *et al.*, 1989). Bacteria become covered with polymerized actin within two hours after entry. They are essentially located within the intricated network of actin cables at the level of the adherence plaques. As incubation progresses, they lose their actin 'coat' and appear followed by a tail of polymerized actin. These tails materialize the movement of bacteria in the intracellular compartment. In some instances, bacteria appear located at the tip of a tail within a protrusion of the cell membrane which allows passage from one cell to the next. A striking aspect of this phenomenon is shown in Fig. 1(*a*). A similar behavior has been described in the case of *Listeria monocytogenes* (Mounier *et al.*, 1990; Tilney & Portnoy, 1989). Both species express a surface factor which induces actin nucleation and subsequent polymerization. This process leads to the formation of a coat. This coat is then polarized at one end of the bacterium thus forming the observed tails. The capacity to spread intracellularly and from cell to cell is easily studied in *vitro* in the plaque assay (Oaks, Wingfield & Formal, 1985). This assay allows screening of mutants impaired in this phenotype. A plaque negative Tn*phoA* mutant of *S. flexneri* has been obtained. This mutant, *icsA*, does not move intracellularly and from cell to cell. It does not produce a 120 kD outer membrane protein and does not induce actin polymerization (Bernardini *et al.*, 1989). The actin nucleation and/or polymerization activity of this protein is currently under study. *icsA* appears to be similar to *virG* (Makino *et al.*, 1986) its sequence has been published (Lett *et al.*, 1989), it is positively regulated at the transcriptional level by the product of *kcp* (Pal *et al.*, 1989) a gene located at 13 min on the chromosome of *S. flexneri* which has been shown, several years ago, to be necessary for the invasive microorganism to elicit a positive Sereny test (Formal *et al.*, 1971).

Invasion assays of human intestinal cells cultivated in vitro

The apical surface of enterocytes displays a particular pattern of differentiation constituted by microvilli which form the brush border. These

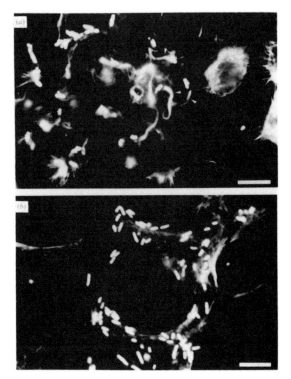

Fig. 1. (*a*) Infection of HeLa cells by *Shigella flexneri*. Double fluorescence labelling by NBD-phalloïdin which specifically stains F-actin filaments, and anti-LPS-rhodamin antiserum which specifically stains the *Shigella flexneri* surface. Several bacteria are seen at the tip of a bright 'comet' of actin. These comets are indicated by small arrowheads. This pattern materializes bacterial movement and corresponds to the formation of protusions which achieve colonization of adjacent cells. This is characteristic of the Ics phenotype. (*b*) Infection of intestinal epithelial cells Caco-2 (grown to confluency) by *Shigella flexneri*. Double fluorescence labelling by NBD-phalloïdin and anti-LPS-rhodamin as in Fig. 1(*a*). Bacteria are seen interacting with the actin filaments that form a ring at the level of the perijunctional area in polarized epithelial cells. This is characteristic of the Olm phenotype. (Bars: 10 μm.)

microvilli contain bundles of polymerized actin filaments cross-linked by fimbrin and villin and laterally connected to the membrane by a calmodulin/ 110 kD protein complex. The latter protein belongs to the family of type 1 myosins (Louvard, 1989; Mooseker & Coleman, 1989). The *in vivo* target therefore appears quite different from the HeLa cell surface that is encountered *in vitro*. If a bacterium needs to penetrate such a system, it has to disorganize the brush border pattern in order to render the membrane–cytoskeleton complex flexible enough and the pool of available actin and actin-binding proteins sufficient for the phagocytic process to occur. When the human colonic continuous cell line Caco-2 (Rousset, 1986) is grown to confluency, shigellae do not invade these differenciated cells that express a brush border (Mounier *et al.*, 1992). They rather bind to the outer edge of the islet and subsequently enter peripheral cells. If the confluent monolayer is treated with EGTA which, through Ca^{2+} chelation, impairs the function of cadherins (Volberg *et al.*, 1986), opens up intercellular junctions, and exposes the laterobasal pole of the cells, bacteria can now invade the epithelial lining quite efficiently. *S. flexneri* therefore does not recognize a receptor on the apical pole of enterocytes but rather a receptor expressed on the laterobasal pole of these cells. Unlike *Salmonella* which penetrates straight through the apical pole of MDCK (Finlay, Gumbiner & Falkow, 1988) and Caco-2 cells (Finlay & Falkow, 1990) shigellae apparently need to reach a subepithelial position to be able to efficiently penetrate enterocytes. Subsequently, colonization of the epithelial lining is obtained by spread along the cylinder of actin cables which is anchored on the perijunctional area of the cells as shown on Fig. 1(*b*) (Vasselon *et al.*, 1992). Then passage occurs from one cell to another due to actin polymerization. This colonization process therefore makes use of both the Olm and the Ics phenotype.

In vivo models

It has recently been demonstrated that the Ics phenotype is important *in vivo*. Macaque monkeys infected orally by SC560, a deletion mutant of *icsA*, developed very limited symptoms of shigellosis as compared with the animals infected by the wild-type strain. Endoscopic examination of the rectum and sigmoïd colon of these animals detected only a limited number of small nodular abscesses and minor ulcerations (Sansonetti *et al.*, 1991).

In spite of many gaps in understanding the pathogenesis of shigellosis, several steps now appear crucial. The entry phenotype is essential. Strains that have lost the expression of this phenotype are avirulent even in the most definitive assays. The major question, however, is where does the entry phenotype take place in the human colon? It has already been demonstrated that bacteria bind to, and invade, the basolateral pole of enterocyte-like Caco-2 cells, thus suggesting that the receptor is not expressed at the level of the brush border. How do shigellae gain access to the laterobasal pole of enterocytes *in vivo*? They may invade colonic crypts where enterocytes are in a less differentiated stage. Another hypothesis, infection via M cells, seems more likely, although not necessarily exclusive. When macaque monkeys are infected by an *icsA* mutant of *S. flexneri*, animals develop very limited symptoms of dysentery (Sansonetti *et al.*, 1991). Colonoscopic examination reveals a limited number of small nodular abscesses or tiny ulcerations. Upon biopsy, these lesions always appear located over lymphoid follicles, thus suggesting the presence of colonic equivalents of Peyer's patches. This mutant points to the initial site of entry of shigellae within the epithelium. This site therefore corresponds to Peyer's patches equivalents which contain many M cells (Bye, Allan & Trier, 1984). These sites may represent the 'Trojan Horse' allowing entry of the invasive microorganism within the epithelium. Bacteria may penetrate into the epithelium via M cells, then spread from one cell to another by expressing their Ics phenotype which allows 'underground colonization'. Subepithelial spread to resident macrophages of the lamina propria may also allow, after lysis of these cells, subsequent 'retrograde' infection of enterocytes by their basolateral pole. This scheme is still hypothetical; however, recent data obtained in the rabbit ligated ileal loop model indicate that M cells actually represent the initial site of entry of shigellae (Wassef, Keren & Mailloux, 1989). Similar areas of the intestine also seem to allow entry of salmonellae and yersiniae (Grützkan *et al.*, 1990; Kohbata, Yokoyama & Yabuuchi, 1986).

ACKNOWLEDGEMENTS

I wish to thank all those who have made this work possible: my collaborators from the 'Unité de Pathogénie Microbienne Moléculaire' as well as our colleagues from the 'Station Centrale de Microscopie Electronique', M. C. Prevost and M. Lesourd, and from the 'Unité de Biologie des Membranes', R. Hellio. I also thank Colette Jacquemin for typing this manuscript.

REFERENCES

Adler, B., Sasakawa, C., Tobe, T., Makino, S., Komatsu, K. & Yoshikawa, M. (1989). A dual transcriptional activation system for the 230 kb plasmid genes coding for virulence associated antigens of *Shigella flexneri*. *Molecular Microbiology*, **3**, 627–35.

Andrews, G. P., Hromockyj, A. E., Coker, C. & Maurelli, A. T. (1991). Two novel virulence loci, *mxiA* and *mxiB*, in *Shigella flexneri* 2a facilitate excretion of invasive plasmid antigens. *Infection and Immunity*, **59**, 1997–2005.

Barak, L. S., Yocum, R. R., Nothnagel, E. A. & Webb, W. W. (1980). Fluorescence staining of the actin cytoskeleton in living cells with 7-nitrobenz-2-oxa-1,3-diazole-phallacidin. *Proceedings of the National Academy of Sciences, USA*, **77**, 980–4.

Baudry, B., Kaczorek, M. & Sansonetti, P. J. (1988). Nucleotide sequence of the invasion plasmid antigen B and C genes (*ipaB* and *ipaC*) of *Shigella flexneri*. *Microbial Pathogenesis*, **4**, 345–57.

Baudry, B., Maurelli, A. T., Clerc, P., Sadoff, J. C. & Sansonetti, P. J. (1987). Localization of plasmid loci necessary for the entry of *Shigella flexneri* into HeLa cells, and characterization of one locus encoding four immunogenic polypeptides. *Journal of General Microbiology*, **133**, 3403–13.

Bernardini, M. L., Mounier, J., d'Hauteville, H., Coquis-Rondon, M. & Sansonetti, P. J. (1989). Identification of *icsA*, a plasmid locus of *Shigella flexneri* that governs intra- and intercellular spread through interaction with F-actin. *Proceedings of the National Academy of Sciences, USA*, **86**, 3867–71.

Bielecki, J., Youngman, P., Connelly, P. & Portnoy, D. A. (1990). *Bacillus subtilis* expressing a hemolysin gene from *Listeria monocytogenes* can grow in mammalian cells. *Nature*, **345**, 175–6.

Buysse, J. M., Stover, C. K., Oaks, E. V., Venkatesan, M. & Kopecko, D. J. (1987). Molecular cloning of invasion plasmid antigen (*ipa*) genes from *Shigella flexneri*: analysis of *ipa* gene products and genetic mapping. *Journal of Bacteriology*, **169**, 2561–9.

Bye, W. A., Allan, C. H. & Trier, J. S. (1984). Structure, distribution and origin of M cells in Peyer's patches of mouse ileum. *Gastroenterology*, **86**, 789–801.

Clerc, P., Baudry, B. & Sansonetti, P. J. (1986). Plasmid-mediated contact hemolytic activity in *Shigella* species: correlation with penetration into HeLa cells. *Annales de l'Institut Pasteur Microbiologie*, 137A(3), 267–78.

Clerc, P. & Sansonetti, P. J. (1987). Entry of *Shigella flexneri* into HeLa cells: evidence for directed phagocytosis involving actin polymerization and myosin accumulation. *Infection and Immunity*, **55**, 2681–8.

Falkow, S. (1991). Bacterial entry into eucaryotic cells. *Cell*, **65**, 1099–102.

Finlay, B. B. & Falkow, S. (1988). Comparison of the invasion strategies used by *Salmonella cholerae-suis*, *Shigella flexneri* and *Yersinia enterocolitica* to enter cultured animal cells: endosome acidification is not required for bacterial invasion or intracellular replication. *Biochimie*, **70**, 10891–9.

Finlay, B. B., Gumbiner, B. & Falkow, S., (1988). Penetration of *Salmonella* through a polarized Madin–Darby canine kidney epithelial cell monolayer. *Journal of Cell Biology*, **107**, 221–30.

Finlay, B. B. (1990). Cell adhesion and invasion mechanisms in microbial pathogenesis. *Current Opinion in Cell Biology*, **2**, 815–20.

Finlay, B. B. & Falkow, S. (1990). *Salmonella* interactions with polarized human intestinal Caco-2 epithelial cells. *Journal of Infectious Diseases*, **162**, 1096–106.

Finlay, B. B., Ruschkowski, S. & Dedhar, S. (1991). Cytoskeletal rearrangements accompanying *Salmonella* entry into epithelial cells. *Journal of Cell Science*, **99**, 283–96.

Fontaine, A., Arondel, J. & Sansonetti, P. J. (1988). Role of Shiga-toxin in the pathogenesis of shigellosis as studied using a Tox- mutant of *Shigella dysenteriae* 1. *Infection and Immunity*, **56**, 3099–109.

Formal, S. B., Gemski, P., Baron, L. S. & LaBrec, E. H. (1971). A chromosomal locus which controls the ability of *Shigella flexneri* to evoke keratoconjunctivitis. *Infection and Immunity*, **3**, 73–9.

Gaillard, J. L., Berche, P., Mounier, J., Richard, S. & Sansonetti, P. J. (1987). *In vitro* model of penetration and intracellular growth of *Listeria monocytogenes* in the human enterocyte-like cell line CaCo2. *Infection and Immunity*, **55**, 2822–9.

Grützkau, A., Hanski, C, Hahn, H. & Riecken, E. O. (1990). Involvement of M cells in the bacterial invasion of Peyer's patches: a common mechanism shared by *Yersinia enterocolitica* and other enteroinvasive bacteria. *Gut*, **31**, 1011–15.

Hale, T. L. & Bonventre, P. F. (1979). *Shigella* infection of Henle intestinal epithelial cells: role of the bacteria. *Infection and Immunity*, **24**, 879–86.

Hale, T. L., Morris, R. E. & Bonventre, P. F. (1979). *Shigella* infection of Henle intestinal epithelial cells: role of the host cell. *Infection and Immunity*, **24**, 887–94.

Hale, T. L., Oaks, E. V. & Formal, S. B. (1985). Identification and antigenic characterization of virulence-associated, plasmid-coded proteins of *Shigella* spp. and enteroinvasive *Escherichia coli*. *Infection and Immunity*, **50**, 620–9.

Hromockyj, A. E. & Maurelli, A. T. (1989). Identification of *Shigella* invasion genes by construction of temperature-regulated *inv::lacZ* operon fusions. *Infection and Immunity*, **57**, 2963–70.

Isberg, R. R. (1989). Mammalian cell adhesion functions and cellular penetration of enteropathogenic *Yersinia* species. *Molecular Microbiology*, **3**, 1449–53.

Isberg, R. R. & Leong, J. M. (1990). Multiple beta-1 chain integrins are receptors for

Invasin, a protein that promotes bacterial penetration into mammalian cells. *Cell*, **60**, 861–71.

Isberg, R. R. (1991). Discrimination between intracellular uptake and surface adhesion of bacterial pathogens. *Science*, **252**, 934–8.

Katariou, S., Metz, P., Hof, H. & Goebel, W. (1987). Tn916-induced mutations in the hemolysin determinant affecting virulence of *Listeria monocytogenes*. *Journal of Bacteriology*, **169**, 1291–7.

Kohbata, S., Yokoyama, H. & Yabuuchi, E. (1986). Cytopathogenic effect of *Salmonella typhi* GIFU 10007 on M cells of murine ileal Peyer's patches in ligated ileal loops: an ultrastructural study. *Microbiology Immunology*, **30**, 1225–37.

Koster, F., Levin, J., Walker, L., Tung, K. S. K., Gilman, R. H., Rahaman, M. M., Majid, M. A., Islam, S. & Williams, R. C. (1978). Hemolytic-uremic syndrome after shigellosis. Relation to endotoxemia and circulating immune complexes. *New England Journal of Medicine*, **298**, 927–33.

LaBrec, E. H., Schneider, H., Magnani, T. J. & Formal, S. B. (1964). Epithelial cell penetration as an essential step in pathogenesis of bacillary dysentery. *Journal of Bacteriology*, **88**, 1503–18.

Lett, M. C., Sasakawa, C., Okada, N., Sakai, T., Makino, S., Yamada, M., Komatsu, K. & Yoshikawa, M. (1989). *virG*, a plasmid coded virulence gene of *Shigella flexneri*: identification of the *VirG* protein and determination of the complete coding sequence. *Journal of Bacteriology*, **171**, 353–9.

Louvard, D. (1989). The function of the major cytoskeletal components of the brush border. *Current Opinion in Cell Biology*, **1**, 51–7.

Makino, S., Sasakawa, C., Kamata, K., Kurata, T. & Yoshikawa, M. (1986). A genetic determinant required for continuous reinfection of adjacent cells on large plasmid in *Shigella flexneri* 2a. *Cell*, **46**, 551–5.

Maurelli, A. T., Baudry, B., d'Hauteville, H. & Sansonetti, P. J. (1985). Cloning of plasmid DNA sequences involved in invasion of HeLa cells by *Shigella flexneri*. *Infection and Immunity*, **49**, 164–71.

Mills, J. A., Buysse, J. M. & Oaks, E. V. (1988). *Shigella flexneri* invasion plasmid antigens IpaB and C: epitope location and characterization by monoclonal antibodies. *Infection and Immunity*, **56**, 2933–41.

Mooseker, M. S. & Coleman, T. R. (1989). The 110-kD protein-calmodulin complex of the intestinal microvillus (brush border myosin I) is a mechanoenzyme. *Journal of Cell Biology*, **108**, 2395–400.

Mounier, J., Ryter, A., Coquis-Rondon, M. & Sansonetti, P. J. (1990). Intracellular and cell to cell spread of *Listeria monocytogenes* involves interaction with F-actin in the enterocyte-like cell line Caco-2. *Infection and Immunity*, **58**, 1048–58.

Mounier, J., Vasselon, T., Hellio, R., Lesourd, M. & Sansonetti, P. J. (1992). *Shigella flexneri* enters human colonic Caco-2 epithelial cell through their basolateral pole. *Infection and Immunity*, in press.

Oaks, E. V., Hale, T. L. & Formal, S. B. (1986). Serum immune response to *Shigella* protein antigens in Rhesus monkeys and humans infected with *Shigella*. *Infection and Immunity*, **53**, 57–63.

Oaks, E. V., Wingfield, M. E. & Formal, S. B. (1985). Plaque formation by *Shigella flexneri*. *Infection and Immunity*, **48**, 124–9.

O'Brien, A. D. & Holmes, R. K. (1987). Shiga and Shiga-like toxins. *Microbiological Review*, **51**, 206–20.

Ogawa, H., Nakamura, A. & Nakaya, R. (1968). Cinematographic studies of tissue cell cultures infected with *Shigella flexneri*. *Japanese Journal of Medical Science and Biology*, **21**, 259–73.

Pal, T., Newland, J. W., Tall, D. & Formal, S. B. (1989). Intracellular spread of

Shigella flexneri associated with the kcpA locus and a 140-kilodalton protein. *Infection and Immunity*, **57**, 477–86.

Portnoy, D. A., Jacks, P. S. & Hinrichs, D. J. (1988). Role of hemolysin for the intracellular growth of *Listeria monocytogenes*. *Journal of Experimental Medicine*, **167**, 1459–71.

Relman, D. A., Tuomanen, E., Falkow, S., Golenbock, D. T., Saukkonen, K. & Wright, S. (1990). Recognition of a bacterial adhesin by an eukaryotic integrin: CR3 (alphaMbeta2, CD11b/CD18) binds filamentous hemaglutinin of *Bordetella pertussis*. *Cell*, **61**, 1375–82.

Rousset, M. (1986). The human colon carcinoma cell lines HT29 and Caco-2: two *in vitro* models for the study of intestinal differentiation. *Biochimie*, **68**, 1035–40.

Ruoslahti, E. & Pierschbacher, M. D. (1987). New perspectives in cell adhesion: RGD and integrins. *Science*, **238**, 491–7.

Sakai, T., Sasakawa, C., Makino, S. & Yoshikawa, M. (1986). DNA sequence and product analysis of the *virF* locus responsible for Congo red binding and cell invasion in *Shigella flexneri*. *Infection and Immunity*, **54**, 395–402.

Sansonetti, P. J., Kopecko, D. J. & Formal, S. B. (1982). Involvement of a plasmid in the invasive ability of *Shigella flexneri*. *Infection and Immunity*, **35**, 852–60.

Sansonetti, P. J., Hale, T. L., Dammin, G. I., Kapper, C., Collins, H. H. & Formal, S. B. (1983). Alterations in the pathogenesis of *Escherichia coli* K12 after transfer of plasmid and chromosomal genes from *Shigella flexneri*. *Infection and Immunity*, **39**, 1392–402.

Sansonetti, P. J., Ryter, A., Clerc, P., Maurelli, A. T. & Mounier, J. (1986). Multiplication of *Shigella flexneri* within HeLa cells: lysis of the phagocytic vacuole and plasmid mediated contact hemolysis. *Infection and Immunity*, **51**, 461–9.

Sansonetti, P. J., Arondel, J., Fontaine, A., d'Hauteville, H. & Bernardini, M. L. (1991). *Omp B* (osmo-regulation) and *icsA* (cell to cell spread) mutants of *Shigella flexneri*: vaccine candidates and probes to study pathogenesis of shigellosis. *Vaccine*, **9**, 416–22.

Sasakawa, C., Kamata, K., Sakai, T., Makino, S., Yamada, M., Okada, N., Yoshikawa, M., (1988). Virulence-associated genetic regions comprising 31 kilobases of the 230-kilobase plasmid in *Shigella flexneri* 2a. *Journal of Bacteriology*, **170**, 2480–4.

Sasakawa, C., Adler, B., Tobe, T., Okada, N., Nagai, S., Komatsu, K. & Yoshikawa, M. (1989). Functional organization and nucleotide sequence of virulence region-2 on the large virulence plasmid of *Shigella flexneri* 2a. *Molecular Microbiology*, **170**, 2480–4.

Sereny, B. (1957). Experimental keratoconjunctivitis shigellosa. *Acta Microbiologica Academica Scientifica Hungarica*, **4**, 367–7.

Sheterline, P., Rickard, J. E. & Richards, R. C. (1984). Fc receptor directed phagocytic stimuli induce transient actin assembly at an early stage of phagocytosis in neutrophil leucocytes. *European Journal of Cell Biology*, **34**, 80–7.

Small, P. L. C., Isberg, R. R. & Falkow, S. (1987). Comparison of the ability of enteroinvasive *Escherichia coli, Yersinia* pseudotuberculosis and *Yersinia* enterocolitica to enter and replicate within HEp-2 cells. *Infection and Immunity*, **55**, 1674–9.

Stendahl, O. I., Hartwig, J. H., Brotschi, E. A. & Stossel, T. P. (1980). Distribution of actin-binding protein and myosin in macrophages during spreading and phagocytosis. *Journal of Cell Biology*, **84**, 215–24.

Takeuchi, A., Spring, H., Labrec, E. H. & Formal, S. B. (1965). Experimental bacillary dysentery: an electron microscopic study of the response of intestinal mucosa to bacterial invasion. *American Journal of Pathology*, **4**, 1011–44.

Takeuchi, A., Formal, S. B. & Spring, H. (1968). Experimental acute colitis in the rhesus monkey following peroral infection with *Shigella flexneri*. *American Journal of Pathology*, **52**, 503–29.

Tanenbaum, S. W. (1978). *Cytochalasins: Biochemical and Cell Biological aspects*. North-Holland Publishing Co. Amsterdam.

Tilney, L. G. & Portnoy, D. A. (1989). Actin filaments and the growth, movement, and spread of the intracellular bacterial parasite, *Listeria monocytogenes*. *Journal of Cellular Biology*, **109**, 1597–608.

Vasselon, T., Mounier, J., Prevost, M. C., Hellio, R. & Sansonetti, P. J. (1991). A stress fiber-based movement of *Shigella flexneri* within cells. *Infection and Immunity*, **59**, 1723–32.

Vasselon, T., Mounier, J., Hellio, R. & Sansonetti, P. J. (1992). Movements along actin filaments of the perijunctional area and *de novo* polymerization of cellular actin are required for *Shigella flexneri* colonization of epithelial Caco-2 cell monolayer. *Infection and Immunity*, in press.

Volberg, T., Geiger, B., Kartenbeck, J. & Franke, W. W. (1986). Changes in membrane microfilaments interaction in intercellular *adherens* junctions upon removal of extracellular Ca^{2+} ions. *Journal of Cell Biology*, **102**, 1832–42.

Wassef, J. S., Keren, D. F. & Mailloux, J. L. (1989). Role of M cells in initial bacterial uptake and in ulcer formation in the rabbit intestinal loop model in shigellosis. *Infection and Immunity*, **57**, 858–63.

Watanabe, H., Arakawa, E., Ito, K., Kato, J. I. & Nakamura, A. (1990). Genetic analysis of an invasion region by use of a Tn3-*lac* transposon and identification of a second positive regulator gene, *inv* E, for cell invasion of *Shigella sonnei*: significant homology of *inv* E with *par* B of phage P1. *Journal of Bacteriology*, **172**, 619–29.

ANTIGENIC VARIATION IN BACTERIAL PATHOGENS

BRIAN D. ROBERTSON AND THOMAS F. MEYER

Max-Planck-Institut für Biologie, Abt. Infektionsbiologie, Spemannstrasse 34, D7400 Tübingen, Germany

INTRODUCTION

The mechanisms which control gene expression in bacteria fall into two broad categories. In one category, genes can be regulated in response to signals such as pH, temperature and the presence or absence of particular nutrients or chemicals, which allows the organism to achieve an optimal level of gene expression for any given environment or stage of development, and usually leads to a phenotypically homogeneous population. The second category covers genetic alterations which are not *per se* a response to signals, but occur independently and provide a source of variation. The mechanisms in this category have been described as 'random mechanisms' (DiRita & Mekalanos, 1989), because they result in a phenotypically heterogeneous population. From another point of view these mechanisms have also been described as 'programmed rearrangements' (Borst & Greaves, 1987) because the genome is organised in such a way that certain rearrangements are likely to occur. Mindful of these definitions this chapter will concentrate on the category of random mechanisms, which has been widely studied for pathogenic bacteria. These are of particular importance because most of the structures studied are essential either for the colonization of the host, or for the survival of the pathogen within that host. The ability to vary the immunogenicity of important structures and to 'fine tune' the specificities of receptors or adhesins, is of considerable evolutionary advantage to the organism. This cannot be achieved by non-random means, since neither the variation of ligands encountered, nor the target of the immune response can be predicted in advance. Such alterations in gene expression usually only occur in response to environmental stimuli.

DNA rearrangements occur in a wide variety of organisms including viruses, bacteria and eukaryotes. At the most extreme this rearrangement involves the shuffling of entire chromosomes with a concurrent reassortment of genes. In the case of Influenza Virus this results in the presentation to the immune system of an antigenically novel virus, leading to pandemics (Skehel & Wiley, 1986). Bacterial plasmid and chromosomal DNA can be transferred during conjugation, and this may allow exchange of the genetic material available for the expression of a variety of functions such as virulence and antibiotic resistance (for examples see the chapter by

Achtman & Hakenbeck, this volume). Extensive genetic rearrangements occur in the mammalian immune system and are crucial for maintaining the diversity of both antibodies and the T cell receptor (Yancopoulos & Alt, 1986). In order to avoid this extraordinarily diverse immune response, many pathogens alter the immunogenicity of the surface molecules they present to the immune system. Viruses do not use 'programmed rearrangement' as described by Borst & Greaves (1987). Instead variation is achieved simply by the fact that the polymerase does not possess any proof-reading function, and therefore mistakes occurring during replication cannot be corrected. Selective pressure to maintain essential genes is provided by functional constraints on their gene products. The Human Immunodeficiency Virus (HIV) envelope protein carries a neutralization epitope in the V3 region, which does not seem to be functionally constrained. Neutralization is avoided, however, because this region is hypervariable and the epitope undergoes antigenic variation, resulting from point mutations (Goudsmit, Back & Nara, 1991). Neutralization epitopes in the capsid protein VP1 of Foot and Mouth Disease Virus are similarly hypervariable as a result of single non-silent base pair mutations (Mateu et al, 1989). An intensively studied eukaryotic pathogen is the African trypanosome (for review see Borst, 1991), which employs gene rearrangements to modify its surface. From a repertoire of more than 1000 silent variant surface glycoprotein (VSG) genes the parasite transfers copies to one of many telomeric expression sites by duplicate transposition. Alternatively, it uses a different expression site containing another VSG. An immunogenically different surface is thereby presented to the host immune system.

   This chapter will cover the mechanisms which underly random phase and antigenic variation in pathogenic bacteria, concentrating on some of the newer discoveries in the field. Detailed reviews of other mechanisms have already been published (Borst & Greaves, 1987; Miller, Mekalanos & Falkow, 1989; DiRita & Mekalanos, 1989; Berg & Howe, 1989).

GENERAL HOMOLOGOUS RECOMBINATION

Methods of antigenic variation utilizing homologous recombination are widely distributed and are dependent on the *recA* function. The Gram-negative pathogen *Neisseria gonorrhoeae* expresses on its surface type 4 pili that undergo both phase and antigenic variation (reviewed by Meyer, Gibbs & Haas, 1990). Sequence analysis of pilin transcripts from variants of the strain MS11 expressing different pilin types, suggested that the variation might result from intragenic recombination events within the pilin expression locus *pilE* (Meyer, Mlawer & So, 1982). Further analysis demonstrated that this recombinatorial exchange involved the transfer of variable minicassette sequences from a repertoire of more than 10 silent *pilS* loci, which lack the amino-terminal invariant domain, to the expression locus

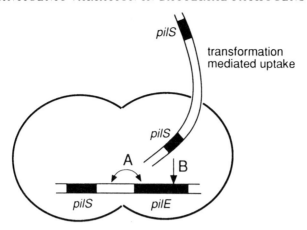

Fig. 1. The two pathways used by *N. gonorrhoeae* for pilin antigenic variation. The intra-genomic pathway (A) involves recombinatorial exchange between *pilS* loci and the expression site *pilE*. The intergenomic pathway (B) involves the take up of chromosomal DNA released from lysed cells by transformation. *pilS* genes carried by the incoming DNA subsequently recombine with *pilE* sequences leading to pilin antigenic variation.

(Haas & Meyer, 1986; Segal *et al.*, 1986; Swanson *et al.*, 1986). The other pathway leading to recombination and variation utilizes intergenomic rather than intragenomic exchange (Fig. 1). This is transformation-mediated recombination in which *Neisseria* cells take up exogenous homologous DNA from lysed bacteria, which can then be used to replace the pilin expression gene (Seifert *et al.*, 1988; Gibbs *et al.*, 1989). This occurs at a frequency of 10^{-3} per cell division. This reassortment of pilin sequences by recombination not only leads to the production of antigenically different pili, but is also responsible for their phase variation. There are two commonly observed groups of non-piliated variants. The production of soluble or S-pilin occurs when propilin is processed at position 40 rather than position 1, resulting in a pilin that is incompatible with pilus assembly and is instead secreted into the medium. This alternative cleavage seems to be pilin sequence-dependent. The second group of non-piliated variants produce very long or L-pilin molecules. These cannot form pili but the pilin is instead found in the periplasm or outer membrane of the cell. L-variants result from unequal recombination between *pilS* and *pilE* leading to the presence of multiple tandem copies of *pilS* in the expression site. These intervening sequences are usually in frame and lead to the observed long unstable pilin molecules. These variants can sometimes revert to the piliated phenotype when the tandem copy within the *pilE* gene is lost by site-specific deletion.

The Gram-negative spirochaete *Borrelia hermsii*, which causes relapsing fever, also undergoes antigenic variation by a mechanism not dissimilar to that used by *N. gonorrhoeae* and trypanosomes. *B. hermsii* expresses lipoproteins called variable major proteins (Vmp) on its surface, which are thought to be encoded by at least 27 variant genes, carried on linear plasmids. The majority of *Vmp* genes exist as promoterless silent copies which are expressed after transposition to a single telomeric expression site on another linear plasmid (Plasterk, Simon & Barbour, 1991), leading to duplication of the silent copy and loss of the original active gene. It is not clear if this occurs by a single or a double crossover event. Recombination between silent gene copies and the expression locus leads to expression of new Vmps, at a frequency of 10^{-4} to 10^{-3}. The process is subject to a higher level of control, since there is some order to the serotypes appearing, although the mechanism behind this is unclear.

SPECIALIZED SITE-SPECIFIC RECOMBINATION

Phase variation of surface structures in a number of organisms also occurs by homologous recombination, but in a site-specific manner via the inversion of a control element. Examples include the flagella of *Salmonella typhimurium* (reviewed by Glasgow, Hughes & Simon, 1989) and type 1 mannose-sensitive fimbriae of *Escherichia coli* (Fig. 2). Upstream of the gene *fimA* encoding the major fimbrial subunit in *E. coli* is a consensus binding site for integration host factor (IHF) (Dorman & Higgins, 1987; Eisenstein *et al.*, 1987) and a 314 bp invertible fragment containing the promoter (Abraham *et al.*, 1985). In one orientation the promoter faces upstream away from *fimA* and fimbrial production is switched OFF. After inversion of the fragment the promoter faces downstream, *fimA* is transcribed, and production of fimbriae is switched ON. *E. coli* strains carrying mutations in genes for either of the IHF subunits are incapable of phase variation, demonstrating that IHF is important for the inversion of the fragment. There is no evidence so far of antigenic variation in any of the fimbrial components.

Fimbrial Expression ON

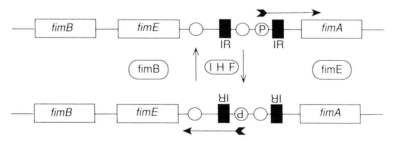

Fimbrial Expression OFF

Fig. 2. Control of type 1 fimbrial expression by inversion of promoter containing element. The promoter element is bounded by 9 bp inverted repeats (IR). There are two binding sites for integration host factor (IHF) shown by open circles. *fimB* and *fimE* direct the inversion of the promoter element to the ON and OFF orientations, respectively.

The type 4 pili of *Moraxella bovis* also undergo a phase switch between two antigenically different pilins, I and Q. This is mediated by the inversion of a 2 kb DNA sequence whose endpoints lie within the coding region of the pilin gene (Marrs *et al.*, 1988). In one orientation the I sequences are adjacent to the promoter and contribute I-specific leader and amino-terminal amino acid sequences. In the other orientation the Q-specific sequences are in the expression locus next to the promoter and provide the Q-specific leader and amino terminal amino acids. Switches to the non-piliated state probably result from imperfect inversion leading to frameshift mutations and either the P^- (no pilin) state or P^- (nonsense pilin phenotype). The amino-terminus of the Q-specific sequence has 58% homology to the *hin* gene product of *Salmonella typhimurium*, which is involved in site-specific inversion, perhaps suggesting a mechanism for this switching.

VARIATION VIA OLIGONUCLEOTIDE REPEATS

Another variation on the homologous recombination theme involves the use of repetitive sequences. This is known to be RecA-independent in two examples: the outer membrane opacity proteins of *N. gonorrhoeae* (reviewed by Meyer *et al.*, 1990), and lipopolysaccharide (LPS) epitopes of *Haemophilus influenzae* (Weiser, Love & Moxon, 1989; Weiser *et al.*, 1990; Maskell *et al.*, 1991; see Moxon & Maskell, this volume). In other cases it is unknown.

Gram-positive streptococci are resistant to phagocytosis by virtue of the M protein which covers the bacterial surface (for reviews see Scott, 1990; Fischetti, 1991). However, opsonization of the bacteria by antibodies directed against it would allow it to be phagocytosed and destroyed. Consequently M protein is highly variable in size with over 80 antigenically different serotypes recognized so far (Fischetti, Jones & Scott, 1985). The serotype M6 has been studied in detail and amino acid sequence analysis revealed the mechanism underlying the observed polymorphisms. The protein consists of repeats of amino acids that can be grouped into four regions. At the amino-terminus, the first region (A) contains five tandem repeats, each of 14 amino acids. The three central repeats are identical, while those at either end diverge slightly. Moving toward the carboxy-terminus the next region (B) contains five repeats of 25 amino acids each, and region C contains two and a half repeats of 42 amino acids each. Finally, at the carboxy-terminus, region D consists of four partial repeats of seven amino acids. This pattern is consistent with the observed serological data, in that the carboxy-terminus is very highly conserved (98% identity) among different serotypes, and amino acid homology decreases approaching the amino-terminus, where the last eleven amino acids at the tip are unique. The relatively conserved carboxy-terminal region is possibly protected from host attack by binding with a human regulatory protein factor H. The only region exposed to antibodies is the A region at the tip of the molecule, which is antigenically highly variable with the potential to avoid opsonization by antibodies. The higher order of the structure of the molecule is, however, conserved throughout, in the form of an alpha-helical coil. M protein size variation is thought to occur by the deletion of repeats from different regions, probably by homologous recombination between tandemly repeated elements within the coding sequence (Hollingshead, Fischetti & Scott, 1987; Jones *et al.*, 1988), although the role of the RecA protein is not clear at this time.

Antigenic diversity in *Mycoplasma hyorhinis* results from combinatorial expression, and phase variation of multiple, size-variant surface lipoproteins (VLPS) (Rosengarten & Wise, 1990, 1991; Yogev *et al.*, 1991). The lipoprotein can be divided into two domains, one of which is conserved and required for membrane insertion, and another divergent external domain with repetitive coding sequences. Size variation is controlled via the loss or gain of repetitive intragenic coding sequences in external domains of the protein. This probably occurs by homologous recombination or slipped-strand mispairing. Phase variation is transcriptionally controlled via homopolymeric repeats (see section on variation via homopolymeric repeats below).

Another recently described example of size variation utilizing repetitive sequences, in which the role of the RecA protein is also unclear, is provided by the major surface antigen of *Anaplasma marginale* (Allred *et al.*, 1990). This protein shows overall homology between isolates but contains a domain with various numbers of tandemly repeated sequences, the number of which correlates with the size of the surface antigen although there is no correlation with antigenic variation in this region. Variation in repeat numbers probably occurs by slipped-strand mispairing or by homologous recombination.

H. influenzae LPS shows a reversible loss of epitopes defined by reactivity with monoclonal antibody and variations in LPS size. Two loci have been identified (*lic*-1 and *lic*-2), that are involved in this variation (Weiser *et al.*, 1989). The *lic*-1 locus is responsible for expression of two epitopes, and contains four genes. The first gene has no known homologies but mediates phase variation via 5′ tandem repeats of CAAT (approximately 30 repeats). Sequence analysis of isogenic variants defined by a monoclonal antibody demonstrated that the number of repeats varied with the expression status. Alterations in the number of repeats, which is RecA independent (Weiser *et al.*, 1989) and probably occurs by slipped-strand mispairing (Levinson & Gutman, 1987), acts as a translational switch between alternative start codons. The expression of the first ORF is thereby altered, which in turn, by some unknown mechanism, influences epitope expression through downstream genes. When there are 29 repeats there is no epitope expression, while with both 30 and 31 repeats there is expression although it is greater with 31 than 30. The mechanism differentiating between these three levels of expression is not clear, although it may be more efficient translation of the first ORF. The *lic-2* locus and another locus *lic-3* also contain CAAT repeats (see also Moxon & Maskell, this volume).

The genes encoding the opacity proteins of *N. gonorrhoeae* include a coding repeat at their 5′ end consisting of repeats of CTCTT, which encode the hydrophobic core of the Opa leader peptide. The number of repeats varies between 6 and 27 and determines the translation frame of the gene, and thus its phase status (Stern *et al.*, 1986). The number of repeats is probably altered by slipped-strand mispairing (Meyer, 1987; Levinson & Gutman, 1987), and is RecA independent (Murphy *et al.*, 1989; Belland, 1991). Eleven *opa* loci have been identified in strain MS11 and form the basis for the antigenic variation, with sometimes four or more proteins being expressed simultaneously. Further variation is provided by recombination events between *opa* genes leading to reassortment of the hypervariable regions.

VARIATION VIA HOMOPOLYMERIC REPEATS

Deletions or additions of nucleotides in homopolymeric sequences are quite commonly used to switch the transcription of genes on and off, probably because such mistakes can occur during DNA replication. The fimbriae of *Bordetella pertussis* undergo phase variation (reviewed by Coote, 1991). This is controlled in two different ways. First, there is a coordinated regulation of the virulence determinants, including the fimbriae, exercised in *trans* by products of the *bvg* locus. Fimbrial phase variation is also controlled at the level of transcription by insertions or deletions in a run of approximately 15 cytidine residues in the promoter region (Stibitz *et al.*, 1989; Willems *et al.*, 1990). This affects transcription by altering the distance between the binding site for an activator protein and the −10 box (there is no −35 box), which in turn alters the efficiency with which the promoter is recognized by RNA polymerase. This mechanism is not entirely independent of coordinate regulation because the activator protein itself is thought to be controlled by the *bvg* locus.

Phase variation of the surface lipoproteins (Vlps) (Rosengarten & Wise, 1990, 1991; Yogev *et al.*, 1992) of *Mycoplasma hyorhinis* is controlled transcriptionally by loss or gain of adenine nucleotides in a polyA tract between the −10 and −35 boxes in the promoter region. The number of adenines controls transcription, perhaps by influencing the interaction with RNA polymerase or other factors, such that with seventeen adenines the gene is transcribed, while with eighteen to twenty adenines transcription is switched off.

The genus *Yersinia* provides us with further examples of this and also provides some clues to the transition that pathogens can make between commensalism and virulence. Both *Y. pestis* which is highly virulent, and *Y. pseudotuberculosis* which is less virulent, carry the *yopA* gene encoding an outer membrane protein (Bolin, Norlander & Wolf-Watz, 1982), and the *inv* gene encoding the invasin (Isberg & Falkow, 1985). However only *Y. pseudotuberculosis* expresses YopA and invasin proteins. The *Y. pestis* *yopA* sequence differs from that of *Y. pseudotuberculosis* by 15 nucleotides, including a single base deletion in a poly A tract which shifts it out of frame (Rosqvist, Skurnik & Wolf-Watz, 1988). Restoration of *Y. pestis* to YopA$^+$ reduces its virulence, and the introduction of mutations into the *Y. pseudotuberculosis yopA* and *inv* genes greatly increases its virulence. Thus mutations in two genes of *Y. pseudotuberculosis* could cause it to become highly virulent, perhaps causing epidemics, although the time scale over which this might occur is not known. While many other pathogens tend to become less virulent as they adapt to their host, ultimately becoming commensal, these *Yersinia* species have maintained the potential to switch between virulence states, perhaps because this aids transmission of the organism, while otherwise maintaining the population in endemic hosts at lower virulence.

N. gonorrhoeae, that most variable of bacteria, also provides an example of phase variation controlled via homopolymeric sequences with the recently described PilC protein (Jonsson, Nyberg & Normark, 1991), which is reportedly involved in pilus assembly (Jonsson *et al.*, 1991). These authors identified two copies of the *pilC* gene (*pilC*1 and 2) on the chromosome of strain MS11. Phase variation of *pilC* consequently controls the phase variation of pili, in addition to whether or not the pilin molecule is competent for pilus assembly. It was also postulated that different pilins might be more efficiently assembled by different PilC proteins. The part of the *pilC* gene encoding the signal peptide contains a polyG tract, and deletions or additions in this region cause frameshifts with respect to the start codon, exercising translational control. However as well as a role in pilus assembly, PilC may have other important functions (T. Rudel & T. F. Meyer, unpublished data).

VARIATION WITHOUT DNA REARRANGEMENTS

The phase variation of the Pap pili of *E. coli* is transcriptionally controlled but does not involve any of the DNA rearrangements described above. Instead it uses a unique mechanism dependent on the methylation status of two Dam methylation sites (GATC 1028 and 1130) situated upstream of *papA*, the gene encoding the major subunit pilin (Blyn *et al.*, 1989; Blyn, Braaten & Low, 1990). The methylation states of these two sites in phase OFF and phase ON cells was determined by restriction enzyme analysis, and demonstrated that GATC$_{1028}$ was unmethylated when expression of pili was ON, and GATC$_{1130}$ was unmethylated when expression of pili was OFF.

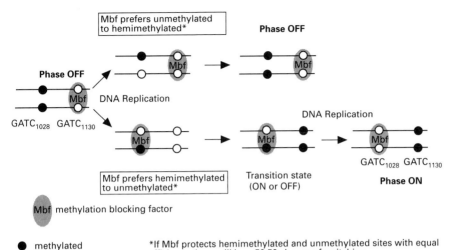

Fig. 3. Model of phase variation of *E. coli* Pap pili as proposed by Blyn *et al* (1990) and Braaten *et al.* (1991). Methylation blocking factor (Mbf) prevents methylation of site $GATC_{1028}$ or $GATC_{1130}$. During DNA replication either the unmethylated and/or hemimethylated site is protected. If Mbf protects the hemimethylated and unmethylated sites with equal efficiency then there will be a 50:50 chance of a switch occurring, assuming both sites cannot be protected simultaneously. Any bias towards either hemimethylated or unmethylated sites will be reflected in a tendency always to switch in the first case, and never to switch in the second. The phase status in the transition state is not known.

The presence of unmethylated GATC sites in *E. coli* is very unusual, but was confirmed using *E. coli* that overexpress Dam and which do not display ON–OFF transitions. This unmethylated state is maintained by a methylation blocking factor (Mbf) (Braaten *et al.*, 1991) which is required for the inhibition of methylation at these sites. In *E. coli* which are mbf^+ both sites are methylated and no Pap pili are expressed. Another gene product encoded by the Pap operon, PapI is also required for the inhibition of methylation at $GATC_{1028}$. The authors propose that PapI and Mbf (or gene products regulated by them) combine to inhibit Dam methylation at $GATC_{1028}$, perhaps by steric hindrance. $GATC_{1130}$ may be protected by Mbf alone. The model explaining phase variation (Blyn *et al.*, 1990) is rather complex and depends on the protection factors having a higher affinity for unmethylated or hemimethylated DNA, versus methylated DNA (Fig. 3). During replication these sites will be protected from Dam methylase. If only unmethylated sites are protected then piliation status will remain fixed. If, however, hemimethylated and unmethylated sites are equally preferred, there will be an equal chance of switching, although the transition from ON to OFF (or vice versa) will require two rounds of DNA replication, with a transition state containing one hemimethylated and one methylated site in between. Whether bacteria are piliated or not in this transition state is unclear. Thereafter daughter cells will either be ON or OFF, or in the transition state. PapB and PapI are also positive regulators of *pap* transcription but it is unknown if they also alter methylation. Pap pilus phase variation is also responsive to environmental changes such as carbon source and temperature. The response to carbon source may be transmitted via cAMP-CRP, which can influence the ON/OFF rate. Low temperature induces ON to OFF transitions and probably turns off pili production outside the host, when it is no longer useful.

CONCLUSIONS

A number of examples have been reviewed of how pathogens alter some of their surface structures in order to evade the host immune response, adapt to the microenvironments they encounter within that host, and to facilitate transmission to the next host. It must be emphasized that the methods here are all examples of random switching, as opposed to regulation, which means that pathogens are able to encounter all manner of unforeseen alterations in their surroundings. However both the rate and direction of switching can be regulated in certain cases. The frequency of the site-specific inversions mediating fimbrial phase variation in *E. coli* is increased by mutations in *osmZ*, the gene product of which strongly influences *in vivo* DNA supercoiling in response to osmotic stress (Higgins *et al.*, 1988; see also Dorman & Ní Bhriain, this symposium). The direction of fimbrial phase variation is controlled by FimB and FimE proteins, which direct the phase switch towards ON and OFF, respectively (Fig. 3), possibly by acting like integrases (Klemm, 1986). Therefore, the otherwise random mechanisms of antigenic variation can be regulated at a higher level, providing the organism with a means of coordinating its survival strategy. Both recombination and especially slipped-strand mispairing provide powerful forces for the rapid evolution of DNA sequences, and consequently gene expression. Antigenic variation not only facilitates escape from the immune response but also allows the optimization of receptor–ligand interactions, and in some cases, such as *Yersinia*, the modulation of virulence potential to suit long-term commensal-like association with the host, or a highly virulent state that is perhaps required for the transmission of the pathogen. The study of antigenic variation and the mechanisms that underly it will continue to provide insight into not only the means by which pathogens survive within the host, but also into the evolution of virulence and the transition from environmental organism to pathogen, and back.

REFERENCES

Abraham, J. M., Freitag, C. S., Clements, J. R. & Eisenstein, B. I. (1985). An invertible element of DNA controls phase variation of type 1 fimbriae of *Escherichia coli*. *Proceedings of the National Academy of Sciences, USA*, **82**, 5724–7.

Allred, D. R., McGuire T. C., Palmer, G. H., Leib, S. R., Harkins, T. M., McElwain, T. F. & Barbet, A. F. (1990). Molecular basis for surface antigen size polymorphisms and conservation of a neutralization-sensitive epitope in *Anaplasma marginale*. *Proceedings of the National Academy of Sciences, USA*, **87**, 3220–4.

Belland, R. J. (1991). H-DNA formation by the coding repeat elements of neisserial *opa* genes. *Molecular Microbiology*, **5**, 2351–60.

Berg, D. E. & Howe, M. M. (1989). *Mobile DNA*. Washington: American Society for Microbiology.

Blyn, L. B., Braaten, B. A., White-Ziegler, C. A., Rolfson, D. H. & Low, D. A. (1989). Phase-variation of pyelonephritis-associated pili in *Escherichia coli*: evidence for transcriptional regulation. *EMBO Journal*, **8**, 613–20.

Blyn, L. B., Braaten, B. A. & Low, D. A. (1990). Regulation of *pap* pilin phase variation by a mechanism involving differential Dam methylation states. *EMBO Journal*, **9**, 4045.

Bolin, I., Norlander, L. & Wolf-Watz, H. (1982). Temperature-inducible outer membrane protein of *Yersinia pseudotuberculosis* and *Yersinia enterocolitica* is associated with the virulence plasmid. *Infection and Immunity*, **37**, 506–12.

Borst, P. & Greaves, D. R. (1987). Programmed rearrangements altering gene expression. *Science*, **235**, 658–67.

Borst, P. (1991). Molecular genetics of antigenic variation. *Immunology Today*, Supplement A: A29–33.

Braaten, B. A., Blyn, L. B., Skinner, B. S. & Low, D. A. (1991). Evidence for a methylation-blocking factor (*mbf*) locus involved in *pap* pilus expression and phase variation in *Escherichia coli*. *Journal of Bacteriology*, **173**, 1789–800.

Coote, J. G. (1991). Antigenic switching and pathogenicity: environmental effects on virulence gene expression in *Bordetella pertussis*. *Journal of General Microbiology*, **137**, 2493–503.

DiRita, V. J. & Mekalanos, J. J. (1989). Genetic regulation of bacterial virulence. *Annual Review of Genetics*, **23**, 455–82.

Dorman, C. J. & Higgins, C. F. (1987). Fimbrial phase variation in *Escherichia coli*: dependence on integration host factor and homologies with other site-specific recombinases. *Journal of Bacteriology*, **169**, 3840–3.

Eisenstein, B. I., Sweet, D. H., Vaughn, V. & Friedman, D. I. (1987). Integration host factor is required for the DNA inversion that controls phase variation in *Escherichia coli*. *Proceedings of the National Academy of Sciences, USA*, **84**, 6506–10.

Fischetti, V. A. (1991). Streptococcal M protein. *Scientific American*, **264** (6), 32–9.

Fischetti, V. A., Jones, K. F. & Scott, J. R. (1985). Size variation of the M protein in group A streptococci. *Journal of Experimental Medicine*, **161**, 1384–401.

Gibbs, C. P., Reimann, B.-Y., Schultz, E., Kaufman, A., Haas, R. & Meyer, T. F. (1989). Reassortment of pilin genes in *Neisseria gonorrhoeae* occurs by two distinct mechanisms. *Nature*, **338**, 651–2.

Glasgow, A. C., Hughes, K. T. & Simon, M. I. (1989). Bacterial DNA inversion systems. In *Mobile DNA*, eds D. E. Berg & M. M. Howe, pp. 637–659. Washington: American Society for Microbiology.

Goudsmit, J., Back, N. K. & Nara, P. L. (1991). Genomic diversity and antigenic variation of HIV-1: links between pathogenesis, epidemiology and vaccine development. *FASEB Journal*, **5**, 2427–36.

Haas, R. & Meyer, T. F. (1986). The repertoire of silent pilus genes in *Neisseria gonorrhoeae*: evidence for gene conversion. *Cell*, **44**, 107–15.

Higgins, C. F., Dorman, C. J., Stirling, D. A., Waddell, L., Booth, I. R., May, G. &

Bremer, E. (1988). A physiological role for DNA supercoiling in the osmotic regulation of gene expression in *S. typhimurium* and *E. coli*. *Cell*, **52**, 569–84.

Hollingshead, S. K., Fischetti, V. A. & Scott, J. R. (1987). Size variation in group A streptococcal M protein is generated by homologous recombination between intragenic repeats. *Molecular and General Genetics*, **207**, 196–203.

Isberg, R. R. & Falkow, S. (1985). A single genetic locus encoded by *Yersinia pseudotuberculosis* permits invasion of cultured mammalian cells by *E. coli* K12. *Nature*, **317**, 262–4.

Jones, K. F., Hollingshead, S. K., Scott, J. R. & Fischetti, V. A. (1988). Spontaneous M6 protein size mutants of group A streptococci display variation in antigenic and opsonogenic epitopes. *Proceedings of the National Academy of Sciences, USA*, **85**, 8271–75.

Jonsson, A.-B., Nyberg, G. & Normark, S. (1991). Phase variation of gonococcal pili by frameshift mutation in *pilC*, a novel gene for pilus assembly. *EMBO Journal*, **10**, 477–88.

Klemm, P. (1986). Two regulatory *fim* genes, *fimB* and *fimE*, control the phase variation of type 1 fimbriae in *Escherichia coli*. *EMBO Journal*, **5**, 1389–93.

Levinson, G. & Gutman, G. A. (1987). Slipped-strand mispairing: a major mechanism for DNA sequence evolution. *Molecular Biology and Evolution*, **4**, 203–21.

Marrs, C. F., Ruehl, W. W., Schoolnik, G. K., Falkow, S. (1988). Pilin gene phase variation of *Moraxella bovis* is due to an inversion of the pilin genes. *Journal of Bacteriology*, **170**, 3032–9.

Maskell, D. J., Szabo, M. J., Butler, P. D., Williams, A. E. & Moxon, E. R. (1991). Molecular analysis of a complex locus from *Haemophilus influenzae* involved in phase-variable lipopolysaccharide biosynthesis. *Molecular Microbiology*, **5**, 1013–22.

Mateu, M. G., Martinez, M. A., Rocha, E., Andru, D., Parejo, J., Irait, E., Sobrino, F. & Domingo, E. (1989). Implications of a quasispecies genome structure: effect of frequent, naturally occurring amino acid substitutions on the antigenicity of foot-and-mouth disease virus. *Proceedings of the National Academy of Sciences, USA*, **86**, 5883–7.

Meyer, T. F. (1987). Molecular basis of surface antigen variation in *Neisseria*. *Trends in Genetics*, **3**, 319–24.

Meyer, T. F., Mlawer, N. & So, M. (1982). Pilus expression in *Neisseria gonorrhoeae* involves chromosomal rearrangement. *Cell*, **30**, 45–52.

Meyer, T. F., Gibbs, C. P. & Haas, R. (1990). Variation and control of protein expression in *Neisseria*. *Annual Review of Microbiology*, **44**, 451–77.

Miller, J. F., Mekalanos, J. J. & Falkow, S. (1989). Coordinate regulation and sensory transduction in the control of bacterial virulence. *Science*, **243**, 916–22.

Murphy, G. L., Connell, T. D., Barritt, D. S., Koomey, M. & Cannon, J. G. (1989). Phase variation of gonococcal protein II: regulation of gene expression by slipped-strand mispairing of a repetitive DNA sequence. *Cell*, **56**, 539–47.

Plasterk, R. H. A., Simon, M. I. & Barbour, A. G. (1991). Transposition of structural genes to an expression sequence on a linear plasmid causes antigenic variation in the bacterium *Borrelia hermsii*. *Nature*, **318**, 257–63.

Rosengarten, R. & Wise, K. (1990). Phenotypic switching in mycoplasmas: phase variation of diverse surface lipoproteins. *Science*, **247**, 315–18.

Rosengarten, R. & Wise, K. (1991). The Vlp system of *Mycoplasma hyorhinis*: combinatorial expression of distinct size variant lipoproteins generating high-frequency surface antigenic variation. *Journal of Bacteriology*, **173**, 4782–3.

Rosqvist, R., Skurnik, M. & Wolf-Watz, H. (1991). Increased virulence of *Yersinia pseudotuberculosis* by two independent mutations. *Nature*, **334**, 522–5.

Scott, J. R. (1990). The M protein of group A *Streptococcus*: evolution and regulation. In *Molecular Basis of Bacterial Pathogenesis, The Bacteria*, Vol. XI, ed. B. H. Iglewski & V. L. Clark, pp. 177–203. New York: Academic Press.

Segal, E., Hagblom, P., Seifert, H. S. & So, M. (1986). Antigenic variation of gonococcal pilus involves assembly of separated silent gene segments. *Proceedings of the National Academy of Sciences, USA*, **83**, 2177–81.

Seifert, H. S., Ajioka, R. D., Marchal, C., Sparling, P. F. & So, M. (1988). DNA transformation leads to pilin antigenic variation in *Neisseria gonorrhoeae*. *Nature*, **336**, 392–5.

Skehel, J. J. & Wiley, D. C. (1986). Antigenic variation in Hong Kong influenza virus haemagglutinins. In *Antigenic Variation in Infectious Diseases*, eds T. H. Birbeck & C. W. Penn, pp. 19–24. Oxford: IRL Press.

Stibitz, S., Aaronson, W., Monack, D., Falkow, S. (1989). Phase variation in *Bordetella pertussis* by frameshift mutation in a gene for a novel two-component system. *Nature*, **338**, 266–9.

Stern, A., Brown, M., Nickel, P. & Meyer, T. F. (1986). Opacity genes in *Neisseria gonorrhoeae*: control of phase and antigenic variation. *Cell*, **47**, 61–71.

Swanson, J., Bergström, S., Robbins, K., Barrera, O., Corwin, D. & Koomey, J. M. (1986). Gene conversion involving the pilin structural gene correlates with pilus$^+$ ⇌ pilus$^-$ changes in *Neisseria gonorrhoeae*. *Cell*, **47**, 267–76.

Weiser, J. N., Love, J. M. & Moxon, E. R. (1989). The molecular mechanism of phase variation of *H. influenzae* lipopolysaccharide. *Cell*, **59**, 657–65.

Weiser, J. N., Maskell, D. J., Butler, P. D., Lindberg, A. A. & Moxon, E. R. (1990). Characterization of repetitive sequences controlling phase variation of *Haemophilus influenzae* lipopolysaccharide. *Journal of Bacteriology*, **172**, 3304–9.

Willems, R., Paul, A., van der Heide, H. G. J., ter Avest, A. R. & Mooi, F. R. (1990). Fimbrial phase variation in *Bordetella pertussis*: a novel mechanism for transcriptional regulation. *EMBO Journal*, **9**, 2803–9.

Yancopoulos, G. D. & Alt, F. W. (1986). Regulation of the assembly and expression of variable-region genes. *Annual Review of Immunology*, **4**, 339–68.

Yogev, D., Rosengarten, R., Watson-McKown, R. & Wise, K. (1991). Molecular basis of *Mycoplasma* surface antigenic variation: a novel set of divergent genes undergo spontaneous mutation of periodic coding regions and 5' regulatory sequences. *EMBO Journal*, in press.

PAP I 2103

HAEMOPHILUS INFLUENZAE LIPOPOLYSACCHARIDE: THE BIOCHEMISTRY AND BIOLOGY OF A VIRULENCE FACTOR

E. RICHARD MOXON AND D. MASKELL

Molecular Infectious Diseases Group, Department of Paediatrics & Institute of Molecular Medicine, John Radcliffe Hospital, Headington, Oxford OX3 9DU, UK

INTRODUCTION

Lipopolysaccharide (LPS) is a macromolecule that is unique to Gram-negative bacteria. It is the major component of the outer leaflet of the cell envelope and forms an integral part of the outer membrane structure (Inouye, 1979). Many structural details of LPS have been defined, and have made it possible to correlate particular domains of the LPS macromolecule with its role in virulence and its potential to act as a target for specific and non-specific host responses. LPS consists of two components that have contrasting chemical and physical properties; a hydrophilic polysaccharide and a hydrophobic lipid (lipid A).

Over the past decade, a considerable body of knowledge has accumulated on the biochemistry and biology of *Haemophilus influenzae* LPS. *H. influenzae* is a common commensal of the human upper respiratory tract to which it is highly adapted; it is not known to colonize any other animal species. It is also occasionally pathogenic and, in young children, the highly virulent serotype b strains are a common cause of life-threatening invasive infections such as meningitis, cellulitis, epiglottitis and pneumonia. Other capsular serotypes (a, c, d, e and f) and non-typeable (NT) strains are less likely to be associated with bacteraemic infection. NT strains are one of the most common causes of otitis media and of fatal acute lower respiratory tract infections, the latter afflicting millions of infants living in socio-economically deprived situations throughout the world. This chapter attempts to summarize knowledge of *H. influenzae* LPS and how this knowledge has increased appreciation of its biological functions.

BIOCHEMISTRY

The LPS of *H. influenzae* comprises lipid A and core oligosaccharide components but lacks a high Mr side-chain polysaccharide (Fig. 1; Flesher & Insel, 1978; Inzana, 1983). Because of the similarity to so-called rough forms

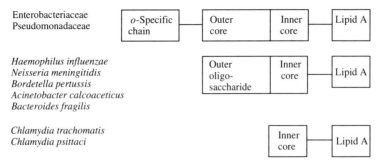

Fig. 1. General architecture of lipopolysaccharides of selected human pathogenic gram-negative bacteria. (Modified after Rietschel *et al.*, 1990.)

of LPS, the term lipooligosaccharide is currently being employed by a few authors. However, it is preferable to retain the original terminology to specify a molecule which has many biological and structural characteristics typical of LPS.

Composition

The lipid A backbone is a GlcN disaccharide substituted by two phosphate groups, one of which is glycosidically linked to the reducing GlcN, the other being linked to the non-reducing GlcN (Helander *et al.*, 1988). This conforms with the composition of the enterobacterial lipid A class (Imoto *et al.*, 1983).

A notable property of the lipid A of *H. influenzae* is the simplicity of its fatty acids, comprising only tetradecanoic and its 3-hydroxylated derivative (Zamze & Moxon, 1987). Some strains of *H. influenzae* have been found to contain small amounts (<5% w/w) of fatty acids 16 : 0 and 16 : 1, most likely representing traces of phospholipid. No 12 : 0 fatty acids have been found. The results are consistent with the taxonomic placement of *H. influenzae* in the family Pasteurellaceae (Erler *et al.*, 1977).

Monoclonal antibody analysis has shown that at least two antigenically distinct components of lipid A exist among typeable and non-typeable strains of *H. influenzae* (Apicella *et al.*, 1985). The epitope responsible for this differentiation is apparently a lipid moiety that can be recovered in the chloroform soluble portion of the lipid A. It is present in the majority of NT and type b strains and is apparently specific for *H. influenzae* since it is absent from other bacterial species tested (Apicella *et al.*, 1985). Variation in lipid A within the same species has been demonstrated previously and is due to variation of the composition of ester-linked fatty acids and substitutions of phosphate groups (Mattsby-Baltzer, Gemski & Alving, 1984).

A chemical analysis of the oligosaccharide from the LPS of a virulent type b strain revealed both similarities to, and some differences from, the

Haemophilus influenzae lipopolysaccharide

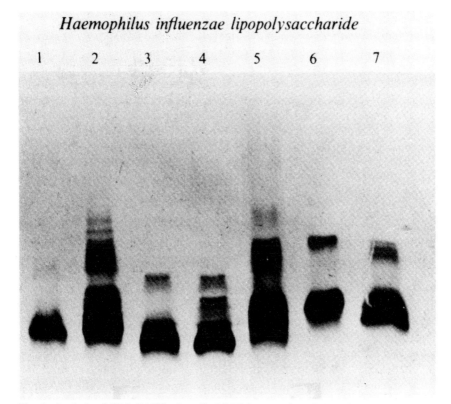

Fig. 2. Analysis of SDS–PAGE of purified LPS from strains of *H. influenzae* of different capsular serotype. Lanes 1–7 show purified LPS (5 μg per lane) from strains:
1: Morgan (type a); 2: Eagan (type b); 3: ATCC9007 (type c); 4: ATCC9008 (type d); 5: ATCC9009 (type e); 6: ATCC9010 (type f); 7: *Salmonella typhimurium* (Ra mutant).

composition of the core region of enterobacterial LPS (Inzana & Anderson, 1985; Flesher & Insel, 1978; Parr & Bryan, 1984). Oligosaccharides from strains representing capsular serotypes a–f are quantitatively similar in composition (Zamze & Moxon, 1987) but the molar ratios and linkages of glucose, galactose, heptose and 2-keto-3-deoxy-*manno*-octulosonic acid (KDO) among different capsular serotypes had quantitative variations. These correlated with antigen-specific reactions with homologous antisera and mobility on SDS–polyacrylamide gel electrophoresis (SDS–PAGE). In general, these differences reflected quantitative variations in either galactose or glucose, or both. SDS–PAGE showed that the LPS from strains of serotype a, c and d were of lower Mr than the LPS from most serotype b strains (Fig. 2; Zamze & Moxon, 1987). A small amount of glucosamine was detected in the oligosaccharides of serotypes b and d. Also, galactosamine was found in a non-capsulate type d variant. Based on the acid hydrolysis (HC1) conditions required for optimum release of galactose and glucose, as compared to heptose, the heptose residues are either extensively phos-phorylated, or are involved in extremely acid-stable linkages.

Studies utilizing Western blot analysis of LPS preparations demonstrated marked antigenic heterogeneity of NT *H. influenzae* (Gupta, Dudas & Apicella, 1986). A series of five cross-adsorbed antisera were developed which were specific for their homologous LPS. Screening of NT strains with the five antisera allowed for the grouping of strains on the basis of their reactivity with the antisera. These studies indicate that strains of NT *H. influenzae* can be characterized on the basis of this heterogeneity (Murphy & Apicella, 1987).

Structure

The most thorough analysis of the *H. influenzae* lipid A to date has been on a mutant, strain Rd:I69 (Zwahlen *et al.*, 1985). This mutant has the most incomplete type of LPS so far identified, being even more defective than the Re chemotype established for Salmonella (Luderitz *et al.*, 1982). The lipid A backbone of this mutant is shown in Fig. 3.

A detailed analysis of the structure of the core oligosaccharides of *H. influenzae* LPS has not yet been achieved. A major technical difficulty impeding progress is that, in addition to the well-recognized microheteroge-neity of LPS and extensive phosphorylation of inner core sugars, there is high-frequency, reversible loss and/or gain of outer core oligosaccharides; some mechanisms implicated in mediating these changes include transcrip-tional regulation, frame-shifts within biosynthetic genes and RNA process-ing (see 'Genetic analysis of LPS variation'). For optimal analysis of oligosaccharides derived from the LPS, mutations in genes mediating phase variation are needed.

Monoclonal antibodies (mabs) raised against core oligosaccharide struc-tures have been helpful and have defined at least two phase-variable structures; the terminal oligosaccharide Galα1-4Galβ (Virji *et al.*, 1990) and Galβ1–4GlcNAc (Mandrell *et al.*, 1992). Both of these structures cross-react with human glycosphingolipids (blood group antigens). Recent evidence indicates that Galβ1-4GlcNAc can be terminally sialylated. Also Galα1–4Galβ is known to be the receptor for PapG, the adhesin of uropathogenic *E. coli* fimbriae (Karlsson *et al.*, 1986) as well as the shiga toxin of *Shigella dysenteriae*.

Antigenic variation

Despite the lack of structural details, mabs raised to specific LPS structures have allowed detailed characterization of the extensive variation and cross-reactivity of *H. influenzae* oligosaccharides. Kimura and Hansen (1986) described spontaneous, high-frequency, reversible acquisition and loss of

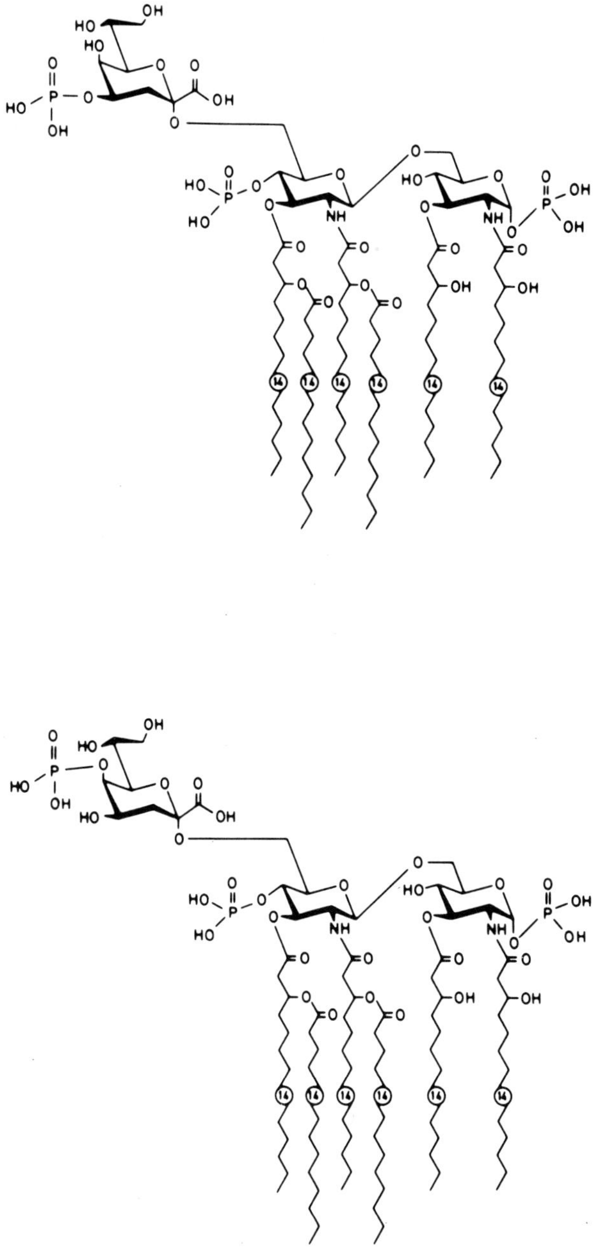

Fig. 3. Structure of the 'deep rough' lipopolysaccharide of *H. influenzae* mutant strain I69. Based on results of Helander *et al.* (1988), the lipopolysaccharide consists of a mixture of two molecular species, one with a dOclA 4-phosphate (upper) and the other with a dOclA 5-phosphate (lower). Numbers in circles refer to the chain length of acyl groups. 14:0(3–OH) posesses the (R) configuration. The fully protonated forms are shown. (Reproduced with permission.)

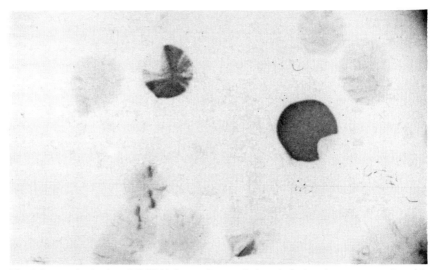

Fig. 4. Sectored colonies of RM7004 due to phase variation. A single colony, non-reactive with *mab* 12D9, was inoculated into supplemented BHI broth, grown to the mid-log phase, and plated on BHI agar. These colonies were immunoblotted with *mab* 12D9 and photographed. (Weiser *et al.*, 1989*a*.)

reactivity with oligosaccharide-specific mabs and that these phenotypic changes were associated with susceptibility to serum killing and virulence for infant rats (see 'Pathogenic effects mediated by core oligosaccharides). This propensity for LPS antigenic variation was typical of serotypable (capsulate) and NT *H. influenzae*, among which extensive cross-reactivity of LPS epitopes exists. Despite the phase-variable reactivity of these mabs, the findings incidentally produced a basis for grouping isolates by the patterns of their antigenicity (Patrick *et al.*, 1987; Kimura *et al.*, 1987). For example, among type b strains, three major antigenic groups were defined based on reactivity with two mabs, designated 12D9 and 4C4. These inter-strain patterns of binding were shown to be consistent with the variable expression of structures present on at least two distinct LPS molecules (Patrick *et al.*, 1989). Subsequently, genes required for 12D9 and 4C4 expression were mapped to separate loci (*lic1* and *lic2*) in type b strains (RM7004 and Eagan) as discussed in 'Genetic analysis of LPS variation' (Weiser *et al.*, 1989*a*, 1990). Intra-strain variations of LPS phenotype were demonstrated for colonies (Kimura & Hansen, 1986) and individual bacteria (Figs 4, 5) (Weiser *et al.*, 1989*a*, Weiser, Love & Moxon, 1989*b*). The frequency with which loss or gain of phase-variable epitopes occurs is usually about 10^{-2} per bacterium per generation, but is highly variable (~ 1–16%).

Weiser *et al.*, (1989*b*) examined the relationship between five phase-variable LPS epitopes. Whereas loss of a single epitope was noted, concurrent loss of several (at least four) epitopes could also occur. Thus, the

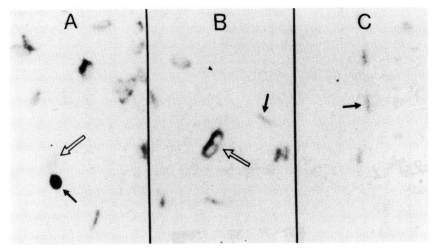

Fig. 5. Reactivity of individual organisms with lipopolysaccharide-specific monoclonal antibody (4C4) by an immunoperoxidase staining technique. Bacteria from a single colony were incubated with the monoclonal antibody followed by peroxidase-conjugated second antibody and stained for peroxidase. They were then counterstained with Giemsa. Antibody reactive cells (open arrows) appear larger with an enhanced outer ring. The counterstain is adsorbed by non-reactive cells (solid arrows) and stains encapsulated organisms more intensely than unencapsulated organisms. (A) RM.7004 (*H. influenzae*, type b); (B) RM.118–1 (*H. influenzae*, capsule deficient); (C) RM.118 (*H. influenzae*, capsule deficient). Magnification, *c.* × 4650.

molecular basis of on/off switching is complex and can involve both independent and co-ordinate switching of epitopes. Until detailed structural analyses are completed, one can only speculate on the extent of the structural variation exhibited by *H. influenzae*. None the less, using the several mabs made available, the theoretical permutations arising from independent on/off switching of just six structures indicates the potential for extensive diversity.

Similar, phase-variable switching of LPS epitopes has been observed in *N. meningitidis, N. gonorrhoeae* and *N. lactamica* but apparently does not occur in the enterobacteriaceae. Interestingly, cross-reactions between LPS structures expressed by *Haemophilus* and *Neisseria* occur. In immunoblotting experiments, common epitopes were present on *N. gonorrhoeae, N. lactamica, N. meningitidis, H. influenzae* (including bio-group aegyptius) and *H. parainfluenzae*. A mab raised against *N. gonorrhoeae* reacted with several strains of *N. gonorrhoea, N. meningitidis* and *H. influenzae* (Virji *et al.*, 1990). In addition, Galα1–4Galβ epitopes expressed by *H. influenzae* were also found on strains of *N. gonorrhoeae*; this structure is a receptor for PapG, the adhesin of uropathogenic *E. coli*. Mutants expressing P-fimbriae (with and without the adhesin) were mixed with *H. influenzae* or *N. meningitidis*; agglutination occurred only in reactions involving recombinant

E. coli expressing the G adhesin and the *N. meningitidis* or *H. influenzae* strains expressing the digalactoside epitope. Thus, there are similarities between the variable Galα1–4Galβ epitopes of *H. influenzae* and *N. gonorrhoeae*, structures that are also shared by human paragloboside and other glycosphingolipids (Mandrell *et al.*, 1992).

GENETICS

Genes required for LPS biosynthesis

Sequences linked to the capsulation locus of *H. influenzae* affect LPS phenotype (Zwahlen, Winkelstein & Moxon, 1983). It is reasonable that steps in the biosynthesis of capsular polysaccharide and LPS may share some common enzymatic or regulatory genes (Jann & Jann, 1990), although this has not been shown in *H. influenzae*. However, when donor DNA from strain Eagan was used to transform a spontaneously derived capsule deficient recipient (strain Rd), transformants were obtained that were of two distinct lipopolysaccharide phenotypes, designated Rdb+:01 and Rdb+:02 (Zwahlen, Rubin & Moxon, 1986). These differences in LPS phenotype depend on whether one or two copies of the capsule genes (duplicated in the donor strain) are recombined into Rd. Rdb+:02 has one copy, and Rdb+:01 two copies of capsule genes. Biochemically, Rdb+:02 has a relative reduction in the amount of galactose in its LPS core and is more virulent for infant rats than is Rdb+:01 (see 'Pathogenic effects mediated by core oligosaccharides').

The mutant, Rd:I69, has a deep rough LPS phenotype and lacks all core sugars except a single phosphorylated KDO. It is surprising that this mutant is viable, but it is able to grow at normal rates *in vitro* despite having the 'roughest' LPS type described to date (Zamze *et al.*, 1987). Because the LPS of this mutant contained only one phosphorylated KDO residue, I69 LPS provides information about the biosynthesis of the lipid A-KDO region since its structure implies that KDO is transferable to lipid A (precursor) as a monosaccharide (Helander *et al.*, 1988). This is at variance with the situation in *E. coli* where two KDO molecules are transferred to lipid IV_A in rapid succession to form $KDO_2–IV_A$. The bifunctional KDO transferase has been cloned from *E. coli* (Raetz, 1990) and it should be interesting to compare the structure of this enzyme with that of the *H. influenzae* homologue.

Recently, Spinola and colleagues (1990) have identified and cloned a locus (EMBLOS-1) involved in the biosynthesis of *H. influenzae* LPS core sugars. When the cloned *H. influenzae* DNA was transfected into an *E. coli* host strain, the LPS core sugars were modified so that there was an addition of an estimated 1.4 kD oligosaccharide species. The *H. influenzae* DNA contains a cluster of at least three loci, designated *isg*-1, *isg*-2 and *isg*-3, whose products act sequentially in LPS biosynthesis. *isg*-1 was required for modification of the *E. coli* core sugars, and therefore may be an enzyme recognizing an *E. coli* core structure. *isg*-3 assembles or exposes a KDO containing epitope recognized by the mab 6E4. The 6E4 reactive, KDO-containing epitope did not phase vary and therefore probably represents a relatively conserved region of the core. A transposon mutation in the cloned *H. influenzae* DNA was made and used to transfect *H. influenzae*; the transformant had an altered LPS that did not react with the mab 6E4, confirming that the cloned genes were indeed involved in LPS biosynthesis (Kwaik *et al.*, 1991).

In contrast to the biosynthetic genes of EMBLOS-1, two loci have been identified, *lic1* and *lic2*, that are required for the expression of phase-variable oligosaccharide epitopes (Weiser *et al.*, 1989*a*; Weiser *et al.*, 1990). The structure of one of these phase-variable epitopes, for which at least one gene in *lic2* is required (Cope *et al.*, 1991), is a terminal Galα1–4Galβ structure. A third locus, *lic3*, contains *galE* (Maskell *et al.*, 1991). When a *H. influenzae galE* mutant was grown in glucose-containing medium in the absence of galactose, the Galα1–4Galβ structure was not expressed. Unlike the organization of the well-described galactose biosynthetic genes (*gal* operon) in *E. coli* (Adhya, 1987), the genes for galactose biosynthesis are arranged in *H. influenzae* such that *galK* (kinase), *galT* (transferase) and *galR* (repressor) map together elsewhere in the chromosome, an unknown distance from *galE*.

Genetic analysis of LPS variation

To analyse the molecular basis of the variable LPS epitope expression, a genomic library from a type b strain (RM7004) was screened for donor clones that could transform a recipient *H. influenzae* strain (Rd) to express novel LPS epitopes (Weiser *et al.*, 1989*a*). Strain Rd does not react with the oligosaccharide-specific mabs (4C4, 5G8, 12D9 or 6A2) described by Patrick *et al.* (1989), whereas RM 7004 shows phase-variable reactivity with each of them. A chromosomal locus cloned from RM7004, designated *lic1*, was shown to be involved in the expression of two of the phase-variable epitopes, 12D9 and 6A2 (Weiser *et al.*, 1989*a*). The original clone also transformed Rd to result in 4C4 expression, but at extremely low frequency. It is now evident that genes required for 4C4 expression in the donor (RM7004) are located within a second and separate locus, *lic2*, which maps close to *lic1*. Mutagenesis of *lic1* in RM7004 was used to generate a series of isogenic mutants with altered mab binding specificities on colony blots; these changes were associated with changes in LPS sugar composition.

Site-specific deletion mutations in *lic1* eliminated expression of two epitopes, 6A2 and 12D9, and confirmed that *lic1* was involved in oligo-saccharide biosynthesis. The nucleotide sequence of *lic1* comprises four open-reading frames (ORFs), *lic1*A, *lic1*B, *lic1*C and *lic1*D, encoding

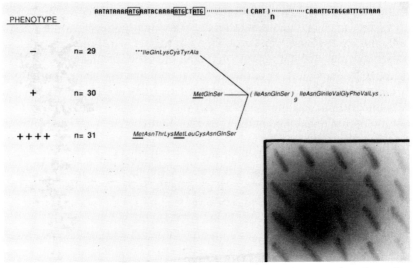

Fig. 6. Proposed molecular mechanism of LPS phase-variation. Phase-variation of *H. influenzae* LPS epitopes expressed by *lic1* is determined by a translational switch. The nucleotide sequence of the 5' end of *lic*A is listed with possible ATG initiation codons boxed. Variation in the number of multiple CAAT repeats (n) shifts the reading frame of *lic*A, altering the translational phase upstream of the repeats. The three possible amino-terminal translational products (n = 29-31) are shown below the nucleotide sequence. Depending upon the number of CAAT units, either one, two, or no initiation codons are in-frame so as to result in variable translation of *lic*A. It is proposed that the three levels of expression observed for *lic1*-determined epitopes, (+), (++++) or (−), correspond to the three translation phases of the 5' end of *lic*A. An immunoblot of a clonal population of streaked-out individual colonies of RM7004 shows the three levels of expression of the 6A2 epitope recognized by *mab* 6A2. Most colonies are (+) with single examples of the (−) and (++++) phenotypes. [Weiser J. N. *et al.*, (1989*b*).]

proteins of 38.5, 32, 24.5 and 30.5 kD (Weiser *et al.*, 1989*b*) and none was homologous with sequences in the EMBL database. Upstream from *lic1*A there are consensus sequences for -35 and -10 binding sites for RNA polymerase, and downstream of *lic1*D there is a stem-loop structure which may be a rho-independent terminator. The first ORF, *lic1*A, is involved in mediating phase variation. At its 5' end, immediately downstream from 3 ATG (start) codons, there is a variable number of tandem repeats of the tetranucleotide 5'-CAAT-3'. Loss or gain of copies of CAAT occur (Weiser *et al.*, 1989*b*), possibly by slipped-strand mispairing (Levinson & Gutman, 1987). It is interesting that the sequence immediately upstream of *lic1*A is relatively AT rich compared with the average AT content characteristic of the *H. influenzae* genome; this might favour denaturation and promote mispairing of the downstream CAAT repeats. Through loss or gain of CAAT repeats, frame shifts may occur with respect to upstream initiation codons resulting in altered translation (Fig. 6). Two of the three possible frames could initiate translation, with the first frame containing two ATGs and the second containing the third ATG. Three levels of expression of the phase-variable 6A2 and 12D9 epitopes in *lic1* were observed in colony blots of RM7004: strong [++++], weak [+] and undetectable [−] so that it seemed reasonable to propose that these different levels of expression correlated with the three possible frames. To investigate the hypothesis that variation in the number of repeats correlated with altered translation of *lic1*A, oligonucleotides flanking the repeat region, on opposite strands, were used in a polymerase chain reaction (PCR) to amplify a fragment of *lic1*A containing the CAAT repeats from spontaneously derived colony variants that were 6A2 [++++] or 6A2 [−]. The most prevalent species of DNA obtained from the 6A2 [−] variant was 184 bp in size corresponding to 29 CAAT repeats, whereas that from the 6A2 [++++] variant was 192 bp corresponding to 31 repeats of CAAT. Of colonies, 97% showed weak [+] expression of the phase-variable epitope corresponding to 30 repeats of CAAT. Thus, although loss or gain of repeats resulted in either strong or undetectable mab binding, 30 repeats is apparently favoured over 29 or 31 (or other equivalent copy numbers of CAAT). The mechanisms exerting a bias in favour of a particular number of repeats (or its associated phenotype) are not understood. These data suggested that variation in the number of CAAT repeats correlates with altered expression of the epitopes dependent on *lic1* and this may be due to altered translation of *lic1*A. As further evidence of the probable role of *lic1*A in determining phase-variation, an insertion mutation was constructed within the *lic1*A open-reading frame. This mutant allowed low level but constitutive expression of the 6A2 and 12D9 epitopes (Weiser *et al.*, 1989*b*). The present model to explain the phase variable expression of epitopes dependent on *lic1* is that spontaneous loss or gain of CAAT repeat units affects the translation of *lic1*A which, in turn, affects the expression of the other three genes of the *lic1* operon. The functions of these *lic1* proteins are completely unknown; they lack homology to glycosyl transferases and, particularly in the case of *lic1*A, may serve a regulatory role.

To investigate the control of the *lic* loci at the level of transcription, RM7004 RNA was probed with radioactive DNA from *lic1*. A transcript of ~3 kb containing message from *lic1*A, *lic1*B and *lic1*C, and terminating between *lic1*C and *lic1*D was observed. Inspection of the DNA in the intergenic region between *lic1*C and *lic1*D suggested that it may form an ordered secondary structure allowing processing, attenuation or termination of the message. A *lacZ* fusion was constructed in-frame with *lic1*D on the *H. influenzae* chromosome. Resultant colonies showed variable, high frequency and reversible switching between β-galactosidase (β-gal) positive or negative colony phenotypes (Szabo *et al.*, 1992). Using single colony PCR, the relationships between the number of CAAT repeats and β-gal expression were as follows:

29 copies of CAAT : out of frame:β-gal[+], or β-gal[−].
30 copies of CAAT : not seen.
31 copies of CAAT : in frame:β-gal[++++], or β-gal[−].

Thus, loss and gain of CAAT in *lic1*A apparently influences the expression of *lic1*D, but only in conjunction with other mechanisms, which in turn must be affecting the antigenic switching. To determine if independent transcription of *lic1*D occurs from promoter sites between *lic1*C and *lic1*D, an omega insertion, containing multiple 'stops', was also introduced into *lic1*C in the presence of the in-frame *lacZ* fusion in the *lic1*D–*lacZ* fusion strain; there was no β-gal expression (unpublished data). Thus, *lic1*D may be transcribed as part of a polycistronic *lic1* message, but then there may be processing and rapid degradation of *lic1*D message. Mutations in *lic1*C which abolish both 6A2 and 12D9 expression, may be having a polar effect on *lic1*D (12D9 epitope); this is consistent with the above model. A 12D9 reactive epitope can be expressed in the absence of 6A2 in some strains of *Haemophilus* (Weiser *et al.*, 1989*a*) but mutation of *lic1*D nullifies expression of both the 12D9 and 6A2 reactive epitopes. Thus, the 12D9 reactive structure is apparently required before the 6A2 epitope can be added.

Additional loci, *lic2* and *lic3*, were indicated when endonuclease digested chromosomal DNA from RM7004 showed three differently sized fragments hybridizing strongly to an oligonucleotide probe with the sequence 5′– CAAT CAAT CAAT CAAT CAAT–3′ (Weiser *et al.*, 1989). Appropriate clones were then obtained from an EMBL3 lambda library and, from these clones, genes within the *lic2* and *lic3* loci were characterized. Within *lic2* of RM7004, there is an open reading frame containing multiple copies of the tetramer 5′–CAAT–3′ at its 5′ end. This *lic2* ORF showed no homology to any known sequences in an extended data base. Mutations within this ORF resulted in loss of expression of the oligosaccharide epitopes recognized by the mabs 4C4, 5G8 and A1. An identical gene (*lex1*) has been found in another *H. influenzae* strain (Cope *et al.*, 1991). Downstream from *lex1*, separated by several nucleotides, is the 3′ end of an open reading frame that would be transcribed in the opposite direction to *lex1* that has homology to the *E. coli* gene *ksg*A. Thus, in contrast to *lic1*, in which three and possibly four genes of a transcriptional unit are involved in LPS biosynthesis, only one open reading frame in *lic2* has been implicated.

The *lic3* locus was identified and characterized by probing the chromosome of RM7004 with an oligonucleotide (CAAT)$_5$ and then sequencing an appropriate clone obtained from an EMBL3 lambda library (Maskell *et al.*, 1991). *lic3* comprises four ORFs and is probably transcribed as a single message that undergoes rapid processing. The sequences between each ORF are in each case capable of forming stem loop structures. Primer extension analyses indicated two start sites for the mRNA (unpublished).

The first gene, the function of which is unknown, has multiple repeats of CAAT and there are two ATGs situated 1 and 15 nucleotides upstream of the first CAAT. The (A + T)% content immediately upstream of the CAAT repeats is 75% and, as noted for *lic1*, this may also facilitate denaturation of the DNA upstream of the CAAT repeats with slipped-strand mispairing leading to frequent frame-shifts. A *lacZ* fusion has been constructed downstream of the CAAT repeats in *lic3*A on the *H. influenzae* chromosome (Szabo *et al.*, 1992). Three levels of expression of this fusion were observed on plates containing Xgal as substrate: blue (+ + + +), bull's-eye (+) and white (−). PCR on single colonies, followed by direct sequencing of the PCR products, was used to assay whether *lacZ* expression was regulated by the CAAT repeats. High level expression (blue) always correlated with useage of the ATG one bp upstream of the repeats, while medium expression (bull's-eyes) could arise from usage of either of the ATGs in *lic3*A. White colonies were derived from utilization of any of the three possible frames. Thus, while different levels of expression depended to some extent on CAAT repeats, other (possibly transcriptional) mechanisms were also operating. Phase variation of *lacZ* expression was observed on sequential platings of the bacteria. This depended on the CAAT repeats changing in number as well as on a second mechanism. To assess whether a second mechanism was working, a fusion was constructed between the ATG required for high level expression and *lacZ*, with concomitant deletion of the CAAT repeats, on the *H. influenzae* chromosome. On plating on Xgal, bull's-eye and white colonies were observed and these phenotypes could phase vary, strongly supporting the existence of a mechanism independent of the CAAT repeats for regulation of the expression of *lic3*A.

The ATG of the second ORF (*lic3*B) is situated 172 bases from the TAG stop codon of *lic3*A. *lic3*B codes for a protein that has 56% amino acid identity with UDP-galactose-4-epimerase (*GalE*). This gene complemented a *galE* mutation in *Salmonella typhimurium*. A mutation which deletes *lic3*B of RM7004 was constructed and its phenotype was typical of a *galE* mutant (Maskell *et al.*, 1991). When compared to wild-type RM7004, the *galE* mutant failed to grow in a defined medium containing galactose, whereas in the absence of galactose, but with glucose and glycerol as carbon source, the mutant failed to express the digalactoside epitope recognized by mab 4C4. Interestingly, if both glucose and galactose were added, growth occurred but was slow compared to wild-type RM7004, suggesting that the regulation of galactose biosynthesis is, as in *E. coli*, subject to the glucose effect (Adhya, 1987). In *E. coli* and other Gram-negative bacteria, galactose metabolism is effected by *galE*, *galT* and *galK* which make up an operon under the control of two alternative promoters that are positively and negatively regulated by the cAMP receptor protein complex and which may be repressed by the galactose repressor *(GalR)*. cAMP levels are modulated by the concentration of available glucose such that, in its presence, galactose utilization is shut down, the 'glucose effect' (Adhya, 1987). *galE* is key to maintaining the balance between the levels of nucleotide sugars UDP-glucose and UDP-galactose that are substrates for variably expressed LPS oligosaccharide antigens. As noted previously, it is predominantly variations in the molar ratios and linkages of glucose and galactose that mediate the inter-strain and intra-strain heterogeneity of *H. influenzae* LPS. Furthermore, these variations are critical determinants of the organism's pathogenic potential (see 'Pathogenic effects mediated by core oligosaccharides'). Given that *galE* maps in a different location to *galT, galK* and *galR*, and considering the differences in the regulation of galactose biosynthesis in *H. influenzae* as compared to *E. coli*, it is interesting that the third *lic3* ORF, a highly hydrophobic protein suggesting a membrane location, shares homology with *ampG* of *Enterobacter cloacae* and *E. coli* (S. Normark: personal communication). *ampG* is involved in the signal transduction process by which β-lactam antibiotics induce expression of the chromosomal β-lactamase gene (Korfmann & Sanders, 1989). Thus, although the role of the *ampG* homologue in *H. influenzae* has not been defined, and no alteration(s) in phenotype of mutations in *lic3*C has been recognized to date (the mutations are compatible with normal growth *in vitro*), it is reasonable to speculate that it may be involved in transducing changes in the availability of exogenous glucose, galactose or other substrates to modulate LPS biosynthesis. Indeed, the most distal gene, *lic3*D, has 71% identity with adk of *E. coli* whose product (adenylate kinase) is a key enzyme involved in the control of cell energy levels (Glaser, Nulty & Vagelos, 1975); it would seem advantageous to the bacterium for LPS biosynthesis and cell energy balance to be co-ordinately regulated.

VIRULENCE

The role of *H. influenzae* LPS in pathogenicity has been studied through experimental infection of infant rats, organ cultures and cultured human cell lines. It is now clear that *H. influenzae* LPS can enhance bacterial survival in the nasopharynx, facilitate invasion of organisms across cellular barriers, promote intravascular survival of bacteria and result in microbial damage to tissues, including those of the respiratory epithelium and meninges. The broad spectrum of pathophysiological effects mediated by LPS is not surprising given its amphipathic properties. In general, fatty acyl groups on lipid A mediate 'classical endotoxin' effects that cause tissue damage and the oligosaccharide core structures are involved in facilitating bacterial survival in the host.

Pathogenic effects attributable to lipid A

Following the preliminary observations on *H. influenzae* LPS that indicated its typical endotoxic properties (Flesher & Insel, 1978; Denny, 1974),

Johnson studied organ cultures of guinea-pig and rat trachea and demonstrated that purified *H. influenzae* LPS caused loss of ciliary activity and disruption of ciliated epithelial cells (Johnson & Inzana, 1986). These observations were consistent with injury to respiratory epithelium observed in organ cultures of human respiratory mucosa (Farley *et al.*, 1986; Read *et al.*, 1991). Thus, both LPS as well as other low molecular weight factors (probably glycopeptides) may compromise the mucociliary clearance mechanisms (for review see Moxon & Wilson, 1991). As a result, *H. influenzae* remain attached to pooled mucus and possibly epithelial cells in the respiratory tract layer allowing the organisms more time to replicate and make surface contact with areas of damaged epithelium (Read *et al.*, 1991).

Intracisternal instillation of purified *H. influenzae* type b LPS (as little as 20 pg) into rabbits induced a brisk inflammatory reaction in the CSF as evidenced by increases in inflammatory cells and the concentration of serum proteins as well as the formation of a miningeal exudate (Syrogiannopoulos *et al.*, 1988). Since concentrations of up to 10^8/ml bacteria are found in the CSF in bacterial meningitis, these findings appear biologically relevant to the pathophysiology of bacterial induced CNS injury in natural infection. Although the mechanism by which LPS induces inflammation is somewhat conjectural, a role for inflammatory mediators such as IL-1 and TNF is favoured (Beutler *et al.*, 1986; Camussi *et al.*, 1991). Also, LPS may activate complement with subsequent release of complement-derived chemotactic signals, e.g. C5 (Ernst *et al.*, 1984).

Pathogenic effects mediated by core oligosaccharides

An essential role for LPS in pathogenicity was also suggested by the isolation of a deep rough, avirulent mutant (I69) of a type b strain (Zwahlen *et al.*, 1985). Subsequently, core oligosaccharide phase variants (Kimura & Hansen, 1986) were shown to differ in virulence and a comparison of genetically related transformants, Rdb+:01 and Rdb+:02, following intranasal inoculation of rats showed that the Rdb+:02 variant was significantly more virulent (Zwahlen *et al.*, 1986). Since the differences in the magnitude of bacteraemia following intranasal inoculation of Rdb+:01 and Rdb+:02 were also observed following systemic (intraperitoneal) inoculation, the altered LPS apparently facilitated intravascular survival.

Using an immunoperoxidase staining technique, the LPS phenotype of organisms have been looked at *in vivo* by examining smears of CSF obtained from individuals having an established meningitis. In a typical result, one or more mabs bound to the majority (>99%) of organisms, but occasional bacteria did not bind these mabs. These findings were consistent with the occurrence of phase variation *in vivo*. When organisms isolated from the CSF of the same individual were grown *in vitro*, mab reactivity was absent in the majority of organisms, indicating differences in the pattern of epitope expression under *in vivo* and *in vitro* growth conditions (Weiser *et al.*, 1989*b*; Mertsola *et al.*, 1991). More recently, the cloning of genes required for the expression of the phase-variable oligosaccharide structures has enabled the construction of defined mutants, and the unambiguous demonstration of the role of LPS in the pathogenesis of experimental bacteraemia and meningitis (Weiser *et al.*, 1989*a*; Weiser *et al.*, 1990; Cope *et al.*, 1990; Maskell *et al.*, 1991).

To investigate the potential contribution of the expression of particular oligosaccharide epitopes in the pathogenesis of experimental *H. influenzae* bacteraemia in infant rats, deletion mutations were introduced into both *lic1* and *lic2*. This double mutant lacked expression of phase-variable oligosaccharide structures that were characteristic of the parent strain RM7004 as defined by loss of reactivity with mabs. The virulence of this mutant was compared to that of the wild-type strain following intranasal challenge of 5-day old rats. Both the mutant and wild-type strain colonized efficiently and nasopharyngeal washings showed similar numbers of organisms. In contrast, there was a significant reduction in the incidence of bacteraemia among rats infected with the mutant. However, the mutant strain was similar in virulence, as judged by both the incidence and magnitude of bacteraemia following systemic (intra-peritoneal) challenge (Weiser *et al.*, 1990). These results indicated that LPS is involved in the translocation of *H. influenzae* from the nasopharynx to the blood stream, a stage in the pathogenesis which has proved difficult to investigate and for which defined mutants have previously not been identified. These findings are interesting in the light of previous experiments which showed that prior exposure of the nasopharyngeal mucosa of infant rats to *H. influenzae* LPS increased the incidence of bacteraemia following subsequent challenge, despite the absence of any evidence of inflammation or damage to the nasopharyngeal mucous membrane (Kaplan *et al.*, 1988).

In addition, major changes to the LPS oligosaccharide can be effected by mutating genes that are essential for the biosynthesis of core sugars. Deletion mutants have been generated in *H. influenzae galE* and *galK* genes. The double *galE galK* mutant and the single *galE* mutant are unable to synthesize any galactose-containing LPS structures, while the *galK* single mutant should still be able to synthesize galactose-containing structures. After intraperitoneal inoculation of 10^2 cfu into infant rats, wild-type and *galK H. influenzae* both lead to high level bacteraemia (10^5 cfu per 10 μl blood), whereas *galE* and *galE galK* mutants do not induce detectable bacteraemia. In one experiment, only two out of six infant rats given 1.1 × 10^6 cfu of the *galE* or the *galE galK* mutants developed bacteraemia. *galE* mutants are susceptible to lysis when grown in the presence of galactose, so the avirulence of the *galE* mutant may have been due to galactose-induced lysis *in vivo*. However, this possibility has been discounted, since moving the *galK* mutation into the *galE* background removes this galactose sensitivity.

The degree of attenuation in the *gal* mutants is similar to that seen for capsule mutants, indicating that LPS is of major importance in the virulence of *H. influenzae*.

Serum-factor induced resistance to bacteriolysis

In common with other gram-negative bacteria, *H. influenzae* type b can be killed by bactericidal antibody and complement *in vitro*. However, Shaw *et al.*, (1976) made the interesting observation that organisms obtained from the blood of rats with *H. influenzae* type b bacteraemia were more resistant to bactericidal killing than the same strain grown in conventional media. Further studies showed that the serum-factor(s) involved was of low M_r (less than 500) and was present in blood and nasopharyngeal secretions (Rubin & Moxon, 1985). Indeed, the phenotypic conversion of most *H. influenzae* type b strains to the relatively serum-resistant phenotype could be effected by incubation in a mixture of glucose, lactate, urea and biocarbonate (GLUB) (Kuratana & Anderson, 1991). The relative resistance conferred by GLUB affected complement-mediated killing by antibodies to LPS; involvement of outer membrane proteins was also implicated (Inzana & Anderson, 1985; Kuratana *et al.*, 1990). Thus, *H. influenzae* type b organisms with the relatively resistant (res) phenotype apparently contained fourfold more LPS of similar electrophoretic mobility when compared to relatively sensitive (sen) organisms that had not been subjected to the phenotype shift. The shift could be blocked by exposing the cells to chloramphenicol or puromycin. Why the increase in LPS results in a relatively greater resistance to antibody-dependent killing remains unexplained. No convincing changes in structure have been correlated with the shift to the res phenotype. However, the importance of the observation resides in the fact that the phenotype of *H. influenzae* type b organisms *in vivo* is different from that displayed by organisms grown in standard laboratory media, and that the virulence of res organisms, as shown by experiments in infant rats, is greater than that of the sen organisms (Rubin and Moxon, 1985).

CONCLUSIONS

Over the past decade, there have been substantial gains in knowledge of *H. influenzae* LPS. Together with the data for other bacteria, especially the classical studies on the enterobacteriaceae, a more complete profile of the role of *H. influenzae* LPS in membrane physiology and pathogenicity is emerging. To pick on just a few examples; LPS molecules of *H. influenzae* have only a single KDO unit in the core as compared to the three units characteristically found in, for example, *Salmonella typhimurium*; the I69 mutant of *H. influenzae* indicated that the minimal structure, lipid A-KDO-P is compatible with rapid, relatively healthy growth of *H. influenzae in vitro*. The extent of the variation in core oligosaccharide structure within a clonal population, and the complex mechanisms that have evolved to mediate this diversity, are also provocative. However, this variation has rendered more difficult the task of defining the LPS structure; the identification of genes required for variable LPS biosynthesis represents an important step in facilitating structural analysis, thereby lessening a major deficit in existing information. These mutants are also useful in defining more clearly the important role of *H. influenzae* LPS in virulence.

Finally it is intriguing that *Haemophilus* and *Neisseria* share cross-reactive core oligosaccharide epitopes, and both express structures that arc cross-reactive with human tissue-derived glycosylation residues. Whether the bacterial structures represent convergent or divergent evolution is not clear. In both cases, given that *Neisseria* and *Haemophilus* are highly adapted to humans, it is likely that epitope switching may be an adaptive mechanism facilitating their bacterial survival, perhaps because of the constraints imposed on host immune responses directed to self antigens. Alternatively, when organisms are transmitted from one individual to another, selection of particular LPS structures from a compendium of variants might be advantageous in microbial host cell interactions given the polymorphic nature of the carbohydrate structures on host cells. There is now a firm basis for a molecular analysis of the inter-relationships of structure and function of *H. influenzae* LPS.

ACKNOWLEDGEMENTS

This review is dedicated to the memory of Dr Peter Butler, Research Scientist in the Molecular Infectious Diseases Group 1987–91, who died from an inoperable brain tumour on the 18th September 1991. In addition, we wish to record our appreciation of the contributions of Drs L. Rubin and A. Zwahlen who were post-doctoral scientists in ERM's laboratory at the Johns Hospital, Baltimore, Maryland, and Mrs M. Deadman, Dr N. High, Dr P. Langford, Mrs J. Love, Dr J. Manning, Miss M. Szabo, Dr M. Virji, Dr J. Weiser, Dr A. Williams and Dr S. Zamze, now or formerly of the Molecular Infectious Diseases Group, Institute of Molecular Medicine, Department of Paediatrics, University of Oxford. Work in ERM's laboratory on *H. influenzae* has been supported by the National Institutes of Health, USA; the Medical Research Council, UK; the National Meningitis Trust, UK and The Wellcome Trust, UK.

We also wish to thank Mrs Sheila Hayes for her help in preparing the manuscript.

REFERENCES

Adhya, S. (1987). The galactose operon. In *Escherichia coli* and *Salmonella typhimurium. Cellular and Molecular Biology*, F. C. Neidhart, J. L. Ingraham, K. Brooks Low, B. Magasanik, M. Schaechter and H. E. Umbarger, eds, pp. 1503–12. Washington, DC: American Society for Microbiology.

Apicella, M. A., Dudas, K. C., Campagnari, A., Rice, P., Mylotte, J. M. & Murphy, T. F. (1985). Antigenic heterogeneity of lipid A of *Haemophilus influenzae. Infection and Immunity*, **50**, 9–14.

Berk, S. L., Holtsclaw, S. A., Wiener, S. L. & Smith, J. K. (1982). Nontypable *Haemophilus influenzae* in the elderly. *Archives of Internal Medicine*, **142**, 537–9.

Beutler, B., Krochin, N., Milsark, I. W., Luedke, C. & Cerami, A. (1986). Control of cachectin (tumor necrosis factor) synthesis: mechanisms of endotoxin resistance. *Science*, **232**, 977–80.

Butler, P. D. & Moxon, E. R. (1990). A physical map of the genome of *Haemophilus influenzae* type b. *Journal of General Microbiology*, **136**, 2333–42.

Camussi, G., Albano, E., Tetta, C. & Bussolino, F. (1991). The molecular action of tumor necrosis factor-α. *European Journal of Biochemistry*, **202**, 3–14.

Cope, L. D., Yogev, R., Mertsola, J., Argyle, J. C., McCracken, (Jr) G. H. & Hansen, E. J. (1990). Effect of mutations in lipo-oligosaccharide biosynthisis genes on virulence of *Haemophilus influenzae* type b. *Infection and Immunity*, **58**, 2343–51.

Cope, L. D., Yogev, R., Mertsola, J., Latimer, J. L., Hanson, M. S., McCracken (Jr) G. H. & Hansen, E. J. (1991). Molecular cloning of a gene involved in lipooligosaccharide biosynthesis and virulence expression by *Haemophilus influenzae* type b. *Molecular Microbiology*, **5**, 1113–24.

Denny F. W. (1974). Effect of a toxin produced by *Haemophilus influenzae* on ciliated respiratory epithelium. *Journal of Infectious Diseases*, **129**, 93–100.

Erler, W., Feist, H. & Flossmann, K.-D. (1977). Die bindungsverhältmisse der fettsäuren im lipida der lipopolysaccharide aus *Pasteureusa multocida. Archives of Experimental Veterinary Medicine*, **31**, 203–9.

Ernst, J. D., Hartiala, K. T., Goldstein, I. M. & Sande, M. A. (1984). Complement (C5)-derived chemotactic activity accounts for accumulation of polymorphonuclear leukocytes in cerebrospinal fluid of rabbits with pneumococcal meningitis. *Infection and Immunity*, **46**, 81–6.

Farley, M. M., Stephens, D. S., Mulks, M. H., Cooper, M. D., Bricker, J. V., Mirra, S. S. & Wright, A. (1986). Pathogenesis of IgA1 protease-producing and non-producing *Haemophilus influenzae* in human nasopharyngeal organ cultures. *Journal of Infectious Diseases*, **154**, 752–9.

Flesher, A. R. & Insel, R. A. (1978). Characterization of lipopolysaccharide of *H. influenzae. Journal of Infectious Diseases*, **138**, 719–30.

Glaser, M., Nulty, W. & Vagelos, P. R. (1975). Role of adenylate kinase in the regulation of macromolecular biosynthesis in a putative mutant of *Escherichia coli* defective in membrane phospholipid biosynthesis. *Journal of Bacteriology*, **123**, 128–36.

Gupta, M. R., Dudas, K. C. & Apicella, M. A. (1986). Antigenic heterogeneity of LPS from non-typeable *Haemophilus influenzae* [abstract no 283]. In *Programme and Abstracts of the 25th Interscience Conference on Antimicrobial Agents and Chemotherapy*. Washington, DC: American Society for Microbiology.

Helander, I. M., Lindner, B., Brade, H., Altmann, K., Lindberg, A. A., Rietschel, E. Th. & Zahringer, U. (1988). Chemical structure of the lipopolysaccharide of *Haemophilus influenzae* strain I-69 Rd$^-$/b+: description of a novel deep-rough chemotype. *European Journal of Biochemistry*, **177**, 483–92.

Imoto, M., Kusumoto, S., Shiba, T., Naoki, H., Iwashita, T., Rietschel, E. T., Wollenweber, H.-W., Galanos, C. & Luderitz, O. (1983). Chemical structure of *E. coli* lipid A: linkage site of acyl groups in the disaccharide backbone. *Tetrahedron Letters*, **24**, 4017–20.

Inouye, M. (ed.). (1979). *Bacterial Outer Membranes, Biogenesis and Functions*. New York: Wiley and Sons.

Inzana, T. J. (1983). Electrophoretic heterogeneity and interstrain variation of the lipopolysaccharide of *H. influenzae*. *Journal of Infectious Diseases*, **148**, 492–9.

Inzana, T. J. & Anderson, P. (1985). Serum factor-dependent resistance of *Haemophilus influenzae* type b to antibody to lipopolysaccharide. *Journal of Infectious Diseases*, **151**, 869–77.

Jacobs, D. M. & Morrison, D. C. (1977). Inhibition of the mitogenic response to lipopolysaccharide (LPS) in mouse spleen cells by polymyxin B. *Journal of Immunology*, **118**, 21–7.

Jann, B. & Jann, K. (1990). Structure and biosynthesis of the capsular antigens of *Escherichia coli*. In *Bacterial Capsules*, K. Jann and B. Jann, eds, pp. 19–42, Berlin: Springer-Verlag.

Johnson, A. P. & Inzana, T. J. (1986). Loss of ciliary activity in organ cultures of rat trachea treated with lipo-oligosaccharide isolated from *Haemophilus influenzae*. *Journal of Medical Microbiology*, **22**, 265–8.

Kaplan, S. L., Hawkins, E. P., Inzana, T. J., Patrick, C. C. & Mason, Jr. E. O. (1988). Contribution of *Haemophilus influenzae* type b lipopolysaccharide to pathogenesis of infection. *Microbial Pathogenesis*, **5**, 55–62.

Karlsson, K. A., Bock, K., Stromberg, N. & Teneberg, S. (1986). Fine dissection of binding epitopes on carbohydrate receptors for microbiological ligands. In *Protein–Carbohydrate Interactions*, D. L. Lark, ed., pp. 207–13, London: Academic Press.

Kimura, A. & Hansen, E. J. (1986). Antigenic and phenotypic variations of *Haemophilus influenzae* type b lipopolysaccharide and their relationship to virulence. *Infection and Immunity*, **51**, 69–79.

Kimura, A., Patrick, C. C., Miller, E. E., Cope, L. D., McCracken, Jr. G. H. & Hansen, E. J. (1987). *Haemophilus influenzae* type b lipo-oligosaccharide: stability of expression and association with virulence. *Infection and Immunity*, **55**, 1979–86.

Korfmann, G. & Sanders, C. C. (1989). *ampG* is essential for high level expression of *AmpC* β-lactamase in *Enterobacter cloacae*. *Antimicrobial Agents and Chemotherapy*, **33**, 1946–51.

Kuratana, M., Hansen, E. J. & Anderson, P. (1990). Multiple mechanisms in serum factor-induced resistance of *Haemophilus influenzae* type b antibody. *Infection and Immunity*, **58**, 914–17.

Kuratana, M. & Anderson, P. (1991). Host metabolites that phenotypically increase the resistance of *Haemophilus influenzae* type b to clearance mechanisms. *Journal of Infectious Diseases*, **163**, 1073–9.

Kwaik, Y. A., McLaughlin, R. E., Apicella, M. A. & Spinola, S. M. (1991). Analysis of *Haemophilus influenzae* type b lipo-oligosaccharide-synthesis genes that assemble or expose a 2-keto-3-deoxyoctulosonic acid epitope. *Molecular Microbiology*, **5**, 2475–80.

Levinson, G. & Gutman, G. A. (1987). Slipped-strand mispairing: a major mechanism for DNA sequence evolution. *Molecular Biology of Evolution*, **4**, 203–21.

Luderitz, O., Freudenberg, M. A., Galanos, Ch., Lehmann, V., Rietschel, E. T. & Shaw, D. H. (1982). Lipopolysaccharides of gram-negative bacteria. In *Current*

Topics in Membranes and Transport, vol 17: *Membrane Lipids of Prokaryotes*, S. Razin and S. Rottem, eds, pp. 79–151. New York: Academic Press.

Lund, B., Lindberg, F., Marklund, B. I. & Normark, S. (1987). The PapG protein is the α-D-galactopyranosyl-(1–4)-β-D-galactopyranose-binding adhesin of uropathogenic *Escherichia coli*. *Proceedings of the National Academy of Sciences, USA*, **84**, 5898–902.

Mandrell, R. E., McLaughlin, R., Kwaik, Y. A., Lesse, A., Yamasaki, R., Gibson, B., Spinola, S. M. & Apicella, M. A. (1992). Lipo-oligosaccharides of some *Haemophilus* species mimic human glycosphingolipids and are sialylated. *Infection and Immunity* (in press).

Maskell, D. J., Szabo, M. J., Butler, P. D., Williams, A. E. & Moxon, E. R. (1991). Molecular analysis of a complex locus from *Haemophilus influenzae* involved in phase-variable lipopolysaccharide biosynthesis. *Molecular Microbiology*, **5**, 1013–22.

Mattsby-Baltzer, I., Gemski, P. & Alving, C. R. (1984). Heterogeneity of lipid A: comparison of lipid A types from different gram-negative bacteria. *Journal of Bacteriology*, **159**, 900–4.

Mertsola, J., Cope, L. D., Saez-Llorens, X., Ramilo, O., Kennedy, W., McCracken, Jr, G. H. & Hansen, E. J. (1991). *In vivo* and *in vitro* expression of *Haemophilus influenzae* type b lipooligosaccharide epitopes. *Journal of Infectious Diseases*, **164**, 555–63.

Moxon, E. R. & Wilson, R. (1991). The role of *Haemophilus influenzae* in the pathogenesis of pneumonia. *Reviews of Infectious Diseases*, **13**, S518–27.

Munford, R. S. & Hall, C. L. (1986). Detoxification of bacterial lipopolysaccharides (endotoxins) by a human neutrophil enzyme. *Science*, **234**, 203–5.

Murphy, T. F. & Apicella, M. A. (1987). Nontypeable *Haemophilus influenzae*: a review of clinical aspects, surface antigens, and the human immune response to infection. *Reviews of Infectious Diseases*, **9**, 1–15.

Musher, D. M., Kubitschek, K. R., Crennan, J. & Baughn, R. E. (1983). Pneumonia and acute febrile tracheobronchitis due to *Haemophilus influenzae*. *Annals of Internal Medicine*, **99**, 444–50.

Parr, T. R. & Bryan, L. E. (1984). Lipopolysaccharide composition of three strains of *Haemophilus influenzae*. *Canadian Journal of Microbiology*, **30**, 1184–7.

Patrick, C. C., Kimura, A., Jackson, M. A., Hermanstorfer, L., Hood, A., McCracken, (Jr) G. H. & Hansen, E. J. (1987). Antigenic characterization of the oligosaccharide portion of the lipooligosaccharide of nontypeable *Haemophilus influenzae*. *Infection and Immunity*, **55**, 2902–11.

Patrick, C. C., Pelzel, S. E., Miller, E. E., Haanes-Fritz, E., Rudolf, J. D., Gulig, P. A., McCracken (Jr) G. H. & Hansen, E. J. (1989). Antigenic evidence for simultaneous expression of two different lipooligosaccharides by some strains of *Haemophilus influenzae* type b. *Infection and Immunity*, **57**, 1971–8.

Raetz, C. R. H. (1990). The biochemistry of endotoxins. *Annual Review of Biochemistry*, **59**, 129–70.

Read, R. C., Wilson, R., Rutman, A., Lund, V., Todd, H. C., Brain, A. P. R., Jeffery, P. K. & Cole, P. J. (1991). Interaction of non-typeable *Haemophilus influenzae* with human respiratory mucosa *in vitro*. *Journal of Infectious Diseases*, **163**, 549–58.

Rietschel, E. Th., Brade, L., Holst, O., Kulshin, V. A., Lindner, B., Moran, A. P., Schade, U. F., Zahringer, U., Brade, H. (1990). Molecular structure of bacterial endotoxin in relation to bioactivity. In *Cellular and Molecular Aspects of Endotoxin Reactions*, A. Noscomy, J. J. Spitzer and E. J. Ziegler, eds, pp. 15. Elsevier Science.

Rubin L. G. & Moxon E. R. (1985). The effect of serum-factor induced resistance to somatic antibodies on the virulence of *Haemophilus influenzae* type b. *Journal of General Microbiology*, **131**, 515–20.

Shaw, S., Smith, A. L., Anderson, P. & Smith, D. H. (1976). The paradox of *Haemophilus influenzae* type b bacteremia in the presence of serum bactericidal activity. *Journal of Clinical Investigation*, **58**, 1019–29.

Spinola, S. M., Kwaik, Y. A., Lesse, A. J., Campagnari, A. A. & Apicella, M. A. (1990). Cloning and expression in *Escherichia coli* of a *Haemophilus influenzae* type b lipooligosaccharide synthesis gene(s) that encode a 2-keto-3-deoxyoctulosonic acid epitope. *Infection and Immunity*, **58**, 1558–64.

Syrogiannopoulos G. A., Hansen E. J., Erwin A. L., Munford R. S., Rutledge J. Reisch J. S. & McCracken (Jr) G. H. (1988). *Haemophilus influenzae* type b lipooligosaccharide induces meningeal inflammation. *Journal of Infectious Diseases*, **157**, 237–44.

Szabo, M., Maskell, D., Butler, P., Love, J. & Moxon, E. R. (1992). The role of repetitive DNA in the regulation of genes involved in LPS biosynthesis in *Haemophilus influenzae*: use of chromosomal gene fusions. (Unpublished).

Virji, M., Weiser, J. N., Lindberg, A. A. & Moxon, E. R. (1990). Antigenic similarities in lipopolysaccharides of *Haemophilus* and *Neisseria* and expression of a digalactoside structure also present on human cells. *Microbial Pathogenesis*, **9**, 441–50.

Weiser, J. N., Lindberg, A. A., Manning, E. J., Hansen, E. J. & Moxon, E. R. (1989*a*). Identification of a chromosomal locus for expression of lipopolysaccharide epitopes in *Haemophilus influenzae*. *Infection and Immunity*, **57**, 3045–52.

Weiser, J. N., Love, J. M. & Moxon, E. R. (1989*b*). The molecular mechanism of phase variation of *H. influenzae* lipopolysaccharide. *Cell*, **59**, 657–65.

Weiser, J. N., Maskell, D. J., Butler, P. D., Lindberg, A. A. & Moxon, E. R. (1990). Characterization of repetitive sequences controlling phase-variation of *Haemophilus influenzae* lipopolysaccharide. *Journal of Bacteriology*, **172**, 3304–9.

Wispelwey, B., Hansen, E. J. & Scheld, M. (1989). *Haemophilus influenzae* outer membrane vesicle-induced blood–brain barrier permeability during experimental meningitis. *Infection and Immunity*, **57**, 2559–62.

Zamze, S. E. & Moxon, E. R. (1987). Composition of the lipopolysaccharide from different capsular serotype strains of *Haemophilus influenzae*. *Journal of General Microbiology*, **133**, 1443–51.

Zamze, S. E., Ferguson, M. A. J., Moxon, E. R., Dwek, R. A. & Rademacher, T. W. (1987). Identification of phosphorylated 3-deoxy-*manno*-octulosonic acid as a component of *Haemophilus influenzae* lipopolysaccharide. *Biochemical Journal*, **245**, 583–7.

Zwahlen, A., Winkelstein, J. A. & Moxon, E. R. (1983). Surface determinants of *Haemophilus influenzae* pathogenicity: comparative virulence of capsular transformants in normal and complement-depleted rats. *Journal of Infectious Diseases*, **148**, 385–94.

Zwahlen, A., Rubin, L. G., Connelly, C. J., Inzana, T. J. & Moxon, E. R. (1985). Alteration of the cell wall of *Haemophilus influenzae* type b by transformation with cloned DNA: association with attenuated virulence. *Journal of Infectious Diseases*, **152**, 485–92.

Zwahlen, A., Rubin, L. G. & Moxon, E. R. (1986). Contribution of lipopolysaccharide to pathogenicity of *Haemophilus influenzae*: comparative virulence of genetically-related strains in rats. *Microbial Pathogenesis*, **1**, 465–73.

LIFE WITHIN PHAGOCYTIC CELLS

STEFAN H. E. KAUFMANN AND INGE E. A. FLESCH

Department of Immunology, University of Ulm, Albert-Einstein-Allee 11, D-7900 Ulm, Germany

INTRODUCTION

During their evolutionary struggle, microbial pathogens have identified several strategies that allow them stably to infect their target organisms. One successful step in such a strategy is the identification of niches that allow survival and even replication relatively unaffected by host defence mechanisms. Such habitats can be extracellular and intracellular. Intracellular niches seem to be particularly useful for long-term survival of microbial pathogens in the face of an ongoing immune response because many diseases caused by intracellular microorganisms take a chronic course (Hahn & Kaufmann, 1981; Kaufmann & Reddehase, 1989).

Their capacity to engulf microbial organisms at a high rate, and to kill and degrade them subsequently, distinguishes tissue macrophages and blood monocytes as well as polymorphonuclear neutrophilic granulocytes from the rest of the body's cells. These so-called professional phagocytes are a life-threatening setting for microbial invaders. Yet, even this entity of the body has not escaped microbial colonization, and a subgroup of intracellular bacteria has established strategies that allow their survival within these important effector cells (Moulder, 1985).

Macrophages and granulocytes are of different cell lineage and one distinguishing feature is their lifespan which, for granulocytes, is around a day and for mononuclear phagocytes lasts for several weeks. It is, therefore, not surprising that macrophages, rather than granulocytes, are abused as a safe habitat. Bacteria which have chosen macrophages as their preferred habitat include the following pathogens: *Listeria monocytogenes, Mycobacterium tuberculosis, Mycobacterium leprae, Legionella pneumophila*, and *Salmonella typhimurium* (Hahn & Kaufmann, 1981).

In the following, currently available data will be summarized with respect to: (1) the antimicrobial arsenal potentially available to mononuclear phagocytes, (2) microbial strategies allowing survival inside macrophages, (3) the means by which T-cells transform macrophages from habitat to terminator of microbial parasites, and (4) the potentially harmful sequelae of this otherwise essential step towards protection. Finally, an attempt will be made to discuss these aspects in the context of disease development *in situ*.

THE ANTIMICROBIAL POTENTIAL OF MACROPHAGES AND MICROBIAL EVASION STRATEGIES

Phagocytosis of pathogens by macrophages induces a series of events which usually result in the death of the microorganism:

1. generation of reactive oxygen intermediates (ROI) (Andrew, Jackett & Lowrie, 1985; Babior, 1984);
2. production of reactive nitrogen intermediates (RNI) (Hibbs *et al.*, 1988; Liew & Cox, 1990; Marletta, 1989);
3. acidification of the phagosome and phagosome–lysosome fusion (Horwitz, 1988; Moulder, 1985);
4. limitation of intracellular iron (Alford, King & Campbell, 1991; Byrd & Horwitz, 1991);
5. release of defensins (Ganz, Selsted & Lehrer, 1988).

Phagocytosis of bacteria by mononuclear phagocytes induces the activation of the plasma membrane bound enzyme NAD(P)H-oxidase and the production of ROI (O_2^-, H_2O_2, OH^-, $O_2^{\frac{1}{2}}$) (Babior, 1984). The evidence for the importance of ROI in the defence of intracellular bacteria is not complete. *M. tuberculosis*, *M. leprae* and *Leishmania donovani* enter macrophages via C3b receptors without inducing an oxidative burst (Horwitz, 1988; Wright & Silverstein, 1983). Phenolic glycolipid-1 of *M. leprae*, and lipophosphoglycan of *L. donovani*, are scavengers of ROI (Chan *et al.*, 1989). *In vitro*, exogeneously added superoxide dismutase, catalase and other ROI scavengers fail to alter the capacity of interferon-γ (IFN-γ)-activated bone marrow macrophages to inhibit intracellular growth of *M. bovis* (Flesch & Kaufmann, 1988). These findings suggest that ROI are of minor importance for antimycobacterial activities of macrophages.

A number of microorganisms, including *L. major*, *Cryptococcus neoformans*, *Schistosoma mansoni*, *M. tuberculosis* and *M. bovis*, have been shown to be susceptible to nitrogen oxides ($NO^{\cdot}$, NO_2^-, NO_3^-) derived from the amino acid L-arginine (Liew & Cox, 1990; Marletta, 1989). *In vitro*, mycobacterial infection of IFN-γ-stimulated bone marrow macrophages induces the production of RNI. Inhibition of RNI production by the L-arginine analogue N^G-monomethyl-L-arginine abrogates RNI production as well as growth inhibition of *M. bovis* (Flesch & Kaufmann, 1991). Similar results have been obtained in the leishmanial system (Liew & Cox, 1990; Green, Nacy & Meltzer, 1991). In contrast to ROI, RNI are not only produced by professional phagocytes but also by a variety of other cells including fibroblasts (Amber *et al.*, 1988), endothelial cells (Palmer, Ashton & Monocada, 1988), hepatocytes (Curran *et al.*, 1989) and cerebellar neurons (Garthwaite, Charles & Chess-Williams, 1988). This might imply that RNI represent a basic mechanism of local resistance against pathogenic invaders.

Soon after phagocytosis of particulate material by macrophages, the content of the phagosome is acidified and phagosome–lysosome fusion is induced. Lysosomal enzymes degrade the phagocytosed material. To evade the microbicidal activities of lysosomal enzymes, *M. tuberculosis* inhibits phagosome acidification and phagosome–lysosome fusion, probably by secretion of ammonium chloride (Hart, D'Arcy & Young, 1991). Interestingly, opsonization of bacteria with a specific antiserum reverses inhibition of phagosome–lysosome fusion (Lowrie & Andrew, 1988). In an electron microscopical study, Buchmeier & Hefron (1991) showed that a virulent strain of *S. typhimurium* could also inhibit phagosome–lysosome fusion and appears to preferentially divide within unfused phagocytic vesicles. Another way for an intracellular pathogen to escape from phagosome–lysosome fusion is by evasion into the cytoplasm. Pathogenic listeriae produce listeriolysin, an enzyme which allows disruption of the phagosomal membrane and evasion into the cytoplasm (Gaillard, Berche & Sansonetti, 1986). Leishmania and some strains of salmonellae are resistant to the action of lysosomal enzymes and replicate in the phagolysosome (Kaufmann & Reddehase, 1989).

Intracellular multiplication of a number of bacterial pathogens including *L. monocytogenes*, *L. pneumophila* and the malaria parasite is iron dependent. The major source of iron is iron-saturated transferrin which is endocytosed via transferrin receptors, and ferritin, an intracellular iron storage protein. It has been shown that IFN-γ-activated human monocytes inhibit intracellular multiplication of *L. pneumophila* by limitation of intracellular iron. Limitation of iron is achieved by downregulation of transferrin receptors on the cell surface and decrease of the intracellular concentration of ferritin (Byrd & Horwitz, 1991). In the mouse system, it has been shown that iron is necessary to support macrophage listericidal mechanisms. However, higher concentrations of iron favour intracellular growth of listeriae (Alford *et al.*, 1991).

Defensins are proteins with antimicrobial potential in the lysosomes of neutrophils and alveolar macrophages. They are active against fungi and certain gram-positive bacteria including *L. monocytogenes*. Defensins may play an important role in host-defence mechanisms of the lung. Recently, it has been found that resistance to defensins seems to be encoded by a pag gene product (Miller, Kukral & Mekalanos, 1989).

EXPRESSION OF HEAT SHOCK PROTEINS (HSP)

Microbes as well as mammalian cells, when stressed by a variety of insults, produce increased levels of heat shock proteins (hsp) (for review see Kaufmann, 1990). Such insults not only include increased temperature but also more physiological assaults such as contact with ROI and iron depletion with which microbial pathogens are confronted after phagocytosis by macrophages. It has been shown recently that *S. typhimurium* produces abundant hsp levels inside murine macrophages (Buchmeier & Hefron, 1990). Furthermore, *S. typhimurium* mutants with deficient *hsp* genes are more susceptible to killing by activated macrophages (Fields *et al.*, 1986) and virulence *in vivo* is markedly reduced in *hsp*-deficient mutants of *S. typhimurium* (Johnson *et al.*, 1991). Conversely, *in vitro* exposure of *S. typhimurium* to sublethal H_2O_2 concentrations not only causes increased hsp levels but also resistance to otherwise lethal concentrations of ROI [Christman *et al.*, 1985). Taken together, these findings point to a potential role as virulence factors of hsp.

MACROPHAGE ACTIVATION BY INTERLEUKINS

The activation of macrophage effector functions is regulated by a complex lymphokine network with positive and negative signals. Production of lymphokines is a major function of helper T-cells (T_H cells). T_H cells are CD4-positive and recognize antigenic peptide in association with major histocompatibility complex (MHC) class II molecules. In the mouse, T_H cells segregate into two subpopulations, so-called T_{H1} cells, which produce interleukin 2 (IL-2) and IFN-γ and so-called T_{H2} cells which secrete interleukin 4 (IL-4), interleukin 5 (IL-5), interleukin 6 (IL-6), and interleukin 10 (IL-10). Both subsets produce interleukin 3 (IL-3) (Mosmann & Coffman, 1987). IFN-γ has been identified as a major macrophage activating factor (Adams & Hamilton, 1984). *In vivo*, recombinant IFN-γ protects mice against *L. monocytogenes* in a local, as well as in a systemic, infection model (Kiderlen, Kaufmann & Lohmann-Matthes, 1984). In the leishmania system, some strains develop T_{H1} responses, and IFN-γ production induces resolution of infection. *In vitro*, IFN-γ induces growth inhibition of *M. bovis* and *M. tuberculosis* in murine bone marrow macrophages (Flesch & Kaufmann, 1987). This antimycobacterial activity could be enhanced significantly by the macrophage product tumor necrosis factor-α (TNF-α) (Flesch & Kaufmann, 1990*a*).

IL-4 has also been shown to express macrophage activating activity. It enhances tumor cytotoxicity, antigen presentation and Ia antigen expression by murine peritoneal macrophages (Crawford *et al.*, 1987; Zlotnik *et al.*, 1987). It also induces tuberculostatic macrophage functions, provided that the macrophages are infected with mycobacteria before stimulation with IL-4 (Flesch & Kaufmann, 1990*a*). In the leishmania system, some mouse strains develop a T_{H2} response, and IL-4 production leads to exacerbation of infection (Heinzel *et al.*, 1989).

The pleiotropic mediator, IL-6, is not only produced by T-cells but also by fibroblasts, epithelial cells and monocytes/macrophages. The finding that body fluids of patients with local acute bacterial infections contain elevated levels of IL-6 indicates a role in the host response to infectious agents (Helfgott *et al.*, 1989). Similar to IL-4, IL-6 activates tuberculostatic functions in infected macrophages (Flesch & Kaufmann, 1990*b*). Since TNF-α and IL-6 are produced by macrophages themselves, they might control macrophage antibacterial function at the autocrine level. Another cytokine which is produced by macrophages themselves is transforming growth factor-β (TGF-β). Macrophages exposed to TGF-β fail to produce ROI and RNI and are not able to kill intracellular parasites after activation (Nelson *et al.*, 1991).

ROLE OF CYTOLYTIC T-CELLS

Cytolytic T-cells are CD8$^+$ and recognize antigenic peptides in association with MHC class I molecules. Originally, CD8$^+$ T-cells were thought to be exclusively important for the defence of viral infections. Meanwhile, it has been shown that CD8$^+$ MHC class I-restricted T-cells also play a role in intracellular bacterial infections (Kaufmann, 1988). CD8$^+$ T-lymphocytes, isolated from mice immunized with *L. monocytogenes*, express cytolytic activity towards *L. monocytogenes*-infected bone marrow macrophages (Kaufmann, Hug & DeLibero, 1986). CD8$^+$ T-cells from mice inoculated with *M. bovis* specifically lyse macrophages infected with *M. bovis*, and lysis of infected macrophages leads to growth inhibition of mycobacteria (De-Libero, Flesch & Kaufmann, 1988).

In addition, CD8$^+$ T-cells with specificity to *L. monocytogenes*, *M. leprae*, or *M. tuberculosis* can also produce IFN-γ after restimulation in vitro (DeLibero *et al.*, 1988; Kaufmann, 1988). This indicates that CD8$^+$ T-cells may contribute to protection by lysis of infected target cells as well as by IFN-γ production.

WHICH T-CELL MECHANISMS ARE OF RELEVANCE?

Microbial survival strategies have a decisive influence on the type of immune response. Not only are intracellular pathogens protected from antibody attack but the means which allow pathogens to survive inside macrophages also determine the elicited T-cell type and function. Those microbes which survive in the phagosomal compartment will preferentially induce MHC class II-restricted CD4 T lymphocytes. Bacteria which evade into the cytoplasm can induce the MHC class-I-restricted CD8 T-cell. The latter situation has been studied carefully in the case of *L. monocytogenes* (Cossart & Mengaud, 1989). But also pathogens which remain in the phagosomal compartment such as *M. tuberculosis* can induce CD8 T-cells (Kaufmann, 1988). Perhaps during chronic infection some microbial products pass the phagosomal membrane and thus end up in the cytoplasm.

While the contribution of T-cell-derived interleukins in antibacterial infection is certain, the role of cytolytic T-cell functions remains controversial. *In vitro* studies indicate that lysis of infected macrophages by T-lymphocytes results in mycobacterial growth inhibition (DeLibero *et al.*, 1988). If bacteria infect cells which do not express MHC class II molecules or which do not become microbicidal after lymphokine stimulation, this might be an important defence mechanism. Bacteria are released into the extra-cellular space and become accessible for more efficient effector cells such as blood-borne monocytes.

On the other hand, host cell destruction can contribute to tissue damage and hence to pathogenesis. For example, *L. monocytogenes*, besides pro-fessional phagocytes, inhabits hepatocytes and recent data suggest that hepatocyte destruction is a major pathogenic feature of experimental listeriosis of mice (Sasaki *et al.*, 1990). Schwann cells are a preferred habitat of *M. leprae*, and damage of peripheral nerves is the major pathogenic characteristic during all stages of leprosy. *In vitro* studies indicate that Schwann cells presenting *M. leprae* antigens are attacked and destroyed by cytolytic T-cells (Steinhoff & Kaufmann, 1988). Although both CD4 and CD8 T-cells with specificity for microbial pathogens express cytolytic activity, the genetic restriction of CD4 T cells will focus them on infected macrophages only, whereas broad MHC class I expression will allow CD8 T-cells to attack almost every host cell.

THE *IN VIVO* SITUATION: GRANULOMA FORMATION

Although *in vitro* studies allow analyses of many aspects of protection against, and pathogenesis of infections with, intracellular bacteria, the importance of the *in vivo* setting should not be neglected. *In vivo*, granulo-mas are formed at the site of microbial growth, and it is this microenviron-ment where the combat between intracellular bacteria and host defence takes place. Granulomas are protective since they allow containment of microbes at discrete foci. On the other hand, they harm the host by impairing the physiological functions of affected organs. It appears that, at the early phase of granuloma formation, TNF is of central importance (Kindler *et al.*, 1989). Probably, TNF secretion is induced in macrophages by certain microbial products. Furthermore, IL-4 has been shown to attract blood monocytes and to cause the formation of giant cells (McInnes & Rennick, 1988). Hence, this interleukin seems to play a central role in the sustainment of a productive granuloma.

Even the localization of granulomas seems to be of pivotal importance. *M. tuberculosis* is strictly aerobic and will only grow at high pO values. The centre of a calcified granuloma has a low pO value and, therefore, does not provide an attractive habitat for tubercle bacilli (Kaufmann, 1988). Thus, tissue sites with high pO values will favour microbial growth whereas those with low pO value will be inhibitory. The apical site of the lung is the typical example for good oxygen supply and, indeed, granulomas at this site represent the preferred focus of mycobacterial persistence.

CONCLUDING REMARKS

In this brief discourse an attempt was made to dissect the mechanisms underlying the relationship between host and intracellular pathogens. The discussion of such a complex matter must be incomplete. Intracellular bacterial infections typically cause chronic diseases, and this clinical feature can be taken as strong indicator for the longevity of the relationship. Hence, bacteria and host must have developed ways that allow them to live with each other. Why then do so many people die of intracellular bacterial infections, and why then is *M. tuberculosis*, the paradigm of intracellular pathogens, the number 1 killer among all infectious agents? The figure of 3 million deaths per year is horrifying; however, taking into account that one-third to half of the world population (1.6–2.4 billion people) is infected with this agent, a frequency of about 1 death/650 infected individuals can be assumed. Even the figure of infected individuals, i.e. the 60 million people who failed to establish a protective relationship, results in a frequency of approximately 1 diseased individual/35 infected ones. Therefore, in the vast majority of cases, the relationship between *M. tuberculosis* and man has resulted in a labile, but tared, balance.

ACKNOWLEDGEMENTS

Financial support from SFB 322, EC-India Science and Technology Co-operation Program, UNDP/World Bank/WHO special Program for Research and Training in Tropical Diseases, WHO as part of its program for Vaccine Development, German Leprosy Relief Association, and Landes-schwerpunkt is gratefully acknowledged. Many thanks to R. Mahmoudi for her secretarial help.

REFRENCES

Adams, D. O. & Hamilton, T. A. (1984). The cell biology of macrophage activation. *Annual Review of Immunology*, **2**, 283–318.

Alford, C. E., King, Jr., T. E. & Campbell, P. A. (1991). The role of transferrin, transferrin receptors, and iron in macrophage listericidal activity. *Journal of Experimental Medicine*, **174**, 459–66.

Amber, I. J., Hibbs, Jr., J. B., Taintor, R. R. & Vavrin, Z. (1988). Cytokines induce an L-arginine-dependent effector system in nonmacrophage cells. *Journal of Leukocyte Biology*, **44**, 58–65.

Andrew, P. W., Jackett, P. S. & Lowrie, D. B. (1985). Killing and degradation of microorganisms by macrophages. In *Mononuclear Phagocytes: Physiology and Pathology*, eds R. T. Dean & W. Jessop, pp. 311–35. Amsterdam: Elsevier Biomedical Press.

Babior, B. M. (1984). Oxidants from phagocytes: agents of defense and destruction. *Blood*, **64**, 959–66.

Buchmeier, N. A. & Hefron, F. (1990). Induction of Salmonella stress proteins upon infection of macrophages. *Science*, **248**, 730–2.

Buchmeier, N. A. & Hefron, F. (1991). Inhibition of macrophage phagolysosome fusion by *Salmonella typhimurium*. *Infection and Immunity*, **59**, 2232–8.

Byrd, T. F. & Horwitz, M. A. (1991). Chloroquine inhibits the intracellular multiplication of *Legionella pneumophila* by limiting the availability of iron. *Journal of Clinical Investigation*, **88**, 351–7.

Chan, J., Fujiwara, T., Brennan, P. & McNeil, M. (1989). Microbial glycolipids: possible virulence factors that scavenge oxygen radicals. *Proceedings of the National Academy of Sciences, USA*, **86**, 2453–7.

Crawford, R. M., Finbloom, D. S., Ohara, J., Paul, W. E. & Meltzer, M. S. (1987). B cell stimulatory factor-1 (interleukin 4) activates macrophages for increased tumoricidal activity and expression of Ia antigens. *Journal of Immunology*, **139**, 135–41.

Christman, M. F., Morgan, R. W., Jacobson, F. S. & Ames, B. N. (1985). Positive control of a regulon for defenses against oxidative stress and some heat shock proteins in *Salmonella typhimurium*. *Cell*, **41**, 753–62.

Cossart, P. & Mengaud, J. (1989). *Listeria monocytogenes*. A model system for the molecular study of intracellular parasitism. *Molecular Biology and Medicine*, **6**, 464–74.

Curran, R. D., Billiar, T. R., Stuehr, D. J., Hofmann, K. & Simmons, R. L. (1989). Hepatocytes produce nitrogen oxides from L-arginine in response to inflammatory products from Kupffer cells. *Journal of Experimental Medicine*, **170**, 1769–74.

DeLibero, G., Flesch, I. & Kaufmann, S. H. E. (1988). Mycobacteria-reactive Lyt2$^+$ T cell lines. *European Journal of Immunology*, **18**, 59–66.

Fields, P. I., Swanson, R. V., Haidaris, C. G. & Hefron, F. (1986). Mutants of *Salmonella typhimurium* that cannot survive within the macrophage are avirulent. *Proceedings of the National Academy of Sciences, USA*, **83**, 5189–93.

Flesch, I. E. A. & Kaufmann, S. H. E. (1987). Mycobacterial growth inhibition by interferon-γ-activated bone marrow macrophages and differential susceptibility among strains of *Mycobacterium tuberculosis*. *Journal of Immunology*, **138**, 4408–13.

Flesch, I. E. A. & Kaufmann, S. H. E. (1988). Attempts to characterize the mechanisms involved in mycobacterial growth inhibition by gamma-interferon-activated bone marrow macrophages. *Infection and Immunity*, **56**, 1464–9.

Flesch, I. E. A. & Kaufmann, S. H. E. (1990a). Activation of tuberculostatic macrophage functions by gamma interferon, interleukin-4 and tumor necrosis factor. *Infection and Immunity*, **58**, 2675–7.

Flesch, I. E. A. & Kaufmann, S. H. E. (1990b). Stimulation of anti-bacterial macrophage activities by B-cell stimulatory factor 2 (interleukin-6). *Infection and Immunity*, **58**, 269–71.

Flesch, I. E. A. & Kaufmann, S. H. E. (1991). Mechanisms involved in mycobacterial growth inhibition by gamma interferon-activated bone marrow macrophages: role of reactive nitrogen intermediates. *Infection and Immunity*, **58**, 3213–18.

Gaillard, J. L., Berche, P. & Sansonetti, P. (1986). Transposon mutagenesis as a tool to study the role of hemolysin in the virulence of *Listeria monocytogenes*. *Infection and Immunity*, **52**, 50–5.

Ganz, T., Selsted, M. E. & Lehrer, R. I. (1988). Defensins: antimicrobial/cytotoxic peptides of phagocytes. In *Bacteria–Host Cell Interaction*, ed. M. A. Horwitz, pp. 3–14. New York: Alan R. Liss, Inc.

Garthwaite, J., Charles, S. L. & Chess-Williams, R. (1988). Endothelium-derived relaxing factor release on activation of NMDA receptors suggests role as intracellular messenger in brain. *Nature (London)*, **336**, 385–8.

Green, S. J., Nacy, C. A. & Meltzer, M. S. (1991). Cytokine-induced synthesis of nitrogen oxides in macrophages: a protective host response to Leishmania and other intracellular pathogens. *Journal of Leukocyte Biology*, **50**, 93–103.

Hahn, H. & Kaufmann, S. H. E. (1981). Role of cell mediated immunity in bacterial infections. *Review of Infectious Diseases*, **3**, 1221–50.

Hart, P. D'Arcy, & Young, M. R. (1991). Ammonium chloride, an inhibitor of phagosome–lysosome fusion in macrophages, concurrently induces phagosome–endosome fusion; and opens a novel pathway: studies of a pathogenic mycobacterium and a nonpathogenic yeast. *Journal of Experimental Medicine*, **174**, 881–9.

Heinzel, F. P., Sadick, M. D., Holaday, B. J., Coffman, R. L. & Locksley, R. M. (1989). Reciprocal expression of interferon-gamma or IL-4 during the resolution or progression of murine leishmaniasis. Evidence for expansion of distinct helper T cell subsets. *Journal of Experimental Medicine*, **169**, 59–72.

Helfgott, D. C., Tatter, S. B., Santhanam, U., Clarick, R. H., Bhardwaj, N., May, L. T. & Sehgal, P. B. (1989). Multiple forms of IFN-β2/IL-6 in serum and body fluids during acute bacterial infection. *Journal of Immunology*, **142**, 948–53.

Hibbs, Jr., J. B., Taintor, R. R., Vavrin, Z. & Rachlin, E. M. (1988). Nitric oxide: a cytotoxic activated macrophage effector molecule. *Biochemical and Biophysical Research Communications*, **1987**, 87–94.

Horwitz, M. A. (1988). Intracellular parasitism. *Current Opinion in Immunology*, **1**, 41–6.

Johnson, K., Charles, I., Dougan, G., Pickard, D., O'Gaora, P., Costa, G., Au, T., Miller, I. & Hormaeche, C. (1991). The role of a stress-response protein in *Salmonella typhimurium* virulence. *Molecular Microbiology*, **5**, 401–5.

Kaufmann, S. H. E., Hug, E. & DeLibero, G. (1986) *Listeria monocytogenes*-reactive T lymphocyte clones with cytolytic activity against infected target cells. *Journal of Experimental Medicine*, **164**, 363–8.

Kaufmann, S. H. E. (1988). CD8$^+$ T lymphocytes in intracellular microbial infections. *Immunology Today*, **9**, 168–73.

Kaufmann, S. H. E. & Reddehase, M. J. (1989). Infection of phagocytic cells. *Current Opinion in Immunology*, **2**, 353–9.

Kaufmann, S. H. E. (1990). Heat shock proteins and the immune response. *Immunology Today*, **11**, 129–36.

Kaufmann, S. H. E. (1991). Heat-shock proteins and pathogenesis of bacterial infections. *Springer Seminars in Immunopathology*, **13**, 25–36.

Kiderlen, A. J., Kaufmann, S. H. E. & Lohmann-Matthes, M. L. (1984). Protection of mice against the intracellular bacterium *L. monocytogenes* by recombinant immune interferon. *European Journal of Immunology*, **14**, 964–7.

Kindler, V., Sappino, A.-P., Grau, G. E., Pignet, P.-F. & Vassalli, P. (1989). The inducing role of tumor necrosis factor in the development of bactericidal granulomas during BCG infection. *Cell*, **56**, 731–40.

Liew, F. Y. & Cox, F. E. G. (1990). Non-specific defence mechanism: the role of nitric oxide. *Immunology Today*, **12**, A17–21.

Lowrie, D. B. & Andrew, P. W. (1988). Macrophage antimycobacterial mechanisms. *British Medical Bulletin*, **44**, 624–34.

Marletta, M. A. (1989). Nitric oxide: biosynthesis and biological significance. *Trends in Biochemical Sciences*, **14**, 488–92.

McInnes, A. & Rennick, D. M. (1988). Interleukin 4 induces cultured monocytes/macrophages to form giant multinucleated cells. *Journal of Experimental Medicine*, **167**, 598–611.

Miller, S. I., Kukral, A. M. & Mekalanos, J. J. (1989). A two-component regulatory system (phoP, phoQ) controls *Salmonella typhimurium* virulence. *Proceedings of the National Academy of Sciences, USA*, **86**, 5054–8.

Mosmann, T. R. & Coffman, R. L. (1987). Two types of mouse helper T cell clones. Implications for immune regulation. *Immunology Today*, **8**, 223–7.

Moulder, J. W. (1986). Comparative biology of intracellular parasitism. *Microbiological Reviews*, **49**, 298–337.

Nelson, B. J., Ralph, P., Green, S. J. & Nacy, C. A. (1991). Differential susceptibility of activated macrophage cytotoxic effector reactions to the suppressive effects of TGF-β. *Journal of Immunology*, **146**, 1849–57.

Palmer, R. M. J., Ashton, D. S. & Monocada, S. (1988). Vascular endothelial cells synthesize nitric oxide from L-arginine. *Nature (London)*, **333**, 664–6.

Sasaki, T., Mieno, M., Udono, H., Yamaguchi, K., Usui, T., Hara, K., Shiku, H. & Nakayama, E. (1990). Roles of CD4$^+$ and CD8$^+$ and CD8$^+$ cells, and the effect of administration of recombinant murine interferon in listerial infection. *Journal of Experimental Medicine*, **171**, 1141–54.

Steinhoff, U. & Kaufmann, S. H. E. (1988). Specific lysis by CD8$^+$ T cells of Schwann cells expressing *Mycobacterium leprae* antigens. *European Journal of Immunology*, **18**, 969–72.

Wright, S. D. & Silverstein, S. C. (1983). Receptors for C3b and C3bi promote phagocytosis but not the release of toxic oxygen from human phagocytes. *Journal of Experimental Medicine*, **158**, 2016–23.

Zlotnik, A., Fischer, M., Roehm, N. & Zipori, D. (1987). Evidence for effects of interleukin 4 (B cell stimulatory factor-1) on macrophages: enhancement of antigen presenting ability of bone marrow-derived macrophages. *Journal of Immunology*, **138**, 4275–9.

CHRONIC INFECTIONS, LATENCY AND THE CARRIER STATE

C. W. PENN

School of Biological Sciences, University of Birmingham, UK

INTRODUCTION

Knowledge of pathogenicity has developed in response to the most pressing needs for understanding, tempered by the technical difficulty with which studies could be conducted and the scope for their meaningful interpretation. Towards the end of the last century, the threat posed and the fear induced by many acute diseases, for example, diphtheria or typhoid, was extreme, and efforts to understand aetiology and epidemiology and at least in some cases to devise methods for treatment or prophylaxis were quite successful. This process has continued, and in most of the major bacterial infections which involve, for example, clear pathogenic mechanisms associated with bacterial toxins, understanding is now good.

In the case of chronic infections, often showing phases of latency, understanding has generally been slower to develop. This stems in part from the technical difficulty of meaningful experimental approaches: for example, a long-term interaction between bacterium and host will, almost by definition be a relatively inactive process, and damage to the host will have to be moderate if the interaction is to be sustainable by the host over a long period; therefore it will inevitably be difficult to explore experimental animal models, since the pathology which we wish to reproduce may be slight and poorly defined, and experiments will be long term. Equally, the precise nature of the interaction or near equilibrium between host and parasite which often exists in chronic infection, and which usually involves finely tuned immune responses and their regulation, may be difficult to reproduce in experimental models of infection. Furthermore, the long lifetime of experiments itself slows progress, and additional technical difficulties beset the experimenter: there may be problems maintaining experimental animals free of infection or illness for the long periods during which pathological changes occur. Experimental animals themselves have a limited natural life. Naturally, since research has, to a large extent, been driven by the need to devise methods of treatment or prevention of disease, those experiments that have been done often address the question 'how does immunity operate and how can it be manipulated to eliminate these organisms', rather than 'what are the mechanisms of long term survival of these organisms'. For all of these reasons, progress in this area has been slow, and often it is easier to speculate than to give firm answers to questions of mechanisms of disease, as the following discussions will show.

As examples a wide variety of infections and organisms will be included, but inevitably those the author is most familiar with – the spirochaetal infections – will receive most attention. It is hoped, nevertheless, that the principles involved will be of wider interest among students of pathogenicity.

DEFINITIONS: ESSENTIAL ELEMENTS OF CHRONIC, LATENT AND CARRIER STATES

It is important to have clear definitions of the types of infectious process under consideration: there is overlap between them, but each has a precise meaning. Chronic infection clearly has a long time-span as its main feature ('chronic' meaning lasting rather than the increasing popular usage to mean 'severe'); but what do we really mean by infection? In view of the resemblance and poorly defined boundary between chronic infection with a pathogen and long-term colonization by a commensal (see below), a clear definition of *infection* is required. A dictionary definition (Oxford) is 'the action or process of affecting injuriously', and the concept will be adhered to that *chronic infection* implies a generally small but nevertheless definite degree of ongoing damage to the host. *Latency* is essentially a suspended state: a period of inactivity but with the retained capacity for resumption of pathogenic activity. It becomes awkward to refer to *latent infection* – which by the above definition implies ongoing damage – is referred to since latent stages in disease are periods when essentially no damage occurs. The term *latent state* or simply *latency* should be used, and defined as a period of persistence of a pathogen in the host without concurrent damage, but with the potential for reactivation or resurgence resulting in active pathogenic processes. Finally, the *carrier state* may encompass either chronic, low-grade infection or latency or indeed the presence in a host individual of a potentially pathogenic organism with neither past nor concurrent nor latent pathogenesis in that individual; the essential element is the ability to transmit infection to other susceptible individuals, and normally carriers are distinguished by the lack of evidence of disease in themselves.

PATHOGENICITY: A MULTI-DIMENSIONAL CONTINUUM IN BOTH HOST AND PARASITE

A simplistic view of pathogenicity would be that there are highly pathogenic and weakly pathogenic organisms, and that the most pathogenic cause the most severe disease while non-pathogens do not cause disease: this 'norm' is represented by the simple 45 degree line through the origin in Fig. 1. In reality, however, there is a much more complex relationship between pathogen and host, for two main reasons. First, host factors are not

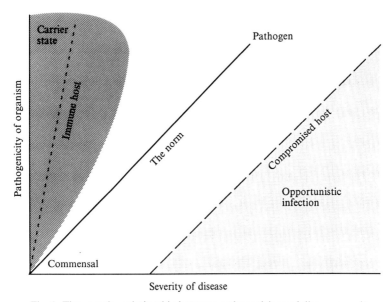

Fig. 1. The complex relationship between pathogenicity and disease severity.

constant: for example, immune status is a crucial influence on pathogenicity, and a solidly immune host is able to downgrade the most highly pathogenic organisms to the equivalent of a harmless commensal. Thus, in Fig. 1, immunity reduces the severity of disease caused by pathogens: the curve is steeper. However, host defences can not only be strengthened by immune responses but also weakened by a variety of circumstances, and both immunospecific and non-specific host defences may be compromised; in this case the whole curve is moved to the right. The severity of disease, whether caused by highly pathogenic or weakly pathogenic organisms, is increased. It is also seen that organisms which are normally commensal (or indeed some which may normally be saprophytes, i.e. environmental organisms not associated at all with the host) may cause disease in the compromised host.

How can we accommodate chronic disease, latency and the carrier state in this continuum? While there will be exceptions for particular combinations of organism and host status, some useful generalizations can be made. First, in both latency and the carrier state, disease severity is very low or absent: this is also the part of the continuum where immunity is effective, and it could be predicted that a degree of host immunity might often be involved in maintenance of these relationships. It is more difficult to place chronic disease in such a two-dimensional graphical representation; duration of infection becomes another dimension. The most meaningful attempt that can be made is to define the zone of chronic infection as the area of both low pathogenicity and low levels of damage: while some damage does occur – by definition – the organisms are not generally highly pathogenic (i.e. acutely damaging), since severe and immediate damage could not equate with survival of a host over prolonged periods, but the resulting disease can be moderately or even extremely severe in its long-term consequences. Again, immune status is a crucial factor, and its influence can act in opposite directions – either to protect or damage the host. As in latency or the carrier state, a degree of immunity may help the host to sustain a chronic infection while limiting the degree of damage that occurs. In some cases, however, immunity is only partially effective and the pathogen is able to persist rather than being totally eliminated from the body. Another complication is the ability of the immune response itself in some circumstances to damage the host, thus increasing rather than decreasing the severity of disease. Overall, chronic infection differs from acute infection primarily in its time-span and in the generally lower level of host damage involved.

There is one more important point to be made from Fig. 1. At the lower end of the range of disease severity, there are a number of quite different sets of circumstances which may obtain, allowing an intimate relationship between host and microorganism without serious damage resulting to the host. Among these is the commensal relationship: a long-term association between host and bacterium, in which presumably the bacterium has a mechanism for avoiding elimination by the host immune system, and does not cause any significant level of host damage. The means by which commensals establish and maintain an equilibrium relationship with the host, without harm to either, might usefully be examined to see whether there are any hints here to the mechanisms of maintenance of the carrier state and latency. It could be argued that the only difference between commensalism and chronic or latent pathogenicity is that the first does no damage to the host at any stage of the interaction.

ESSENTIAL ELEMENTS OF CHRONIC INFECTIONS, LATENCY AND
THE CARRIER STATE

Axiomatically, there appear to be a number of elements which must be present, or conditions which must be fulfilled, if these relationships are to be established.

1. There must be a failure of the immune system to eliminate the organisms over a long time-span.
2. The organisms must be limited in their rate of multiplication. They may either have an intrinsically slow rate of growth, or be constrained by conditions in the host environment, or adopt a state of temporary dormancy, otherwise by its nature an exponential increase in number (characteristic of microbial growth) cannot be sustained in a long-term equilibrium relationship with the host. The only exception to this requirement is that, on mucosal surfaces, microbial numbers may be

regulated by their loss to the system, e.g. by flow of secretions containing organisms to the external environment.

3. The degree of damage inflicted on the host by the organism must be minimal if the host is to survive over a prolonged period.

MECHANISMS OF AVOIDANCE OF THE IMMUNE SYSTEM

This is probably the most researched aspect of mechanisms involved in sustaining a long-term relationship between pathogen and host, and is of fundamental interest to immunologists as well as to microbiologists. The microbiological aspects of these phenomena will be focused on.

The normal commensal flora and its relationship with the host

The nature of the normal flora has been reviewed in some detail by a number of authors (Savage, 1977; Hentges *et al.*, 1984; Drasar and Hill, 1985; Van der Waaij, 1989 and this volume), and need not be re-examined here save to comment that a complete picture of many of the complex communities that inhabit the skin, mouth, and upper respiratory, gastrointestinal and uro-genital tracts of man is still not available. Often these organisms have very specialized growth requirements and may be interdependent on one another, thus defying attempts to culture them axenically. Their presence may be known only because they can be seen microscopically; recent advances in characterization of such communities by molecular biological means based on amplification and sequencing of ribosomal RNA genes (Weisberg *et al.*, 1991) are providing the tools to learn more of their diversity and character.

Most commensals inhabit mucosal sites, bathed in external secretions; it is known that the immune system is active here, particularly secretory IgA antibody. How do these organisms resist elimination by immune mechanisms? It may be that they are not, in general, very antigenic, but it is not really known because, being harmless, they have not been much studied. Another possibility is that they turn over constantly: as the host mounts an antibody response to a resident antigenic type, it is eliminated and the niche filled by a closely related (i.e. physiologically and in other respects fitted for the niche) but antigenically distinct organism. The numerous known sero-types of e.g. streptococci or enterobacteria such as *E. coli* would be good candidates for such processes, and many true commensals might be similarly variable. An ingenious suggestion is that the turnover of types might eventually slow down if ultimately there is selection of antigenic types of organism which are not highly antigenic to an individual host – perhaps there are microbial similarities to self-antigens such as blood group substances which would not be recognized by the immune system (Shedlofsky & Freter,

1974). Such a selection could also be dependent on MHC polymorphism and the accompanying restriction of immune responses. Already then there are two possible mechanisms which commensals may use to avoid the immune response – reduced antigenicity and antigenic variability, which could equally be used by chronic or latent pathogens; further study of this aspect of the host–commensal relationship is needed. In addition, it is clear that commensals do not damage the host significantly – clearly they cannot be elaborating potent toxins, but it is also likely that structural molecules such as lipopolysaccharide will exhibit low toxicity in comparison with those of some acute pathogens.

Suppression of host immunity

As understanding of the immune system, particularly the cellular immune mechanisms and their regulation, has developed during recent decades, it has been attractive to consider whether microorganisms which are able to survive in the body despite a functioning immune system might possess the means actively to suppress the immune response against themselves – or indeed more generally against unrelated antigens. An obvious candidate for such a mechanism was lepromatous leprosy, in which cellular immune responses were known to be weak and microbial multiplication was inadequately checked; evidence for specific suppression of cellular immunity, which might be mediated by stimulation of suppressor T cells, was indeed found (Bloom & Mehra, 1984). More recently however, it has proved difficult to establish the molecular and cellular details of such suppressive mechanisms as clearly as has been achieved with T-helper and cytotoxic T cells (Schwarz, 1989), and Young et al. (1990) have emphasized the paucity of understanding of the ineffectual immune responses in chronic mycobacterial infection. Speculation in this field continues, and Cohen and Young (1991) postulate that immune responses in lepromatous leprosy may be regulated analogously to those (also incompletely understood) against microbially cross-reactive self-antigens. An example of the latter is the class of highly conserved heat shock proteins (see below) to which immune responses in most individuals might be limited to a muted, non-destructive level insufficient to cause serious autoimmune damage – but, at the same time, inadequate to eliminate the self-cross-reactive microorganism. Such a mechanism would partially parallel the maintenance of tolerance to self exhibited by the immune system, and, until self-tolerance is fully understood, complete understanding of these interactions with microbial antigens will be lacking.

Another example of a chronic infection in which evidence for suppression of immune responses has been sought is syphilis. While several investigators have claimed during recent decades that cellular immune responses in both man and the rabbit model were suppressed during active syphilis infection, the evidence was confused and contradictory (for review see Young & Penn, 1990) and suppression has not been further substantiated.

Before leaving the subject of interference with immune responses it should, of course, be remembered that the HIV virus provides the best understood example, albeit by a mechanism not of active or specific suppression but rather by disabling activation of the immune response. Perhaps this paradigm should be kept in mind in considering other immunorefractory infections.

Minimal immunogenicity of organisms

A priori it may be postulated that, if a microorganism is, for whatever reason, a poor antigen, its ability to avoid immune responses should be enhanced. There are two ways in which an organism could be considered a poor antigen: it may fail altogether to stimulate a good immune response; or it may do so, but antibodies or cellular immunospecific receptors may not be able to bind to the surface of the organism (the effective antigens may be confined to its interior). The particular mechanism for poor immunogenicity considered under this heading is the expression of a poorly antigenic microbial surface. While there are several examples of such surface structures which are often quoted as pathogenic attributes, they do not all relate to chronic infection. For example, capsules which consist of poorly immunogenic acidic polysaccharides containing hyaluronic acid as in groups A and C streptococci, or sialic acids as in groups B and C meningococci (especially in group B where the alpha-2,8 conformation is present) and *E. coli* K1, do not stimulate good antibody responses in infection. This is believed to be because the polysaccharide structures resemble those of host connective tissue and cell surface components and are therefore tolerated by the immune system. Among organisms which cause chronic infections, there are, however, no obvious examples of poorly immunogenic surface polysaccharides.

If the classic examples of chronic pathogens are considered, an intriguing emphasis on unusual, lipid-rich bacterial surfaces emerges. In mycobacteria, it has proved difficult to define exactly the molecular architecture of the bacterial surface (Gaylord and Brennan, 1987), but a recent review (Young *et al.*, 1990) suggests that an 'outer membrane-like' structure composed of cell wall mycolates and 'capsular' lipids as well as complex glycolipids – termed lipopolysaccharide but not to be confused with the lipopolysaccharides of Gram-negatives – might form the basic surface structure. In addition, mycobacteria have antigenic proteins of various types associated with their surfaces: some are apparently embedded in the cytoplasmic membrane, some are actively secreted, extracellular proteins, while a third class are believed to be acylated at the amino-terminus by a well-understood mechanism (Wu & Tokunaga, 1986) involving cleavage by specific signal peptidases at a consensus amino acid sequence, resulting in glyceride acylation at an N-terminal cysteine residue. In this class is included a putative nutrient-binding protein which shows sequence homology with the PhoS protein of *E. coli* – which in the latter organism is a periplasmic protein which serves to bind periplasmic phosphate and transfer it to a cytoplasmic membrane permease for uptake into the cell. It is suggested that acylation may enable such proteins in Gram-positive bacteria (Gilson *et al.*, 1988) to remain associated with cell wall lipid structures by hydrophobic interaction so that they are able to carry out surface-located binding and translocation functions (Young *et al.*, 1990). Acylation is not seen on the homologous periplasmic nutrient-binding proteins of Gram-negatives, presumably because they are physically constrained within the periplasm, where their function is needed, by the outer membrane.

Remarkable parallels exist between the above surface structure of mycobacteria and that postulated for another classical chronic pathogen, *Treponema pallidum*. While the structure of spirochaetes basically resembles that of Gram-negative bacteria in their possession of an outer membrane, *T. pallidum* (although by no means spirochaetes generally) differs significantly from the *E. coli* paradigm in the composition and structure of its outer membrane. Several lines of evidence (Penn, Cockayne & Bailey, 1985) suggest that there is no classical Gram-negative-type lipopolysaccharide, nor is there a significant content of protein exposed on the outer surface which appears to be antigenically inert. Detergent phase partition experiments also supported the supposition that integral membrane proteins were sparse in the *T. pallidum* outer membrane (Radolf *et al.*, 1988). The most likely major component of the exposed surface therefore appears to be lipid – the first putative parallel with the mycobacterial surface; so far there are no indications as to the likely molecular composition of treponemal surface lipids. More recent evidence (Radolf, Norgard & Schultz, 1989; Walker *et al.*, 1989) confirms by direct electron microscopic observation of freeze-fractured membranes that protein particles are indeed unusually sparse in the outer membrane of *T. pallidum* (although not in other species of spirochaetes examined). Despite this, there are, in fact, several abundant and antigenically potent proteins associated with the *T. pallidum* surface (i.e. easily removed by simple detergent extraction), albeit of unknown configuration relative to the membrane structure. The second striking parallel with mycobacteria is that these proteins include an unexpectedly large number of lipoproteins (Chamberlain *et al.*, 1989; Schouls *et al.*, 1989), acylated at similar consensus sequences as with the mycobacterial PhoS homologue. However, whereas its unusual outer membrane structure seems unique to *T. pallidum*, the presence of numerous lipoproteins seems to be extended to other spirochaetes including the Lyme disease organism *Borrelia burgdorferi* (Brandt *et al.*, 1990), another chronic disease pathogen. So far the reasons for widespread acylation of surface-associated proteins in spirochaetes can only be speculated on: it has been suggested that hydrophobic tails may enable the proteins to be anchored to membranes without necessarily being embedded in them – so that, for example, a protein associated with the outer membrane of *T. pallidum* could be anchored to it but actually located on the periplasmic surface of the membrane, thus, for example, avoiding exposure on the outer surface of the organism, and hence remaining antigenically cryptic. Whether any such function may attach to lipoproteins of mycobacteria remains entirely unknown. Another property of lipoproteins which might be relevant to interaction of chronic pathogens with the immune system is that their immunogenic properties appear to be enhanced by acylation (Deres *et al.*, 1989). Whether immunogenicity might be modulated in other ways by acylation, perhaps in some circumstances towards tolerogenicity, favourable for chronic pathogens, is a fascinating speculation. Indeed, Vordermeier *et al.*, (1991) observed recently that lymph node cells from mice infected with *M. bovis* BCG did not proliferate in response to a synthetic peptide which included the acylation site of the 38K lipoprotein, suggesting that acylation may not always be stimulatory to cellular immune responses to peptides.

Antigenic variation

There are several well-defined and understood examples of antigenic variation as a factor in the pathogenicity of microorganisms. Again, some definition is needed. Two quite separate categories of antigenic variability exist, and both have been embraced by the term antigenic variation. The first, is the ability of certain pathogens to express, by means of high frequency changes in gene sequence or regulation, novel or alternate antigenic specificities in the face of an immune response, as an evasive strategy. The second category is simply the existence among separate isolates of an organism of a wide variety of antigenic types of a defined antigen. While such heterogeneity could, of course, result from variation of the first type, in many cases it does not, as for example the multiple, and relatively genetically stable, serotypes of enterobacterial O and K antigens.

Antigenic variation of the first type depends on mechanisms for frequent mutation or rearrangement of genes (or changes in their expression) encoding antigens critical to the ability of the organism to evade host immune responses, and several examples have been thoroughly analysed, although again not all are associated with classical chronic infections. Clues to the biological significance of antigenic variation came first from the study of relapsing diseases like spirochaetal relapsing fever and African trypanosomiasis, in which it was clear that resurgences of disease over a prolonged period correlated with emergence of new antigenic types of the infecting organism – a clear mechanism for extending the period of infection in the face of an immune response. Less obviously involved in chronicity is one of the examples of variability best understood at the molecular level – that of *Neisseria gonorrhoeae*, in which outer membrane 'Opa' surface proteins which confer colony opacity properties and are involved in interactions with host cells, as well as pili (fimbriae) with an adhesive role, are both subject to high frequency variation (see chapter by Robertson and Meyer, this volume). While gonococcal infection is in some circumstances chronic in nature, it is primarily an acutely damaging disease, and it may be that the variability has evolved to allow the organism to colonize diverse anatomical sites and transfer successfully from one individual to another, perhaps with different cell surface characteristics as adhesion substrates, rather than to evade immune responses. The inability of the organism to survive outside the human host dictates that it be very successful in establishing colonization in newly infected individuals.

In chronic infection, antigenic variation in arthropod-borne relapsing fever species of *Borrelia* has been well explored. *B. hermsii* has been investigated in great detail by Barbour and colleagues (Barbour & Hayes, 1986; Barbour 1990), and shown to vary in abundant, immunodominant surface 'variable major proteins' (VMPs). These are encoded on unusual linear plasmids, and frequent variation is brought about by the presence of multiple and diverse copies of the VMP gene, of which normally only one is in an expressing site with appropriate expression signals. To effect antigenic change, a recombination event takes place between the VMP sequence in the expression site and one of the different sequences in a non-expressing site. Curiously, despite the close relationships between spirochaetes and the fact that other spirochaetal diseases such as syphilis and Lyme disease are also chronic in nature and in the latter case also arthropod-borne, there is no evidence except in the relapsing fever organisms for antigenic variation as a determinant of persistence in the host. Indeed, an interesting speculation about *T. pallidum* is that it might be exceptionally stable genetically. Several studies have shown that isolates obtained at widely different geographical locations and times are remarkably similar. Could it be that this pathogen, which seems to depend on its surface *not* being recognized by the immune system, would be positively disadvantaged by a mechanism for genetic change which might be likely to make it *more* recognizable rather than less so? A recent observation which might support such a conclusion is the failure to demonstrate the presence of a RecA-like protein in *T. pallidum* (Stamm, Parrish & Gherardini, 1991), thus potentially enhancing its genetic stability.

Binding of host components

A model for the concept that chronic pathogens might bind host macromolecules to their surface in order to appear host-like to the immune system is the apparent use of this strategy by schistosomes. There is, however, little evidence that bacteria employ such a strategy, or indeed whether acquisition of host proteins is common. Among chronic pathogens the best example appears to be that *T. pallidum* is able to bind to its surface a diverse range of plasma proteins including albumin, complement, alpha-2 macroglobulin, transferrin and immunoglobulins (Alderete & Baseman, 1979), class 1 MHC proteins (Marchitto, Kindt & Norgard, 1986), and fibronectin, laminin and collagen (Peterson, Baseman & Alderete, 1983; Fitzgerald & Repesh, 1985; Steiner & Sell, 1985). There is no evidence, however, that this leads to failure of recognition by, or enhanced resistance to, the immune system.

Localization in privileged sites

Immunologically privileged sites are sites where the function of the immune system is impaired because of the nature or structure of the tissues. The full range of anti-microbial effects of the immune system requires access by both humoral and cellular factors ultimately from the bloodstream. Barriers such as the blood–brain barrier, or locations which are poorly supplied by the vascular system, may lead to inadequate expression of host defence. Examples are the cerebrospinal fluid, which is normally deficient in humoral effectors such as antibody and complement, the interior of the eye, and locations where physicochemical conditions will not support antimicrobial action of defence systems. These might include the gall bladder, parts of the kidney where osmotic or ionic conditions are not physiological, and abscesses (including tubercules in which mycobacteria persist for prolonged periods) in which pH, absence of oxygen and other factors may inhibit the action of complement or phagocytes. In a sense also an intracellular location could be considered privileged – since, provided the cell concerned was unable itself to kill the pathogen, protection could be afforded against other more harmful cells or humoral factors. In general, as far as is known, sequestration of chronic pathogens in privileged sites does not appear to be common; an exception is the harbouring of *Salmonella* spp. in the gall bladder. In some examples of chronic and latent infection, the location of persistent organisms are not known: for example, in latent syphilis, the lack of cultivability of the organism makes it impossible to detect the very small numbers which are presumed to persist at this stage of disease. The case of intracellular sequestration is considered separately below.

AVOIDANCE OF NON-SPECIFIC DEFENCES

Chronic pathogens not only avoid elimination by the immune system, but also like any other pathogen must be resistant to, and preferably should not activate, the antimicrobial actions of non-specific host defences, particularly complement and phagocytes. Evidence for microbial strategies to avoid non-specific defences is, however, sparse among chronic pathogens. It can be assumed that in the early stages of syphilis, for example, when organisms are multiplying at the site of infection but are not stimulating the recruitment of polymorphonuclear cells of the acute inflammatory response, they do not activate complement and hence chemotactic factors for inflammatory cells are not released (Penn, 1987). Presumably the surface of *T. pallidum* does not activate the alternative complement pathway: evidence for this is difficult to obtain owing to inability to culture the organism *in vitro* or to obtain organisms from *in vivo* sources which have not been exposed to host factors during infection.

Turning to the mycobacteria, their ability to survive within mononuclear phagocytic cells is well known and has been reviewed extensively elsewhere (Lowrie, 1983; Lowrie & Andrew, 1988; Rastogi & David, 1988; Rastogi, 1990; Kaufmann, this volume). Phagocytic cells, especially monocytes, are part of the specific immune system as well as non-specific host defence effectors; often in chronic infections, however, the initial interaction of phagocyte and pathogen is likely to be non-immune in nature. Clearly, intracellular mycobacteria are protected against either further ingestion by phagocytes or attack by humoral factors, as long as their growth is restricted to a level sustainable by the host cell. Such a limitation would appear, however, to be a considerable handicap to widespread colonization of the host, and mechanisms for spread from cell to cell without harm from host defence mechanisms should also be postulated. There is evidence for at least three mechanisms of survival of mycobacteria within mononuclear phagocytes. Phagosome–lysosome fusion may be inhibited, not only by *M. tuberculosis* but also by *M. avium* and *M. leprae*; *M. leprae* appears able to escape from the phagosome into the cytoplasm and there to multiply, at least *in vivo* (Sibley, Franzblau & Krahenbuhl, 1987; Rastogi & David, 1988); and mycobacteria also appear not only to be able to resist the effects of lysosomal acidity and degradative enzymes, but to multiply within the phagolysosome environment where they may even utilize the products of host degradative enzymes as nutrients (Lowrie & Andrew, 1988). Whether or not macrophages are able to kill *M. tuberculosis in vivo* in the immune host is a matter of controversy – it has not proved possible to demonstrate it unequivocally *in vitro*, and an alternative is that elimination of organisms in the immune host results from the action of cytotoxic T cells on infected macrophages (Lowrie, 1990). In this sense then, the intracellular environment may not be as impregnable a refuge as some have thought.

An intracellular location has also found favour as a reservoir of infection in other chronic diseases such as brucellosis (Corbel, 1990), and it may also be speculated that the intracellular location of obligate intracellular pathogens, *Chlamydia*, *Rickettsia* and *Coxiella* similarly may provide protection from the immune system and enhance chronicity. In a sense, this hypothesis cannot be tested because the organisms cannot replicate extracellularly, so comparisons of intracellular and extracellular survival cannot meaningfully be made. Consideration of chronicity or latency among this group of organisms is beyond the scope of this review. Recently, details of intracellular survival mechanisms of *Salmonellae* have been accumulating. Ishibashi and Arai (1990) and Buchmeier and Heffron (1991) have shown that *Salmonellae* too may inhibit phagosome–lysosome fusion; Fields, Groisman & Heffron (1989) showed that resistance to the action of microbicidal peptides obtained from phagocytes was controlled by *phoP*, a regulatory gene with multiple effects in the organism which presumably include regulation of one or more virulence genes, and mutations in the *phoP* regulon have since been shown to attenuate *Salmonella* strains (Miller *et al.*, 1990). Another suggestion (Buchmeier & Heffron, 1990) has been that stress protein expression, including specifically the *S. typhimurium* homologue of the *E. coli* gene *htrA*, may also enhance survival within phagocytes (Johnson *et al.*, 1991). It remains unclear, however, just how far these manifestations of intracellular persistence in *Salmonella* really relate to chronicity of infection; most have parallels in other, less chronic pathogens, and since we do not know the exact cellular pathology of long-term infection or the carrier state in salmonellosis, for example, in the gall bladder, it is difficult to evaluate the relevance of the phenomena described above.

GROWTH OF ORGANISMS AND ITS LIMITATION

It is unlikely to be an accident that some of the most successful chronic pathogens are extremely slow growing: *T. pallidum* is believed to divide *in vivo* (the only meaningful environment in this context) in rabbit testicular tissues about every 30 h (Turner & Hollander, 1957), and, while most mycobacteria have division times of less than 24 h (Wheeler & Ratledge, 1988), estimates of 7 d for *M. lepraemurium* and 12–14 d for *M. leprae in vivo* have been reported (Brown, 1983). Other pathogens with chronic or latent manifestations *in vivo* do not generally show slow growth *in vitro*, for example *Salmonellae* may grow as rapidly as many acutely pathogenic bacteria, but it is likely that even non-sporeforming organisms possess mechanisms, as yet poorly understood, for specialized physiological adaptation to a dormant state, induced, for example, by starvation (Matin *et al.*, 1989). Recent work indicates that a number of regulons may be coordinately controlled, their expression 'superglobally' regulated and dependent on a novel sigma factor, sigma S (for Stationary or Starvation) encoded by the *katF* gene in *E. coli* (Lange and Hengge-Aronis, 1991*a,b*; Matin, 1991). The function of a 'gearbox' promoter class of sequences in genes induced during the shift from exponential to stationary growth phase has also been implicated in such global regulation (Bohannon *et al.*, 1991). Induction of such states is probably crucial to the success of chronic infection and latency if the problem of accommodating exponential growth of vegetative organisms is to be overcome. The question is also pertinent to the persistence of parasitic organisms between hosts, in the external environment, in a non-replicative state, and features in common to organisms in the two situations may emerge; the existence of physiologically atypical, non-culturable but potentially viable and infectious forms of pathogens in environmental habitats has been postulated (Roszak & Colwell, 1987).

There is one situation in which slow growth of long-term infecting organisms is not an *a priori* requirement: organisms colonizing mucosal surfaces are uniquely able to adopt a growth format resembling continuous cultures *in vitro*, since the elimination of surplus organisms with the flow of mucosal secretions or with sloughed off epithelial cells may allow a long-term dynamic equilibrium in which growth of organisms is balanced by their loss from the colonized site. It can be assumed indeed that this is common in the carrier state, which invariably involves organisms carried on mucosal or other external surfaces of the body; hence carried organisms may not have as great a requirement for slow growth or dormancy as those which persist within the body.

HOST DAMAGE AND ITS LIMITATION

One of the preconditions for long-term equilibrium between host and pathogen outlined above was that damage to the host should be moderate or slow to develop. In fact, it has often been pointed out that the most successful pathogens are those which do only limited damage to the host, thus ensuring their persistence and transmission to other susceptible individuals; clearly the longer the infection persists, without serious damage or at the worst death of the host, the greater the probability of transmission. A quick scan through the list of highly damaging microbial toxins shows that they are not associated with chronic diseases. What then are the features of damage in chronic infections? In syphilis, the damage that occurs in primary or secondary stages is not fully understood, let alone in tertiary disease. Meaningful animal models are lacking. In primary infection, there is presumably no damage during the incubation period since there is no visible pathology. The trigger for the appearance of primary lesions is not clear; secondary lesions may involve in part the participatioin of immune complexes; and in tertiary disease the histopathology suggests a role for immunopathological processes (Young & Penn, 1990). Autoimmune phenomena may also be involved, and the possession by *T. pallidum* of a prominent antigenic stress protein of the GroEL family (Hindersson, Knudsen & Axelsen, 1987) may be significant. There is no evidence for seriously damaging toxins of *T. pallidum*. The situation is similar with *Mycobacteria*, where again the significant damage seems mainly to be due to immunopathological host responses (Rook, 1988). It may be that such immunopathological responses tend to be found with chronic pathogens simply because of their chronicity – in many acute bacterial diseases, there may not be a long enough period of interaction between the pathogen and the immune system for such reactions to develop. Without these reactions, some of the chronic infections might, indeed, approach the ideal of the perfect pathogen.

THE CARRIER STATE

The carrier state differs in several respects from the chronic infections mainly considered above. The mucosal site of carriage not only allows less restricted multiplication of the organism as described above, but also allows organisms to develop probably a similar equilibrium with the host as is achieved by commensals. As with commensals however, there is no very clear idea how this equilibrium with the immune system is determined. The other important consideration in examining the carrier state is to ask whether carried organisms are necessarily full pathogenic – their pathogenicity emerging fully only when they are passed on to a susceptible host (or the carrier becomes compromised in some way and succumbs to full-blown disease?) – or whether in the carried state organisms tend to be in a disabled form, and become pathogenic perhaps after mutation to virulence. Evidence from studies of meningococci suggests that carried organisms do indeed have full pathogenic potential since their clonal character resembles that of case isolates (Crowe et al., 1989; Achtman et al., 1991), and conversely nasopharyngeal colonization, without overt disease, by a disease isolate of N. meningitidis has also been reported (Woods & Cannon, 1990). Thus the choice between carriage and disease seems to rest on the exact balance of interactions between pathogen and host in such infections where invasion of the host is a prerequisite for disease, and mucosal colonization per se causes no pathology.

Finally, there is one aspect of carriage on mucosal surfaces which does not appear to have been explored fully: the detailed location and quantitation of carried organisms on mucosal surfaces. Numerous unanswered questions often arise if attempts are made to speculate on the mechanism of transformation from carrier state to overt disease. For example, to consider the meningococcus: are carried meningococci in the nasopharynx adherent to epithelial cells? If so at exactly what anatomical location and in what numbers? What is the nature of the bacterial adhesin and host receptor involved? Do the numbers of carried bacteria increase prior to invasion? What is the mechanism of invasion at a molecular level? And perhaps, most importantly, are the above factors different in carriers who will remain carriers compared with those who will succumb to disease? Very little information seems to be available to answer these questions.

CONCLUSIONS

There remains much to be learned of the molecular mechanisms of maintenance of microorganisms in long-term relationships, close to equilibrium, with host species in chronic infection, latency and the carrier state. These are unique relationships in biology, and, inherent in their nature, is that they are difficult to investigate experimentally.

It is clear that the relationship between these organisms and the immune system is crucial to their long-term persistence; also of great importance is the balance between microbial growth and the tendency towards exponential increase in numbers, against the effectiveness of host factors in their control. Perhaps the most exciting concept among those recently emerging is that a dormant or latent state in 'vegetative' bacterial cells is now seen to have a basis at the level of global regulation of gene expression which might be the key to successful long-term microbial interaction with the host.

REFERENCES

Achtman, M., Wall, R. A., Bopp, M., Kusecek, B., Morelli, G., Saken, E. & Hassan-King, M. (1991). Variation in class 5 protein expression by serogroup A meningococci during a meningitis epidemic. *Journal of Infectious Diseases*, **164**, 375–82.

Alderete, J. F. & Baseman, J. B. (1979). Surface-associated host proteins on virulent *Treponema pallidum*. *Infection and Immunity*, **26**, 1048–56.

Barbour, A. G. (1990). Antigenic variation of a relapsing fever *Borrelia* species. *Annual Review of Microbiology*, **44**, 155–71.

Barbour, A. G. & Hayes, S. F. (1986). Biology of *Borrelia* species. *Microbiological Reviews*, **50**, 381–400.

Bloom, B. R. & Mehra, V. (1984). Immunological unresponsiveness in leprosy. *Immunological Reviews*, **80**, 5–28.

Bohannon, D. E., Connell, N., Keener, J., Tormo, A., Espinosa-Urgel, M., Zambrano, M. M. & Kolter, R. (1991). Stationary-phase-inducible 'gearbox' promoters: differential effects of *katF* mutations and role of sigma-70. *Journal of Bacteriology*, **173**, 4482–92.

Brandt, M. E., Riley, B. S., Radolf, J. D. & Norgard, M. V. (1990). Immunogenic integral membrane proteins of *Borrelia burgdorferi* are lipoproteins. *Infection and Immunity*, **58**, 983–91.

Brown, I. N. (1983). Animal models and immune mechanisms in mycobacterial infection. In *The Biology of the Mycobacteria*, vol. 2, C. Ratledge & J. Stanford, eds, pp. 173–234. London: Academic Press.

Buchmeier, N. A. & Heffron, F. (1990). Induction of *Salmonella* stress proteins upon infection of macrophages. *Science*, **248**, 730–2.

Buchmeier, N. A. & Heffron, F. (1991). Inhibition of macrophage phagosome–lysosome fusion by *Salmonella typhimurium*. *Infection and Immunity*, **59**, 2232–8.

Chamberlain, N. R., Brandt, M. E., Erwin, A. L., Radolf, J. D. & Norgard, M. V. (1989). Major integral membrane protein immunogens of *Treponema pallidum* are proteolipids. *Infection and Immunity*, **57**, 2872–7.

Cohen, I. R. & Young, D. B. (1991). Autoimmunity, microbial immunity and the immunological homunculus. *Immunology Today*, **12**, 105–10.

Corbel, M. J. (1990). *Brucella*. In *Principles of Bacteriology, Virology and*

Immunity, 8th edn, vol. 2, M. T. Parker, B. I. Duerden, eds, pp. 339–53. London: Edward Arnold.

Crowe, B. A., Wall, R. A., Kusecek, B., Neumann, B., Olyhoek, T., Abdillahi, H., Hassan-King, M., Greenwood, B. M., Poolman, J. T. & Achtman, M. (1989). Clonal and variable properties of *Neisseria meningitidis* isolated from cases and carriers during and after an epidemic in the Gambia, West Africa. *Journal of Infectious Diseases*, **159**, 686–700.

Deres, K., Schild, H., Wiesmuller K.-H., Jung, G. & Rammensee H.-G. (1989). *In vivo* priming of virus-specific cytotoxic T lymphocytes with synthetic lipopeptide vaccine. *Nature (London)*, **342**, 561–4.

Drasar, B. S. & Barrow, P. A. (1985). *Intestinal Microbiology*. London: Nelson.

Fields, P. I., Groisman, E. A. & Heffron, F. (1989). A *Salmonella* locus that controls resistance to microbicidal proteins from phagocytic cells. *Science*, **243**, 1059–62.

Fitzgerald, T. J. & Repesh, L. A. (1985). Interactions of fibronectin with *Treponema pallidum*. *Genitourinary Medicine*, **61**, 147–55.

Gaylord, H. & Brennan, P. J. (1987). Leprosy and the leprosy bacillus: recent developments in characterisation of antigens and immunology of the disease. *Annual Review of Microbiology*, **41**, 645–75.

Gilson, E., Alloing, G., Schmidt, T., Claverys, J. P., Dudler, R. & Hofnung, M. (1988). Evidence for high-affinity binding-protein dependent transport systems in Gram-positive bacteria and in *Mycoplasma*. *EMBO Journal*, **7**, 3971–4.

Hentges, D. J. (ed.) (1983). *Human Intestinal Flora in Health and Disease*. New York: Academic Press.

Hindersson, P., Knudsen, J. D. & Axelsen, N. H. (1987). Cloning and expression of *Treponema pallidum* common antigen (Tp-4) in *Escherichia coli*. *Journal of General Microbiology*, **133**, 587–96.

Ishibashi, Y. & Arai, T. (1990). Specific inhibition of phagosome–lysosome fusion in murine macrophages mediated by *Salmonella typhimurium* infection. *FEMS Microbiology Immunology*, **64**, 35–44.

Johnson, K., Charles, I., Dougan, G., Pickard, D., O'Gaora, P., Costa, G., Ali, T., Miller, I. & Hormaeche, C. (1991). The role of a stress-response protein in *Salmonella typhimurium* virulence. *Molecular Microbiology*, **5**, 401–7.

Lange, R. & Hengge-Aronis, R. (1991*a*). Identification of a central regulator of stationary-phase gene expression in *Escherichia coli*. *Molecular Microbiology*, **5**, 49–59.

Lange, R. & Hengge-Aronis, R. (1991*b*). Growth phase-regulated expression of *bolA* and morphology of stationary phase *Escherichia coli* cells are controlled by the novel sigma factor σS. *Journal of Bacteriology*, **173**, 4474–81.

Lowrie, D. B. (1983). Mononuclear phagocyte–mycobacterium interaction. In *The Biology of the Mycobacteria*, vol. 2, C. Ratledge & J. Stanford, eds, pp. 235–78. London: Academic Press.

Lowrie, D. B. (1990). Is macrophage death on the field of battle essential to victory, or a tactical weakness in immunity against tuberculosis? *Clinical and Experimental Immunology*, **80**, 301–3.

Lowrie, D. B. & Andrew, P. W. (1988). Macrophage antibacterial mechanisms. *British Medical Bulletin*, **44**, 624–34.

Marchitto, K. S., Kindt, T. J. & Norgard, M. V. (1986). Monoclonal antibodies directed against major histocompatibility complex antigens bind to the surface of *Treponema pallidum* isolated from infected rabbits or humans. *Cellular Immunology*, **101**, 633–42.

Matin, A. (1991). The molecular basis of carbon-starvation-induced general resistance in *Escherichia coli*. *Molecular Microbiology*, **5**, 3–10.

Matin, A., Auger, E. A., Blum, P. B. & Schultz, J. E. (1989). Genetic basis of starvation survival in nondifferentiating bacteria. *Annual Review of Microbiology*, **43**, 293–316.

Miller, S. I., Pulkinen, W. S., Selsted, M. E. & Mekalanos, J. J. (1990). Characterisation of defensin resistance phenotypes associated with mutations in the *phoP* virulence regulon of *Salmonella typhimurium*. *Infection and Immunity*, **58**, 3706–10.

Penn, C. W. (1987). Pathogenicity and immunobiology of *Treponema pallidum*. *Journal of Medical Microbiology*, **24**, 1–9.

Penn, C. W., Cockayne, A. & Bailey, M. J. (1985). The outer membrane of *Treponema pallidum*: biological significance and biochemical properties. *Journal of General Microbiology*, **131**, 2349–58.

Peterson, K. M., Baseman, J. B. & Alderete, J. F. (1983). *Treponema pallidum* receptor binding proteins interact with fibronectin. *Journal of Experimental Medicine*, **157**, 1958–70.

Radolf, J. D., Chamberlain, N. R., Clausell, A. & Norgard, M. V. (1988). Identification and localization of integral membrane proteins of virulent *Treponema pallidum* subsp. *pallidum* by phase partitioning with the non-ionic detergent Triton X-114. *Infection and Immunity*, **56**, 490–8.

Radolf, J. D., Norgard, M. V. & Schultz, W. W. (1989). Outer membrane ultrastructure explains the limited antigenicity of virulent *Treponema pallidum*. *Proceedings of the National Academy of Sciences of the USA*, **86**, 2051–5.

Rastogi, N. (1990). 5th Forum in Microbiology: organised by N. Rastogi. Killing intracellular mycobacteria: dogmas and realities. *Research in Microbiology*, **141**, 191–270.

Rastogi, N. & David, H. L. (1988). Mechanism of pathogenicity in mycobacteria. *Biochimie*, **70**, 1101–20.

Rook, G. A. W. (1988). Role of activated macrophages in the immunopathology of tuberculosis. *British Medical Bulletin*, **44**, 611–23.

Roszak, D. B. & Colwell, R. R. (1987). Survival strategies of bacteria in the natural environment. *Microbiological Reviews*, **51**, 365–79.

Savage, D. C. (1977). Microbial ecology of the gastrointestinal tract. *Annual Review of Microbiology*, **31**, 107–33.

Schouls, L. M., Mout, R., Dekker, J. & van Embden, J. D. A. (1989). Characterisation of lipid-modified immunogenic proteins of *Treponema pallidum* expressed in *Escherichia coli*. *Microbial Pathogenesis*, **7**, 175–88.

Schwartz, R. H. (1989). Acquisition of immunologic self-tolerance. *Cell*, **57**, 1073–81.

Shedlofsky, S. & Freter, R. (1974). Synergism between ecological and immunological control mechanisms of the intestinal flora. *Journal of Infectious Diseases*, **129**, 296–303.

Sibley, L. D., Franzblau, S. B. & Krahenbuhl, J. L. (1987). Intracellular fate of *Mycobacterium leprae* in normal and activated mouse macrophages. *Infection and Immunity*, **55**, 680–5.

Stamm, L. V., Parrish, E. A. & Gherardini, F. C. (1991). Cloning of the *recA* gene from a free-living leptospire and distribution of RecA-like protein among spirochetes. *Applied and Environmental Microbiology*, **57**, 183–9.

Steiner, B. M. & Sell, S. (1985). Characterisation of the interaction between fibronectin and *Treponema pallidum*. *Current Microbiology*, **12**, 157–66.

Turner, T. B. & Hollander, D. H. (1957). Biology of the Treponematoses. *World Health Organisation Monograph Series* No. 35. Geneva: World Health Organisation.

van der Waaij, D. (1989). The ecology of the human intestine and its consequences for overgrowth by pathogens such as *Clostridium difficile*. *Annual Review of Microbiology*, **43**, 69–87.

Vordermeier, H. M., Harris, D. P., Roman, E., Lathigra, R., Moreno, C. & Ivanyi, J. (1991). Identification of T cell stimulatory peptides from the 38-kDa protein of *Mycobacterium tuberculosis*. *Journal of Immunology*, **147**, 1023–9.

Walker, E. M., Zampighi, G. A., Blanco, D. R., Miller, J. N. & Lovett, M. A. (1989). Demonstration of rare protein in the outer membrane of *Treponema pallidum* subsp. *pallidum* by freeze–fracture analysis. *Journal of Bacteriology*, **171**, 5005–11.

Weisberg, W. G., Barns, S. M., Pelletier, D. A. & Lane, D. J. (1991). 16S ribosomal DNA amplification for phylogenetic study. *Journal of Bacteriology*, **173**, 697–703.

Wheeler, P. R. & Ratledge, C. (1988). Metabolism in *Mycobacterium leprae, M. tuberculosis* and other pathogenic mycobacteria. *British Medical Bulletin*, **44**, 547–61.

Woods, J. P. & Cannon, J. G. (1990). Variation in expression of class 1 and class 5 outer membrane proteins during nasopharyngeal carriage of *Neisseria meningitidis*. *Infection and Immunity*, **58**, 569–72.

Wu, H. C. & Tokunaga, M. (1988). Biogenesis of lipoproteins in bacteria. *Current Topics in Microbiology and Immunology*, **125**, 127–57.

Young, D., Garbe, T., Lathigra, R. & Abou-Zeid, C. (1990). Protein antigens: structure, function and regulation. In *Molecular Biology of the Mycobacteria*, J. McFadden, ed., pp. 1–35. Guildford: Surrey University Press.

Young, H. & Penn, C. W. (1990). Syphilis and related treponematoses. In *Principles of Bacteriology, Virology and Immunity*, 8th edn, vol. 3, G. R. Smith, C. S. F. Easmon, eds, pp. 587–604. London: Edward Arnold.

BACTERIAL ENTEROTOXIN INTERACTIONS

BRENDAN W. WREN

Department of Medical Microbiology, St. Bartholomew's Hospital Medical College, West Smithfield, London EC1A 7BE, UK

INTRODUCTION

Bacterial toxins are a heterologous group of molecules, which, because of their relevance to pathogenicity and vaccine potential, have been the subject of considerable research in the past century. With the advent of animal cell culture analysis, recombinant DNA technology and sophisticated protein analysis, research into bacterial toxins has intensified in more recent times. These studies have culminated in the molecular characterization of several toxins and receptor molecules for toxins, and have provided a detailed knowledge of the mechanism of action of a number of bacterial toxins.

It is not possible to describe all toxins by a common property, but, in general, they cause damage to cells, tissue, or the whole of the host organism, thereby contributing to disease. Cell damage is often mediated through specific interactions of toxins with host-cell receptors as a prelude to the toxic response. Generally, bacterial protein toxins are divided into groups which describe their resultant action (e.g. enterotoxin, neurotoxin, cytotoxin, haemolysin, leukocidin and ciliostatic toxin). However, some toxins exhibit multiple activities (e.g. shiga toxin has neurotoxic, cytotoxic and enterotoxic activities). Alternatively, bacterial protein toxins can be classified into three types according to their mode of action on animal cells. Type I act by binding to a cell-surface protein and induce on this molecule a transmembrane signal. Type II toxins act directly on the cell membrane either by channel formation or by disruption of the lipid bilayer. Type III toxins act by translocating an enzymic component inside the cytosol which modifies an intracellular target molecule by a post-transcriptional modification.

The purpose of this chapter is to summarize present knowledge of the molecular structure of selected examples of type III enterotoxins and their interactions with their respective receptors and substrates. Examples will be drawn primarily from the cholera and related enterotoxins, the shiga and shiga-like toxins, and the toxins from *Clostridium difficile*. Many of the features and mechanisms described for these enterotoxins are relevant to other characterized bacterial toxins. For a comprehensive survey of the literature on bacterial toxins, the reader is referred to Alouf & Freer (1991).

TYPE III ENTEROTOXINS

Most enteric pathogens exert their harmful effects through the production of enterotoxins which cause massive losses of body fluids resulting in diarrhoea, subsequent dehydration and in severe cases, death. Experimentally, enterotoxins can be defined by their ability to cause fluid accumulation in the rabbit ileal loop test. To potentiate disease, most enterotoxins have to interact with specific intestinal cell receptors, which is usually accomplished by binding to specific carbohydrate structures on glycolipids or glycoproteins or both. Type III enterotoxins must possess cell-receptor binding activity, a membrane translocation mechanism, and an enzymic domain for target modification. Enterotoxins that act via this mechanism include three distinct families, namely the cholera toxin (CT) and enterotoxigenic *Escherichia coli* heat-labile toxins (LT-Is and LT-IIs), the shiga toxin (ShT) and shiga-like toxins (SLTs) and the *C. difficile* enterotoxin, toxin A. These three families of enterotoxins bind to different intestinal cell receptors, have distinct enzymic activities and appear to modify different intracellular target molecules. Although the primary structures of all of these toxins have been determined, our knowledge of the overall structure of *C. difficile* toxin A and its mechanism of action remains limited. In contrast, the complete three-dimensional structure has been determined for one of the LT-Is at 2.3 Å resolution and the mechanisms of action of CT, ShT and the LTs are well established. The most recent molecular studies on the structure of these enterotoxins and their interactions with their respective biological receptors and substrates are outlined in the following sections.

CHOLERA TOXIN AND ENTEROTOXIGENIC *E. COLI* HEAT-LABILE TOXINS

Vibrio cholerae produces an enterotoxin, CT, that is structurally, functionally and immunologically similar to the LT-Is produced by enterotoxigenic *E. coli* (Table 1). The best characterized LT-Is include those of strains of human (LT-Ih) and porcine origin (LT-Ip), which are more closely related to each other than to CT. Amino acid homology between LT-Ih and LT-Ip, LT-Ih and CT, and LT-Ip and CT have been estimated at 95, 84 and 79%, respectively. All of these toxins have the same structural AB_5 subunit composition where the A- and B-subunits are quite distinct and can fold and function independently, yet are tightly associated. Biochemical and electron microscopy studies have shown that the binding domain consists of a ring of five non-covalently associated identical B-subunits which surround a central core containing a single enzymatic A-subunit (Fig. 1) (Ribi *et al.*, 1988). The A-subunits of CT and of the LT-Is are part of a family of ADP-ribosylating toxins which include *Pseudomonas aeruginosa* exotoxin A, diphtheria toxin, pertussis toxin and botulinum C2 and C3 toxins (Alouf & Freer, 1991).

Table 1. *Physical and biological properties of cholera and shiga families of toxins and C. difficile toxin A.*

Organism	Disease	Toxin/subunit structure	M_r	Intracellular target	Receptor/carbohydrate group
Vibrio cholerae	Cholera	Cholera toxin A B	85 400 27 400 11 600	Adenylate cyclase	Ganglioside G_{M1} (G_{D1b})
Escherichia coli	Travellers diarrhoea/ Piglet and calf scours	LT-Ih/Ip A B	85 400 27 400 11 600	Adenylate cyclase	Ganglioside G_{M1} (G_{D1b}) (galactoprotein)
Escherichia coli	Diarrhoea in buffalo, cattle and humans?	LT-IIa A B	84 237 27 172 11 413	Adenylate cyclase	Ganglioside ($G_{M1}G_{M2}G_{D1a}G_{D1b}$ $G_{D2b}G_{T1b}G_{Q1b}$)
Escherichia coli	Diarrhoea in buffalo, cattle and humans?	LT-IIb A B	83 714 27 199 11 303	Adenylate cyclase	Ganglioside ($G_{D1a}G_{T1b}$)
Shigella dysenteriae	Bacillary dysentery	Shiga toxin A B	70 675 32 225 7 690	28S rRNA	Glycolipid G_{b3}
Escherichia coli	Haemorrhagic colitis/ Haemolytic uraemic syndrome	SLT-I A B	70 661 32 211 7 690	28S rRNA	Glycolipid G_{b3}
Escherichia coli	Haemorrhagic colitis/ Haemolytic uraemic syndrome	SLT-II A B	72 220 33 135 7 817	28S rRNA	Glycolipid G_{b3}
Escherichia coli	Swine oedema disease	SLT-IIv A B	70 875 33 050 7 565	28S rRNA?	Glycolipid $G_{b4}(G_{b3})$
Clostridium difficile	Pseudomembranous colitis/antibiotic- associated colitis	Toxin A	308 000	?	I, X & Y antigens? Galα1-3Galβ1-4- GlcNAc?

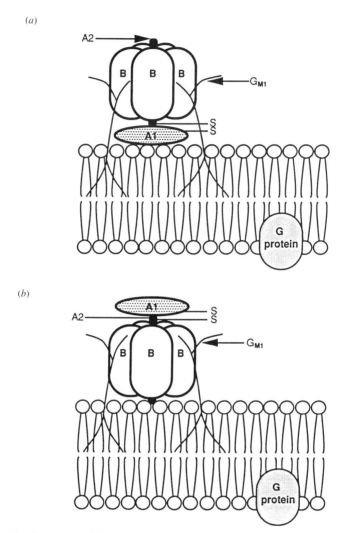

Fig. 1. Subunit structure of cholera toxin and proposed interactions with G_{M1} ganglioside.

Mechanism of action

Both the A- and the B-subunits of CT and LT-Is are synthesized within the respective bacteria as precursor proteins. After translocation and removal of leader peptides, the AB_5 complex is assembled in the periplasm. The final release mechanisms of the toxins from their respective bacterial cells appear to differ. *V. cholerae* strains secrete CT into the surrounding medium whereas enterotoxigenic *E. coli* strains appear to release LT-I as part of an outer membrane fragment. In fact, it has been demonstrated that *V. cholerae* expressing the cloned LT-Ih structural gene, secrete the toxin, whereas in *E. coli* expressing the cloned CT structural gene, both CT and LT-Ih remain cell-associated (Neill, Ivins & Holmes, 1983).

The B-subunits (M_r 11.6 kDa) of CT and the LT-Is are responsible for binding the toxin to epithelial cells which is mediated by specific interactions with G_{M1} ganglioside (Figs. 1 and 2). The B-subunits of CT and the LT-Is also bind ganglioside G_{D1b} (Fig. 3) (Fukuta *et al.*, 1988). In addition, the B-subunits of the LT-Is bind to galactoprotein receptors (Griffiths, Finkelstein & Critchley, 1986). The amino acid difference at residue 95 between the LT-Is and CT has been proposed to partially account for this additional binding specificity (Sixma *et al.*, 1992).

Binding of CT and the LTs to target cell membranes leads to the proteolytic nicking of the A-subunit between residues 192 and 195 and the reduction of the disulphide bridge between residues 187 and 199, which results in two polypeptide fragments, A1 (M_r 22 kD) and A2 (M_r 5.4 kD). The disulphide bond must be reduced to release the enzymatically active A1-fragment. After translocation across the intestinal epithelial cell membrane, the A1-fragment catalyses the ADP-ribosylation of arginine 201 on the $G_{s\alpha}$ regulatory component of adenylate cyclase, thus inducing elevated intracellular levels of cAMP. The accumulation of cAMP is believed to be responsible for the water and electrolyte loss observed in the diarrhoea caused by the toxins. However, the finding that receptor-coupled $G_{o\alpha}$ and $G_{i\alpha}$ are also ADP-ribosylated by CT, suggests this process may be more complex and could involve G proteins other than $G_{s\alpha}$.

Recently, the three-dimensional structure of LT-Ip has been determined from X-ray crystallography (2.3 Å resolution) and computer simulation studies (Sixma *et al.*, 1991, 1992). These results have confirmed many previous speculations on the structure of the toxin, and have provided a detailed knowledge of the interactions of the B-subunits with G_{M1} and the interactions of the A1-subunit with the NAD substrate.

Ganglioside-binding interactions

Fluorescence labelling studies suggested that tryptophan 88 of the B-subunits of LT-Ih and CT might be responsible for the binding of the toxins to G_{M1} oligosaccharide moieties (Moss *et al.*, 1981). Tsuji *et al.* (1985) had also shown that a mutant LT-Ip with aspartic acid in place of glycine 33 was unable to bind to G_{M1} ganglioside and was inactive as a toxin on intact cells. It appears that glycine 33 takes part in a β-turn and any mutation at that position is likely to alter the local backbone conformation. X-ray diffraction studies on LT-Ip suggest that tryptophan 88 is found at the bottom of a small cavity, surrounded by several loops, including one containing glycine 33 from an adjacent B-subunit (Sixma *et al.*, 1991). These observations confirm those made by Iida *et al.* (1989) who suggested that G_{M1} binding requires more than one B-subunit. X-ray diffraction studies on LT-Ih have also

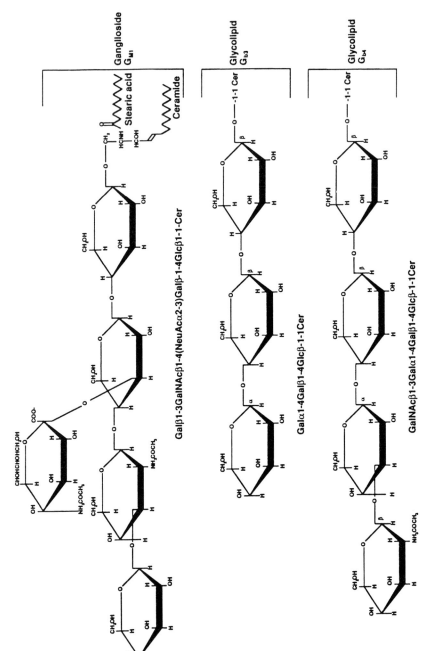

Galβ1-3GalNAcβ1-4(NeuAcα2-3)Galβ-1-4Glcβ1-1-Cer

Galα1-4Galβ1-4Glcβ-1-1Cer

GalNAcβ1-3Galα1-4Galβ1-4Glcβ-1-1Cer

Fig. 2. Structures of G_{M1}, G_{b3} and G_{b4} glycolipids.

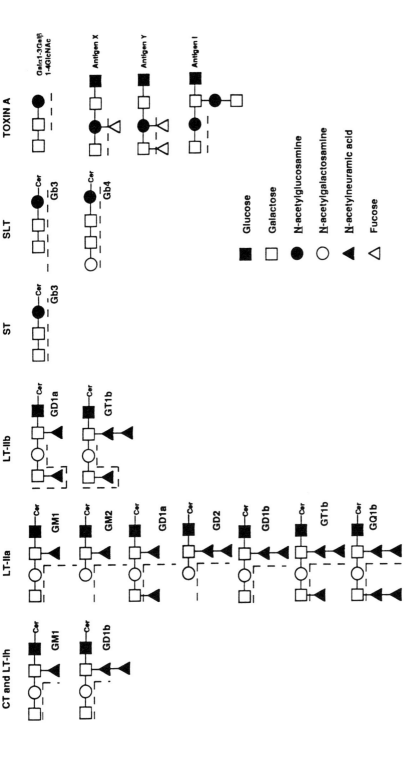

Fig. 3. Structure of terminal carbohydrate groups to which the cholera and shiga families of toxins and *C. difficile* toxin A bind. The probable sugar sequences involved in the binding of each toxin are defined by the dotted line.

confirmed the close proximity (4.8 Å) of the tryptophan 88 to glycine 33 residues from adjacent B-subunits, suggesting that the cavity may well be part of the ganglioside-binding site (Sixma *et al.*, 1991).

Recently, Sixma *et al.* (1992) have determined the three-dimensional structure of LT-Ip complexed with lactose (Gal-β1-4-Glc) by X-ray crystallography at the 2.3 Å resolution. The lactose was found to bind virtually identically with all five B-subunits, interacting mainly through its galactose moiety. Their study showed that several sites on the galactose are engaged in hydrogen bonding with a number of amino acid side-chains from the B-subunit. These include glutamic acid 51, glycine 61, aspartic acid 90 and lysine 91. Sixma *et al.* (1992) also showed that extensive van der Waals contacts exist between the hydrophobic groups of galactose and tryptophan 88. This type of interaction has been observed in other sugar-binding proteins (Vyas *et al.*, 1991).

The molecular mechanism(s) by which the A1-subunit is taken into the cell remain unclear. The B pentamer of CT neither enters the bilayer nor undergoes major conformational change, but does bind with its plane parallel to the membrane (Ribi *et al.*, 1988). A crucial question is which way up are CT and the LT-Is when they bind to the cell; is the A-subunit held close to the cell membrane (Fig. 1(*a*)) or does it protrude away from the B pentamer and cell membrane (Fig. 1(*b*))? Most of the current crystallographic evidence points towards the latter suggestion, with the C-terminus of the A2-subunit initially interacting with the membrane (Sixma *et al.*, 1992).

NAD binding interactions

There are marked similarities in the active site of LT-Ip and that of *P. aeruginosa* exotoxin A (ETA), the protein structure of which has also been determined (Allured *et al.*, 1986). Neither toxin has the Rossman $\beta\alpha\beta\alpha\beta$ fold characteristic of the NAD-binding site of dehydrogenases. This indicates a new type of NAD-binding characteristic of the ADP-ribosyltransferases. ETA, which shares sequence homology with the enzymic domain of diphtheria toxin, ADP ribosylates a diphthamide (modified histidine) residue in elongation factor-2. Despite the significant structural similarity between the LT-Ip A-subunit and ETA (44 residues can be superimposed with a root mean squared difference of 1.5 Å on Cα co-ordinates), only three of the 44 residues are identical (Sixma *et al.*, 1991). These include tyrosine 6, alanine 69 and glutamic acid 112 in the LT-Ip A1-subunit. The latter residue has been shown to be important for LT activity by mutagenesis studies (Tsuji *et al.*, 1990). In ETA the corresponding glutamic acid residue is glutamic acid 553, which is known to be an active-site residue because it can bind NAD. ETA loses activity after mutation of glutamic acid 553 to aspartic acid (Douglas & Collier, 1987). In LT-Ip glutamic acid 112 is hydrogen-bonded to the hydroxyl group of serine 61 which has also been shown to be important for LT-Ip A1-activity by mutagenesis studies (Harford *et al.*, 1989). X-ray crystallography studies on LT-Ip have revealed other residues in the immediate neighbourhood of glutamic acid 112 which may be important for catalysis or recognition. These A1-residues are believed to form an elongated crevice that may form a NAD-binding cleft and include the serines 61 and 114, histidines 44 and 107 and arginine 54, which forms a salt bridge to glutamic acid 112 (Sixma *et al.*, 1991).

Other enterotoxigenic E. coli *heat-labile toxins*

A second group of *E. coli* heat-labile enterotoxins, LT-II, has been described. LT-IIs have similar biological properties to CT and the LT-Is, but do not react with antiserum to CT or LT-Ih in neutralization or immunodiffusion tests (Pickett *et al.*, 1986). Two distinct members of the LT-II family have been described, LT-IIa from an *E. coli* strain isolated from a water buffalo in Thailand and LT-IIb from an *E. coli* strain isolated from cooked meat in Brazil. LT-IIa has been shown to possess ADP-ribosylating activity for the $G_{s\alpha}$ regulatory component of adenylate cyclase, similar to that of CT and the LT-Is (Chang *et al.*, 1987).

The genes encoding both toxins have been sequenced and exhibit 69% DNA sequence homology with each other (Pickett, Weinstein & Holmes, 1987; Pickett *et al.*, 1989). The deduced amino acid sequence of the LT-IIa A1-fragment shows 61% homology to the LT-Ih A1-fragment and 23% homology to the A2-subunit. This suggests that the enzyme active A1-subunits are well conserved between the LT families of toxins and that they may have originated from the same ancestral gene. In contrast, the B-subunits of LT-IIa and LT-IIb show no significant homology with the B-subunits from CT, LT-Ih and LT-Ip, which may explain the different ganglioside affinities of these toxins (Fig. 3) (Fukuta *et al.*, 1988). The deduced amino acid sequences of the LT-IIa and LT-IIb B-subunits show 67% homology. However, despite this relatively high percentage homology, LT-IIa and LT-IIb also have different ganglioside affinities (Fukuta *et al.*, 1988). LT-IIa has affinity for a range of gangliosides which include G_{M1}, whereas LT-IIb has no affinity for G_{M1} but binds to gangliosides such as G_{T1b} and G_{D1a}, which have a terminal sugar sequence of NeuAcα2-3Galβ1-3GalNAc (Fig. 3) (Fukuta *et al.*, 1988).

The high degree of homology, but different ganglioside-binding specificities of these toxins provides a unique opportunity for study of the structural basis of specific carbohydrate binding.

SHIGA TOXINS AND SHIGA-LIKE TOXINS

Shiga toxin (ShT) is produced by *Shigella dysenteriae* type 1, the shigella serotype responsible for the most severe cases of bacillary dysentery (Table

1). Clinical observations have shown that, during outbreaks of bacillary dysentery, a percentage (3 to 53%) of patients will develop sequelae involving the kidneys or the central nervous system. Although ShT is one of the most potent bacterial toxins, its actual role in the pathogenesis of shigellosis is still unclear (Tesh & O'Brien, 1991). In common with CT and the LTs, ShT is composed of an enzymatically active A-subunit (M_r 32.2 kD) in association with a ring of B-subunits (M_r 7.7 kD). Biochemical analysis suggests there are five B-subunits for every A-subunit. However, recent high-resolution X-ray crystallographic analysis has revealed a four-fold symmetry axis within a B-subunit crystal suggesting that the B-subunits exist naturally as tetramers and not pentamers (Hart et al., 1991). Definitive stoichiometry of the toxin awaits X-ray diffraction studies of the cystallized holotoxin.

In the past few years it has been demonstrated that a limited number of E. coli serotypes (predominantly O157:H7) produce toxins that are related to ShT. These E. coli strains were isolated from patients with haemorrhagic colitis, a watery diarrhoea that progresses to bloody diarrhoea, and from patients with haemolytic uraemic syndrome, a disorder characterized by acute renal failure. The shiga-like toxins (SLTs), which are also called Vero toxins, are functionally similar to ShT. For example, they are cytotoxic for Vero and HeLa cells, cause fluid accumulation in the rabbit ileal loop test and, when injected intraperitoneally, cause hind-limb paralysis and death in mice (Tesh & O'Brien, 1991). Two antigenically distinct types of SLTs, SLT-I and SLT-II, have been isolated and characterized from E. coli strains, which may produce one or both of the toxins. SLT-I can be neutralized by polyclonal antisera to ShT, whereas SLT-II cannot be cross-neutralized.

A third major group of SLTs, designated SLT-II variants (SLT-IIv), have been isolated from E. coli strains associated with swine oedema disease (SLT-IIvp) and with human diarrhoeal disease (SLT-IIvh) (Weinstein et al., 1988; Oku et al., 1989). Weaning pigs are mainly affected by swine oedema disease and characteristic features of the disease include convulsions and other neurological disorders. The SLT-IIvs can be neutralized by polyclonal antisera to SLT-II, but unlike ShT, SLT-I and SLT-II, culture filtrates of the SLT-IIvs are not cytotoxic for HeLa cells, but are much more cytotoxic for Vero cells. Also the SLT-IIvs have a much higher 50% lethal dose for BALB/c mice than other SLTs (Samuel et al., 1990).

The nucleotide sequences of ShT, SLT-I, SLT-II and SLT-IIvp have been determined (Weinstein et al., 1988). The deduced amino acid sequences of ShT and SLT-1 are virtually identical, except for a threonine residue at position 45 in the A-subunit of ShT which is replaced with a serine in SLT-I. In contrast, the deduced amino acid sequence from the structural gene for SLT-II shares only 57% homology with ShT and SLT-I. However, SLT-II shares 88% amino acid sequence homology with SLT-IIvp.

Mechanism of action

ShT and the SLTs appear to have the same cytotoxic mechanisms of action. Entry of the toxins into G_{b3}- or G_{b4}-containing cells is thought to be mediated by receptor-dependent endocytosis. The internalised vesicles may fuse with lysosomes and be translocated first to the Golgi apparatus and then into the cytosol (Sandvig et al., 1989). In the process of intracellular translocation, the A-subunit may undergo proteolytic cleavage at a trypsin-sensitive site near the amino terminus. The A-subunit of each of these toxins acts as a specific N-glycosidase which cleaves adenine 4324 from near the 3'-end of the 28S rRNA component of the eucaryotic ribosomal complex. This depurination results in inhibition of peptide elongation by blocking elongation factor-1-dependent aminoacyl-tRNA binding, which in turn causes protein synthesis to cease.

Interestingly, the 28S rRNA N-glycosidase activity of this family of toxins is identical to the mechanism of action of a family of plant toxins which includes ricin. This similarity is one of the few examples in which procaryotic and eucaryotic proteins, which are not directly involved in cellular 'house-keeping' functions, are conserved. Recently, a high resolution crystallographic structure of ricin has been reported, allowing visualization of a cleft in the A-subunit that may contain the active site of the enzyme (Montfort et al., 1987). If the conserved residues between the shiga and ricin toxin families are plotted on the ricin A chain crystal structure, seven amino acids are found to lie in the active site. These include glutamic acid 167, alanine 168, arginine 170, asparagine 202, tryptophan 203, leucine 206 and serine 207 from SLT-I (Hovde et al., 1988). Hovde et al. (1988) have used oligonucleotide-directed mutagenesis to change the glutamic acid 167 to an aspartic acid residue. The specific activity of the SLT-I mutant toxin was decreased by a factor of 1000 compared to the wild-type toxin. They conclude that glutamic acid 167 is an active-site residue in SLT-I and is critical for enzymic activity of the toxin. It is noteworthy that glutamic acid side chains have been shown to be crucial for enzymic activity in the ADP-ribosylating toxins (see above). These include ETA and diphtheria toxin, where conversion of a key glutamic acid residue at the NAD-binding site (residue 553 for ETA and residue 148 for diphtheria toxin) to aspartic acid causes a >100-fold loss of ADP-ribosylation activity (Douglas & Collier, 1987; Tweten, Barbieri & Collier, 1985). Similarly, glutamic acid 112 has been shown to be important for LT-Ip A1-subunit by mutagenesis studies (Tsuji et al., 1990).

There are many questions which remain to be answered regarding the enterotoxic mechanisms of action of the ShT and SLT family of toxins. For example, it is still uncertain if the proposed cytotoxic mechanism of action of ShT and the SLT family of toxins (inhibition of protein synthesis) accounts for the enterotoxic properties of these toxins.

Shiga toxin and Shiga-like toxin receptors

Binding of ShT, SLT-I and SLT-II to mammalian cells is accomplished by the interaction of the B-subunit(s) with a membrane glycolipid of the globoseries, globotriaosylceramide (G_{b3}) (Fig. 2). This glycolipid appears to be the functional receptor on Vero cells, HeLa cells and rabbit microvillus membranes (Jacewicz et al., 1986). Within the glycolipid molecule, the terminal Gal-α1-4-Gal disaccharide appears to be the minimum determinant necessary for toxin binding. The diverse activities of ShT and the SLTs on many tissues may in part be due to the binding potential of the B-subunit to several Gal-α1-4-Gal containing glycolipids.

The difference in susceptibility of cell lines to the cytotoxicity of the SLT-IIv toxins compared to ShT and the other SLTs is consistent with a difference in receptor glycolipid binding. DeGrandis et al. (1989) have shown that SLT-IIvp binds specifically to GalNAcβ1-3Galα1-4Galβ1-4Glc1-1Cer (G_{b4}) (Fig. 2) and has a much lower affinity for G_{b3}. Recently, Samuel et al. (1990) have found that SLT-II, but not SLT-IIvp, bound to the Galα1-4-Gal-BSA disaccharide receptor analogue, whereas both SLT-II and SLT-IIvp bound to the Galα1-4Galβ1-4Glc-BSA trisaccharide receptor analogue. This suggests that SLT-IIv prefers a larger Galα1-4-Gal-binding region than does SLT-II and, since G_{b4} is more strongly recognized by SLT-IIvp than G_{b3}, it is likely that the additional β1-3-linked GalNAc provides further thermodynamic stabilization of the binding. Samuel et al. (1990) also used ST-II/IIvp hybrid toxins to confirm the binding preference of SLT-IIvp for G_{b4}. These specificities were also important in vivo, as there was a 400-fold difference in mouse lethality between SLT-II and SLT-IIvp (Samuel et al., 1990).

These studies show that, despite the extensive homology between SLT-IIvp and SLT-II (80% amino acid sequence homology compared with 55% between the SLT-II and SLT-I B-subunits), SLT-IIvp does not share the same glycolipid binding specificity. Although SLT-IIvp and SLT-II have a high degree of homology, the isoelectric points of the B subunits of these toxins are very different (pI 10.2 for SLT-IIvp B-subunit versus pI 5.4 for the SLT-II B-subunit). These distinct differences are the result of significant variation in the isoelectric points of individual amino acids at comparable positions in the B-subunits of SLT-IIvp and SLT-II.

CLOSTRIDIUM DIFFICILE TOXINS

The toxigenic nature of *C. difficile* was first described over 50 years ago (Hall & O'Toole, 1935). However, *C. difficile* was not implicated in the aetiology of pseudomembranous colitis and antibiotic-associated colitis until the late 1970's (Larson et al., 1978). *C. difficile* produces two related toxins, toxins A and B, which play a significant role in *C. difficile*-mediated disease (Lyerly, Krivan & Wilkins, 1988; Lyerly et al., 1990a).

Toxin A causes an extensive amount of tissue damage to the gut mucosa which is thought to be responsible for most of the gastrointestinal symptoms associated with the disease (Lyerly *et al.*, 1985). Initially, the toxin damages the villous tips, and this is followed by the disruption of the brush border membrane. In rabbit ileal loop tests toxin A elicits production of a haemorrhagic and viscous fluid which differs from that elicited by CT and the LTs which mediate a 'ricewater' fluid accumulation in the loop assay. Also, in contrast to CT and the LTs, toxin A has no ADP-ribosylating activity and does not appear to be a subunit toxin.

Toxin B is an extremely potent cytotoxin. One pg is sufficient to cause rounding up of a range of mammalian tissue culture cells. The cytotoxicity of toxin A can be equally as potent as toxin B, but only on tissue culture cells which appear to have a high density of the $Gal\alpha 1\text{-}3Gal\beta 1\text{-}4GlcNAc$ trisaccharide (e.g. mouse embryonal carcinoma F9 cells) (Tucker, Carrig & Wilkins, 1990). However, toxin B is not a true enterotoxin and is inactive in the intestine when administered orally, but becomes lethal in combination with sublethal doses of toxin A, suggesting that toxin A initiates tissue damage providing toxin B with access to sensitive tissues (Lyerly *et al.*, 1988).

During the past decade several investigators have worked on the isolation and physical properties of toxins A and B, but there have been a number of discrepancies. These include M_r estimates for the native proteins, the ability of the toxins to dissociate into smaller subunits under denaturing conditions and the relative cytotoxicity and enterotoxicity of the respective toxins. Recent cloning and sequencing studies on both toxin genes have largely resolved these controversies and have provided information on the regions of the toxin A molecule responsible for binding to its carbohydrate receptor (Wren, 1992).

Recombinant DNA studies

Two studies have been reported where the entire *C. difficile* toxin A gene (8.1 kb) has been successfully cloned in *E. coli* (Wren *et al.*, 1987; Phelps *et al.*, 1991). In these studies the recombinant polypeptide exhibited haemorrhagic fluid accumulation in the rabbit ileal loop test, cytotoxicity and haemagglutination. These findings demonstrate that the *in vitro* biological activities exhibited by the native toxin A are functions of a single protein encoded by the 8.1 kb toxin A gene, which are independent of any other *C. difficile* gene products, including toxin B.

A recombinant polypeptide derived from a toxin A subclone, plasmid pCD23, consisting of the C-terminal quarter of the toxin A gene, readily agglutinated rabbit erythrocytes. Antiserum against this recombinant polypeptide has been found to neutralize both the enterotoxic and cytotoxic activities of the toxin (Lyerly *et al.*, 1990*b*). These results suggest that this C-terminal domain of the toxin A gene encodes the binding portion of toxin A for its carbohydrate receptor.

Primary structure of C. difficile *toxins*

The complete nucleotide sequence for the toxin A gene from strain VPI 10463 has been reported (Dove *et al.*, 1990). The coding region of 8,133 bp encodes 2710 amino acids which gives a deduced molecular mass for the toxin of *c*. 308 kD. Using the deduced amino acid sequence, data base searches have failed to show any significant amino acid sequence similarity with other bacterial toxins. However, the C-terminal third of the derived amino acid sequence of toxin A, which contains 38 continuous repeat units, shows homology with a family of streptococcal binding proteins (Wren, 1991) (Fig. 4). All 38 repeat groups, which have been designated I, A, B, C, and D, contain the dipeptide tyrosine-phenylalanine and have the consensus sequence of KAVTGxTIxGxxYYFxxNGx (Dove *et al.*, 1990; Wren, Clayton & Tabaqchali, 1990). The multivalent nature of the repetitive domain of toxin A and the fact that the gene product from clone pCD23, which contains 33 out of the 38 repeats, retains the toxin's haemagglutinating ability, provide further evidence that this portion of the polypeptide is responsible for the carbohydrate-binding capacity of the toxin (Lyerly *et al.*, 1990b).

Recently, the peptide sequence which is likely to be important in the binding of the toxin to its receptor has been further localized. Antiserum raised against the synthetic peptide TIDGKKYYFN, which is the most frequent and highly conserved repeat, cross-reacted with toxin A by immunoblot analysis, neutralized cytotoxic activity and inhibited toxin A-mediated haemagglutination (Wren, Russell & Tabaqchali, 1991). These results suggest that the decapeptide may be located on the surface of toxin A, and may play a crucial role in the binding of the toxin to its carbohydrate receptor.

The complete nucleotide sequence of the toxin B gene has revealed a coding region of 7098 bp which encodes 2366 amino acids with a deduced molecular mass for the toxin of *c*. 270 kDa (Barroso *et al.*, 1990). Analysis of the amino-acid sequence derived from the toxin B gene sequence, reveals that the C-terminal quarter of the toxin B protein contains 24 repeat units, designated I or E (Fig. 4). These repeats are clearly related to the toxin A repeats in size, amino acid sequence and arrangement (Wren, 1991). They have many conserved residues upstream of the tyrosine–phenylalanine dipeptide which was always present.

Another pathogenic clostridial species, *Clostridium sordellii*, also produces two toxins, namely, a haemorrhagic toxin and lethal toxin. They are antigenically related to *C. difficile* toxins A and B respectively, and the genes encoding these toxins hybridize with those encoding toxins A and B,

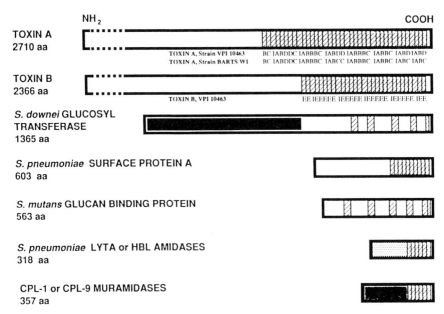

Fig. 4. Arrangement and location of amino acid repeat sequences in *C. difficile* toxins A and B, streptococcal glucan-interactive proteins, *Streptococcus pneumoniae* surface protein A and lytic enzymes from *S. pneumoniae* and its bacteriophages.
Cross hatched areas represent repeat sequences.
Solid black areas represent glucosyltransferase activity.
Dotted lines represent amidase activity.
Dark shading represent muramidase activity.

respectively (Martinez & Wilkins, 1988; Wren *et al.*, 1990). *C. sordellii* causes enteritis and enterotoxaemia in sheep and cattle. However, full molecular characterisation of these toxins remains to be determined.

Structural features of toxin-A binding domain

Comparison of the amino-acid sequences of the related repeat groups from toxins A and B reveals that three amino acids are conserved in almost all cases, the tyrosine-phenylalanine dipeptide and a glycine, ten residues upstream of the dipeptide. It is well recognized that stability can be conferred to the protein–carbohydrate complex in carbohydrate-binding proteins by partial stacking of aromatic amino acids, such as tyrosine and phenylalanine, within the sugar ring structure (Quiocho, 1986; Vyas *et al.*, 1991). Polar residues (e.g. glycine and threonine) may also be involved in hydrogen bonding in protein–polysaccharide complexes (Quiocho, 1986). The turn-promoting glycine residues perhaps confer flexibility to the ligand-binding fold. The repeats are also generally hydrophilic and this, together with the known immunogenicity of the toxin A repeats, suggests that they are probably exposed on the surfaces of the proteins, as one would expect for ligand-binding domains.

Related streptococcal multivalent binding domains

The consensus repeat sequence of KAVTGxTIxGxxYYFxxNGx from the C-terminal regions of toxins A and B is also found in the C-terminal regions of binding proteins important in the virulence of *Streptococcus pneumoniae* and the oral pathogens *Streptococcus mutans* and *Streptococcus downei* (Wren, 1991; Yother & Briles, 1992). These include a glucan-binding protein (GBP) from *S. mutans*, four glucosyltransferases (GTF) from *S. mutans* and *S. downei, S. pneumoniae* surface protein A and four pneumococcus-associated lysins. The glucan-interactive proteins contain a variable number of repeat sequences which are non-tandemly arranged. In contrast, the repeat sequences of the other proteins are tandemly arranged at the C-terminal end of the proteins (Fig. 4). The antiserum raised against the decapeptide TIDGKKYYFN, derived from a toxin A repeat, cross-reacted with the *S. mutans* GBP and GTF, and partially inhibited the ability of these proteins to bind their respective substrates, thus providing immuno-logical and functional evidence of the predicted relationship between the binding portion of toxin A and that of the glucan-binding proteins (Wren *et al.*, 1991). One explanation for the presence of these related repeats, in otherwise unrelated proteins, is that they form binding domains. This family of binding proteins appears to be of modular design, with one module constituting the binding domain located in the C-terminal region and the other module(s) providing enzymic functions.

Toxin A receptor

Toxin A has been shown to bind to brush border membranes from hamsters, which are animals particularly susceptible to the effects of toxin A (Rolfe, 1991). More recently, Pothoulakis *et al.* (1991) have purified a 135 kD glycoprotein from rabbit brush border membranes which appears to be coupled to a G protein. However, the biological receptor to which toxin A binds in humans remains unclear, particularly as most human infants are apparently insensitive to toxin A (Lyerly *et al.*, 1988). Toxin A readily binds the trisaccharide Galα1-3Galβ1-4GlcNAc (Fig. 3), which may account for many of the properties associated with toxin A, including its capacity to agglutinate rabbit erythrocytes, bind bovine thyroglobulin, and provoke a potent cytotoxic effect on cells expressing Galα1-3Galβ1-4GlcNAc (Tucker & Wilkins, 1991). However, despite the fact that hamster and rabbit brush border membranes express Galα1-3Galβ1-4GlcNAc, there is no evidence that human intestinal epithelial cells express the same trisaccharide. Recently, three further carbohydrate antigens, I, X and Y (Fig. 3), have been found to bind toxin A and are present on human intestinal epithelial cells (Tucker, Carrig & Wilkins, 1990). However, at least one of these antigens, antigen Y, is also present on human infant intestinal epithelial cells. These antigens contain the Galβ1-4GlcNAc type 2 core (Fig. 3) and may represent part of the glycosylated portion of the toxin A receptor on human intestinal mucosa.

The receptor and target tissues for toxin B remain to be identified. However, the receptor is likely to be widely found on eucaryotic cells judging from the extreme cytotoxicity of toxin B on a wide variety of cell types.

Conclusions on C. difficile toxins

Toxins A and B appear to have multiple functions which reside on single large polypeptides rather than individual subunits. Unfortunately, the primary structures of toxins A and B have yielded few clues as to their mechanisms of action, since they bear no amino acid sequence similarity to other characterised bacterial toxins. However, the genes encoding toxins A and B do show nucleotide sequence homology to each other over large areas of the regions which do not encode the repeat sequences (45% amino acid homology) which may suggest that they share similar cytotoxic mechanisms of action. Both toxins appear to be internalised by mammalian cells, although the intracellular target(s) and the mechanism(s) of internalisation remain unclear (Lyerly et al., 1988).

The toxins produced by C. difficile evidently lead to disease, but the manner by which this occurs appears more complex than just the direct action on the gut mucosa. The extensive damage which occurs from the action of toxin A and the intense inflammatory response may be due in part to immunopathological damage. It has been reported that toxin A causes granulocytes to release factors that can cause intestinal secretion and enteritis (Triadafilopoulous et al., 1987). It is likely that these factors contribute to the tissue damage associated with toxin A, thus increasing the severity of the disease.

TOXIN INTERACTIONS AND FUTURE STUDIES

The three families of enterotoxins described in this chapter either possess, multiple subunits which constitute the holotoxin (cholera and shiga families of toxins) or repeat units in a single polypeptide (clostridial toxins). The multiplicity of the binding domains appears to be essential for receptor-binding complex formation and may allow the toxin to bind several common (or different) carbohydrate moieties accounting for the specificity and avidity of binding.

The binding of the enterotoxins to specific carbohydrate moieties pro-vides a unique opportunity to study protein–carbohydrate interactions, which are also important in other microbe–host interactions and other biological recognition phenomena. Further insight into protein–carbohydrate complexes can be provided by physicochemical techniques such as fluorimetry, spectrophotometry and nuclear magnetic resonance spectroscopy. These techniques will provide information on the specificity of interaction and also permit the calculation of association constants and other thermodynamic and kinetic parameters. Such data will give a measure of the strength of binding and allow inferences to be made about the chemical groups involved. An understanding of toxin interactions may also have practical benefits in terms of the construction of novel fusion proteins for the delivery of specific therapeutic agents and immunogens of vaccine potential.

In the past few years, a whole range of immunological, tissue culture and molecular biological techniques have been used to characterise bacterial toxins. However, a clearer understanding of the structure–function relationships of most of these toxins must await experiments in which individual segments of the polypeptides are altered and tested for effects on receptor recognition. A detailed knowledge of the residues involved in ligand recognition is also required. Ultimately, three-dimensional structures of toxin–receptor and toxin–substrate complexes will be essential for our understanding of how ligand recognition is achieved.

ACKNOWLEDGEMENTS

Most of the work performed at St. Bartholomew's Hospital has been supported by the Wellcome Trust. I am indebted to Tony Minhas and Susan Colby for criticism of the text, Soad Tabaqchali and other colleagues at St. Bartholomew's Hospital for their help and advice on the *C. difficile* studies.

REFERENCES

Allured, V. S., Collier, R. J., Carroll, S. F. & McKay, D. B. (1986). Structure of exotoxin A of *Pseudomonas aeruginosa* at 3.0-Ångstrom resolution. *Proceedings of National Academy of Sciences of United States of America*, **83**, 1320–4.

Alouf, J. E. & Freer, J. H. (1991). *Sourcebook of bacterial toxins*. London: Academic Press.

Barroso, L., Wang, S.-Z., Phelps, C. J., Johnson, J. L. & Wilkins, T. D. (1990). Nucleotide sequence of *Clostridium difficile* toxin B gene. *Nucleic Acids Research*, **18**, 4004.

Chang, P. P., Moss, P. J., Twiddy, E. M. & Holmes, R. K. (1987). Type II heat-labile enterotoxin of *Escherichia coli* activates adenylate cyclase in human fibroblasts by ADP ribosylation. *Infection and Immunity*, **55**, 1854–8.

DeGrandis, S., Law, H., Brunton, J., Gyles, C. & Lingwood, C. A. (1989). Globotetraosylceramide is recognised by the pig edema disease toxin. *Journal of Biological Chemistry*, **264**, 12520–5.

Douglas, C. M. & Collier, R. J. (1987). Exotoxin A of *Pseudomonas aeruginosa*: substitution of glutamic acid 533 with aspartic acid drastically reduces toxicity and enzymatic activity. *Journal of Bacteriology*, **169**, 4967–71.

Dove, C. H., Wang, S.-Z., Price, S. B., Phelps, C. J., Lyerly, D. L., Wilkins, T. D. & Johnson, J. L. (1988). Molecular characterization of the *Clostridium difficile* toxin A gene. *Infection and Immunity*, **58**, 480–8.

Fukuta, S., Magnani, J. L., Twiddy, E. M., Holmes, R. K. & Ginsburg, V. (1988). Comparison of the carbohydrate-binding specificities of cholera toxin and *Escherichia coli* heat-labile enterotoxins LTh-I, LT-IIa, and LT-IIb. *Infection and Immunity*, **56**, 1748–53.

Griffiths, S. L., Finkelstein, R. A. & Critchley, D. R. (1986). Characterisation of the receptor for cholera toxin and *Escherichia coli* heat-labile enterotoxin in rabbit intestinal brush borders. *Biochemical Journal*, **238**, 313–22.

Hall, I. C. & O'Toole, E. (1935) Intestinal flora in new-born infants, with description of new pathogenic anaerobe, *Bacillus difficile*. American Journal of Diseases of Children, **49**, 390–402.

Harford, S., Dykes, C. W., Hobden, A. N., Read, M. J. & Halliday, I. J. (1989). Inactivation of *Escherichia coli* heat-labile enterotoxin by *in vitro* mutagenesis of the A-subunit gene. *European Journal of Biochemistry*, **183**, 311–6.

Hart, P. J., Monzingo, A. F., Donohue-Rolfe, A., Keusch, G. T., Calderwood, S. B. & Robertus, J. B. (1991). Crystallization of the B-subunit of shiga-like toxin from *Escherichia coli*. *Journal of Molecular Biology*, **218**, 691–4.

Hovde, C. J., Calderwood, S. B., Mekalanos, J. J. & Collier, R. J. (1988). Evidence that glutamic acid 167 is an active-site residue of shiga-like toxin I. *Proceedings of the National Academy of Sciences of United States of America*, **85**, 2568–72.

Iida, T., Tsuji, T., Honda, T., Miwatani, H., Wakabayashi, S., Wada, K. & Matsubara, H. (1989). A single amino acid substitution in B-subunit of *Escherichia coli* enterotoxin affects its oligomer formation. *Journal of Biological Chemistry*, **264**, 14065–70.

Jacewicz, M., Clausen, H., Nudelman, E., Donohue-Rolfe, A. & Keusch, G. T. (1986). Pathogenesis of *Shigella* diarrhoea; isolation of a *Shigella* toxin-binding glycolipid from rabbit jejunum and HeLa cells and its identification as globotriaosylceramide. *Journal of Experimental Medicine*, **163**, 1391–1404.

Larson, H. E., Price, A. B., Honour, P. & Borriello, S. P. (1978). *Clostridium difficile* and the aetiology of pseudomembranous colitis. *Lancet* i: 1063–66.

Lyerly, D. M., Johnson, J. L., Frey, S. M. & Wilkins, T. D. (1990*a*). Vaccination against lethal *Clostridium difficile* enterocolitis with a nontoxic recombinant peptide of toxin A. *Current Microbiology*, **21**, 29–33.

Lyerly, D. M., Johnson, J. L. & Wilkins, T. D. (1990*b*). The toxins of *Clostridium difficile* In *Clinical and Molecular Aspects of Anaerobes*, ed. Borriello, S. P., pp. 137–145. Petersfield, United Kingdom: Wrightson Biomedical Publishing Ltd.

Lyerly, D. M., Krivan, H. C. & Wilkins, T. D. (1988). *Clostridium difficile*: Its disease and toxins. *Clinical Microbiology Reviews*, **1**, 1–18.

Lyerly, D. M., Saum, K., MacDonald, D. & Wilkins, T. D. (1985). Effect of toxins A and B given intragastrically to animals. *Infection and Immunity*, **47**, 349–52.

Martinez, R. D. & Wilkins, T. D. (1988). Purification and characterisation of *Clostridium sordellii* hemorrhagic toxin and cross-reactivity with *Clostridium difficile* toxin A (enterotoxin). *Infection and Immunity*, **56**, 1215–21.

Monfort, W., Villafranca, J. E., Monzingo, A. F., Ernst, S. R., Katzin, B., Rutenber, E., Xuong, N. H., Hamlin, R. & Robertus, J. D. (1987). The three-dimensional structure of ricin at 2.8 Å. *Journal of Biological Chemistry*, **262**, 5398–403.

Moss, J., Osborne, J. C., Fishman, P. H., Nakaya, S. & Robertson, D. C. (1981). *Escherichia coli* heat-labile enterotoxin. *Journal of Biological Chemistry*, **256**, 12861–5.

Neill, R. J., Ivins, B. E. & Holmes, R. K. (1983). Synthesis and secretion of plasmid-coded *Escherichia coli* heat-labile enterotoxin in *Vibrio cholerae*. *Science*, **221**, 289–91.

Oku, Y., Yutsudo, T., Hirayama, T., O'Brien, A. D. & Takeda, Y. (1989). Purification and some properties of a Vero toxin from a human strain of *Escherichia coli* that is immunologically related to shiga-like toxin II (VT2). *Microbial Pathogenesis*, **6**, 113–22.

Phelps, C. J., Lyerly, D. M., Johnson, J. L. & Wilkins, T. D. (1991). Construction and expression of the complete *Clostridium difficile* toxin A gene in *Escherichia coli*. *Infection and Immunity*, **59**, 150–3.

Pickett, C. L., Twiddy, E. M., Belisle, B. W. & Holmes, R. K. (1986). Cloning of genes that encode a new heat-labile enterotoxin of *Escherichia coli*. *Journal of Bacteriology*, **165**, 348–52.

Pickett, C. L., Twiddy, E. M., Coker, C. & Holmes, R. K. (1989). Cloning, nucleotide sequence and hybridisation studies of the type IIb heat-labile enterotoxin gene of *Escherichia coli*. *Journal of Bacteriology*, **171**, 4945–52.

Pickett, C. L., Weinstein, D. L. & Holmes, R. K. (1987). Genetics of type IIa heat-labile enterotoxin of *Escherichia coli*: operon fusions, nucleotide sequence, and hybridisation studies. *Journal of Bacteriology*, **169**, 5180–7.

Pothoulakis, C., LaMont, J. T., Eglow, R., Gao, N., Rubin, J. B., Theoharides, T. & Dickey, B. F. (1991). Characterisation of rabbit ileal receptors for *Clostridium difficile* toxin A. Evidence for a receptor coupled G protein. *Journal of Clinical Investigation*, **88**, 119–25.

Quiocho, F. A. (1986). Carbohydrate-binding proteins: tertiary structures and protein-sugar interactions. *Annual Review of Microbiology*, **55**, 287–315.

Ribi, H. O., Ludwig, D. S., Mercer, L., Schoolnik, G. K. & Kornberg, R. D. (1988). Three-dimensional structure of cholera toxin penetrating a lipid membrane. *Science*, **259**, 1272–6.

Rolfe, R. D. (1991). Binding kinetics of *Clostridium difficile* toxins A and B to intestinal brush border membranes from infant and adult hamsters. *Infection and Immunity*, **59**, 1223–30.

Samuel, J. E., Perera, L. P., Ward, S., O'Brien, A. D., Ginsburg, V. & Krivan, H. C. (1990). Comparison of the glycolipid receptor specificities of shiga-like toxin type II and shiga-like toxin type II variants. *Infection and Immunity*, **58**, 611–18.

Sandvig, K., Olnes, S., Brown, J. E., Petersen, O. W. & Van Deurs, B. (1989). Endocytosis from coated pits of shiga toxin: A glycolipid-binding protein from *Shigella dysenteriae* 1. *Journal of Cell Biology*, **108**, 1331–43.

Sixma, T. K., Pronk, S. E., Kalk, K. H., Wartna, E. S., Van Zanten, B. A. M., Wiltholt, B. & Hol, W. G. J. (1991). Crystal structure of a cholera toxin-related heat-labile enterotoxin from *E. coli*. *Nature*, **351**, 371–7.

Sixma, T. K., Pronk, S. E., Kalk, K. H., Wartna, E. S., Van Zanten, B. A. M., Berghuis, A. M. & Hol, W. G. J. (1992). Lactose binding to heat-labile enterotoxin revealed by X-ray crystallography. *Nature*, **355**, 561–4.

Tesh, V. L. & O'Brien, A. D. (1991). The pathogenic mechanisms of shiga toxin and the shiga-like toxins. *Molecular Microbiology*, **5**, 1817–22.

Tsuji, T., Honda, T., Mitwatani, T., Wakabayashi, S. & Matsubara, H. (1985). Analysis of receptor-binding site in *Escherichia coli* enterotoxin. *Journal of Biological Chemistry*, **260**, 8552–8.

Tsuji, T., Inoue, T., Miyama, A., Okamoto, K., Honda, T. & Mitwatani, T. (1990).

A single amino acid substitution of *Escherichia coli* enterotoxin results in a loss of toxic activity. *Journal of Biological Chemistry*, **265**, 22520–5.

Triadafilopoulous, G., Pothoulakis, C., O'Brien, M. J. & LaMont, J. T. (1987). Differential effects of *Clostridium difficile* toxins A and B on rabbit ileum. *Gastroenterology*, **93**, 273–9.

Tucker, K. D., Carrig, P. E. & Wilkins, T. D. (1990). Toxin A of *Clostridium difficile* is a potent cytotoxin. *Journal of Clinical Microbiology*, **28**, 869–71.

Tucker, K. D. & Wilkins, T. D. (1991). Toxin A of *Clostridium difficile* binds to the human carbohydrate antigens I, X, and Y. *Infection and Immunity*, **59**, 73–8.

Tweten, R. K., Barbieri, J. T. & Collier, R. J. (1985). Diphtheria toxin. Effect of substituting aspartic acid for glutamic acid 148 on ADP-ribosylating activity. *Journal of Biological Chemistry*, **260**, 10392–4.

Vyas, N. K., Vyas, M. N. & Quiocho, F. A. (1991). Comparison of the periplasmic receptors for L-arabinose, D-glucose/D-galactose and D-ribose: Structural and functional similarity. *Journal of Biological Chemistry*, **266**, 5226–37.

Weinstein, D. L., Jackson, M. P., Samuel, J. E., Holmes, R. K. & O'Brien, A. D. (1988). Cloning and sequencing of a shiga-like toxin type II variant from an *Escherichia coli* strain responsible for edema disease of swine. *Journal of Bacteriology*, **170**, 4223–30.

Wren, B. W., Clayton, C. L., Mullany, P. P. & Tabaqchali, S. (1987). Molecular cloning and expression of *Clostridium difficile* toxin A in *Escherichia coli* K12. *Federation of European Biochemical Societies Letters*, **225**, 82–6.

Wren, B. W., Clayton, C. L. & Tabaqchali, S. (1990). Nucleotide sequence of *Clostridium difficile* toxin A gene fragment and detection of toxigenic strains by PCR. *Federation of European Microbiology Societies Letters*, **70**, 1–6.

Wren, B. W., Clayton, C. L., Casteldine, N. B. & Tabaqchali, S. (1990). Identification of toxigenic *Clostridium difficile* strains by using a toxin A gene-specific probe. *Journal of Clinical Microbiology*, **28**, 1808–12.

Wren, B. W. (1991). A family of clostridial and streptococcal ligand-binding proteins with C-terminal repeat sequences. *Molecular Microbiology*, **5**, 797–803.

Wren, B. W., Russell, R. R. & Tabaqchali, S. (1991). Antigenic cross-reactivity and functional inhibition by antibodies to *Clostridium difficile* toxin A, *Streptococcus mutans* glucan-binding protein and a synthetic peptide. *Infection and Immunity*, **59**, 3151–5.

Wren, B. W. (1992). Molecular characterisation of *Clostridium difficile* toxins A and B. *Reviews in Medical Microbiology*, **3**, 21–7.

Yother, J. & Briles D. E. (1992). Structural properties and evolutionary relationship of PspA, a surface protein of *Streptococcus pneumoniae*, as revealed by sequence analysis. *Journal of Bacteriology*, **174**, 601–9.

THE SUPERANTIGENIC ACTIVITIES OF BACTERIAL TOXINS

COLIN R. A. HEWITT, JOHN D. HAYBALL, JONATHAN R. LAMB AND ROBYN E. O'HEHIR

Department of Immunology,
St Mary's Hospital Medical School,
Imperial College of Science, Technology and Medicine,
Norfolk Place,
London W2 1PG, UK

INTRODUCTION

Bacterial toxins have long been recognized for their involvement in the pathogenesis of disease. More recently though, some of these toxins have been shown to possess potent qualities as immunomodulatory molecules. Most notable amongst these qualities are their profound mitogenic effects on T-lymphocytes. Because the polyclonal mechanism of T-cell stimulation by these toxins involves interaction with certain regions of T-cell antigen receptors (TcR), these toxins have been classified as 'superantigens'.

In this article the familiar toxic effects of superantigens are described, the relationship between their structure and function is outlined and the mechanisms of superantigen activity and the qualitative and quantitative consequences of their interaction with the immune system discussed.

BACTERIAL TOXINS AND THE PATHOGENESIS OF DISEASE

Bacterial exotoxins are produced by several genera of bacteria including staphylococci (Bergdoll *et al.*, 1981), streptococci (Cone *et al.*, 1987) and mycoplasma (Cole, Daynes & Ward, 1981). Each of the toxins produced by these organisms has a well-recognized association with a number of disease states ranging from food poisoning and shock, to heart disease and arthritis.

The toxins produced by *Staphylococcus aureus* are probably the best characterized and most completely understood. Thus far seven distinct subtypes of staphylococcal enterotoxin (SE) have been serologically identified and named SEA to SEE (Bergdoll, 1970).

Enterotoxins derived from food contaminated with *S. aureus* are probably responsible for the majority of food poisoning cases. Once ingested, SE cause fever, vomiting and violent diarrhoea, often with severe prostration. It is unclear how much SE is required to induce illness, but human feeding experiments indicate that the ingestion of at least 3.5 μg is required to cause illness in a 70 kg adult. Despite this, investigations surrounding actual outbreaks of food poisoning indicate that as little as 100 ng of SE may produce illness in sensitive individuals (Bergdoll, 1983).

The mechanisms by which SE confer illness are varied but both gastro-intestinal and neural sites are targeted (Bergdoll, 1983). A more recent proposal, of relevance to this article, suggests that massive SE-induced polyclonal stimulation of the cellular immune response induces a burst of cytokine release which may in turn be responsible for many of the symptoms associated with toxin ingestion (Marrack *et al.*, 1990).

SE are also associated with the pathogenesis of diseases other than food poisoning. Certain subtypes of SE for example are commonly identified as the products of isolates from septicaemia cases. In particular, they are implicated in the pathogenesis of severe pseudomembranous enterocolitis associated with infection by antibiotic-resistant staphylococci. Severe diarr-hoea followed by profound shock and fatal circulatory collapse have also been reported in these patients and associated with SE.

A toxin, distantly related to the SE, was identified in the 1980s in *S. aureus* isolates from patients with the then newly recognized toxic shock syndrome (TSS). The toxin identified in such isolates, TSS toxin-1 (TSST-1), is now almost certain to be a causal agent of TSS (Bergdoll *et al.*, 1981), a syndrome associated with tampon use and penetrating wounds. The clinical symptoms of TSS include emesis, diarrhoea, fever, hypotension and leucopenia followed by leucocytosis, renal failure and desquamation of the skin in the presence or absence of cardiac, hepatic and pulmonary dysfunction.

Two additional, less well characterized toxins derived from *S. aureus* are the exfoliating toxins A and B. These toxins are strongly associated with the staphylococcal scalded skin syndrome. Other less well-studied toxins have also been identified as the products of *Streptococcus pyrogenes* and *Myco-plasma arthritidis*, a pathogen of rodents. The pyrogenic exotoxins of *S. pyrogenes* (SPE) are distantly related to the SEs but are strongly implicated in the pathogenesis of scarlet fever and in the induction of heart damage in rheumatic fever (Weeks & Ferretti, 1986; Cone *et al.*, 1987; Goshorn & Schlievert, 1988; Stevens, Tanner & Winship, 1989; Bartter *et al.*, 1988). *M. arthritidis* is a naturally occurring pathogen of rodents responsible for causing a chronic relapsing arthritis associated with conjunctivitis, uveitis, urethritis and paralysis (Cole, Washburn & Taylor-Robinson, 1985). This cluster of symptoms is, however, preceded by a rodent form of toxic shock syndrome caused by the production of a mitogenic toxin (MAM); (Cole *et al.*, 1981) completely unrelated to the SEs. Despite the significance of MAM in the study of bacterial toxicity in animals, no equivalent mycoplasmal toxin has yet been isolated from humans.

STRUCTURAL RELATIONSHIPS BETWEEN BACTERIAL TOXINS

All the SE, (SEA, SEB, SEC1, SEC2, SEC3, SED, SEE, TSST-1) and two of the four SPE (SPEA and SPEC) have been sequenced and characterized as small monomeric proteins of 22 to 28 kD (Huang *et al.*, 1987; Huang & Bergdoll, 1970; Schmidt & Spero, 1983; Bohach & Schlievert 1989; Couch & Betley, 1989; Bayles & Iandolo, 1989; Couch, Soltis & Betley, 1988; Blomster-Hautamaa *et al.*, 1986; Weeks & Ferretti, 1986; Goshorn & Schlievert, 1988). Alignment and comparison of these toxin sequences has allowed the structural and ancestral relationships between each of the toxins to be established.

On the basis of primary amino acid sequence homology the staphylococcal enterotoxins have been aligned and divided into two groups (Marrack & Kappler, 1990). Examination of the amino acid sequences of SEA, SED and SEE demonstrate that these toxins share considerable homology (see Fig. 1), such that the sequences of SEA and SEE are more than 90% homologous, strongly suggesting their derivation from a common ancestral gene. The amino acid sequences of SEB, SEC1, SEC2, and SEC3, are less like those of SEA, SED and SEE, but each toxin still contains stretches of amino acids in common with each other. With the exception of SPEA, which on the grounds of sequence analysis may be derived from a gene related to the SE (Weeks & Ferretti, 1986), the streptococcal exotoxin SPEC and the staphylococcal exfoliating toxins, ETA and ETB, demonstrate as much homology with each other as they do with each of the SE (Marrack & Kappler, 1990). Although TSST-1 shares little homology with any of the SE, SPE or ET, a number of conserved stretches of residues between TSST-1 and the SE can be identified. One particularly notable difference between TSST-1 and the SE, and related to their activity, is the absence of a conserved disulphide loop (Blomster-Hautamaa *et al.*, 1986).

The *M. arthritidis* toxin, MAM, has been sequenced partially and shows no homology to any other bacterial toxin or protein. A peculiar and unique feature of this protein, compared with the other bacterial toxins, is that MAM contains no signal sequence. It has, therefore, been proposed that mature MAM is released only from cells undergoing lytic senescence (Cole & Atkin, 1991).

The minor lymphocyte stimulating (Mls) antigen system of endogenous murine superantigens, although completely unrelated to toxins, warrants discussion here due to their phenotypic similarity to the effects of bacterial toxins on the immune system. Mls antigens have been found recently to be encoded by mouse mammary tumour retroviruses (MMTV) that have integrated into the germline as DNA proviruses (for review see Acha-Orbea and Palmer 1991). Sequence analysis demonstrates that the retroviral Mls antigens and the bacterial toxins have very little homology, and it has thus been proposed that their similar effects on the immune system are due to the

(a)

```
            10         20         30         40         50         60         70         80         90        100        110        120
SEA  SEKSEEINEKDLRKKSELQGTALGNLKQIYYNEKAKTENKESHDQFLQHTILFKGFFTDHSWYNDLLVDFDSKDIVDKYKGKKVDLYGAYYGYQCAGTFPNKTACMYGGVTLHDNNRLTEE

SEE    SEEINEKDLRKKSELQRNALSNLRQIYYNEKAITENKESDDQFLENTLLFKGFFTGHPWYNDLLVDLGSKDATNKYKGKKVDLYGAYYGYQCAGTFPNKTACMYGGVTLHDNNRLTEE

SED     SVKEKELHKKSELSSTALNNMKHSYADKNPIIGENKSTGDQFLENTLLYKKFFTDLINFEDLLINFNSKEMAQHFKSKNVDVYPIRYSINCYGGEIDRTACTYGGVTPHEGNKLKER


           130        140        150        160        170        180        190        200        210        220        230
SEA  KKVPINLWLDGKQNTVPLETVKTNKKNVTVQELDLQARRYLQEKYNLYNSDVFDGKVQRGLIVFHTSTEPSVNYDLFGAQGQYSNTLLRIYRDNKTINSENMHIDIYLYTS

SEE  KKVPINLWIDGKQTTVPIDKVKTSKKEVTVQELDLQARHYLHGKFGLYNSDSFGGKVQRGLIVFHSSEGSTVSYDLFDAQGQYPDTLLRIYRDNKTINSENLHIDLYLYTT

SED  KKIPINLWINGVQKEVSLDKVQTDKKNVTVQELDAQARRYLQKDLKLYNNDTLCGKIQRGKIEFDSSDGSKVSYDLFDVKGDFPEKQLRIYSDNKTLSTEHLHIDIYL
```

(b)

```
            10         20         30         40         50         60         70         80         90        100        110        120
SEB   ESQPDPKPDELHKSSKFTGLMENMKVLYDDNHVSAINVKSIDQFLYFDLIYSIKDTKLGNYDNVRVEFKNKDLADKYKDKYVDVFGANYYYQCYFSKKTNDINSHQTDRKKTCMYGGVTEHN

SEC1  ESQPDFTPDELHKASKFTGLMENMKVLYDDHYVSATKVKSVDKFLAHDLIYNISDKKLKNYDKVKTELLNEGLAKKYKDEVVDVYGSNYYVNCYFSSK-DNV-GKVTGG-KTCMYGGITKHE

SEC2  ESQPDPTPDELHKSSEFTGTMGNMKYLYDDHYVSATKVMSVDKFLAHDLIYNISDKKLKNYDKVKTELLNEDLAKKYKDEVVDVYGSNYYVNCYFSSK-DNV-GKVTGG-KTCMYGGITKHE

SEC3   SQPDFTPDELHKSSEFTGTMGNMKYLYDDHYVSATKVMSVDKFLAHDLIYNISDKKLKNYDKVKTELLNEDLAKKYKDEVVDVYGSNYYVNCYFSSK-DNV-GKVTGG-KTCMYGGITKHE

SPE-A  LQNIYFLYEGDPVTHENVKSVDQLLSHDLIYNVSGP---NYDKLKTELKNQEMATLFKDKNVDIYGVEYYHLCYLC----E-NA---E-RSACIYGGVTNHE


           130        140        150        160        170        180        190        200        210        220        230        240
SEB   GNQLD--KYRSITVRVFEDGKNLLSFDVQTNKKKVTAQELDLTRHYLVKNKKLYEFNNSPYETGYIKFIE-NENSFWYDMMPAPGDKFDQSKYLMMYNDNKMVDSKDVKIEVYLTKKK

SEC1  GNHFDNGNLQNVLIRVYENKRNTISFEVQTDKKSVTAQELDIKARNFLINKKNLYEFNSSPYETGYIKFIENNGNTFWYDMMPAPGDKFDQSKYLMMYNDNKTVDSKSVKIEVHLTTKN

SEC2  GNHFDNGNLQNVLIRVYENKRNTISFEVQTDKKSVTAQELDIKARNFLINKKNLYEFNSSPYETGYIKFIENNGNTFWYDMMPAPGDKFDQSKYLMMYNDNKTVDSKSVKIEVHLTTKN

SEC3  GNHFDNGNLQNVLIRVYENKRNTISFEVQTDKKSVTAQELDIKARNFLINKKNLYEFNSSPYETGYIKFIENNGNTFWYDMMPAPGDKFDQSKYLMMYNDNKTVDSKSVKIEVHLTTKN

SPE-A GNHLEIPK-K-IVVKVSIDGIQSLSFDIETNKKMVTAQELDYVRKYLTDNKQLYTNGPSKYETGYIKFIPKNKESPWFDFPEP-E-FTQSKYLMIYKDNETLDSNTSQIEVLTTK
```

Fig. 1. Amino acid sequence comparison of the two related groups of enterotoxins proposed by Marrack & Kappler (1990). (a) SEA, SED and SEE; (b) SEB, SEC1, SEC2, SEC3 and SPE-A. References for each of the sequences are cited in the text.

convergent evolution of bacterial and viral mechanisms to evade immune destruction.

MANY TOXIN GENES ARE RELATED AND LOCATED ON MOBILE ELEMENTS

Although the bacterial toxins share significant homology in their nucleic and amino acid sequence, the physical location of their genes is diverse. The *SEA, SPEA* and *SPEB* genes are all encoded by different bacteriophages (Betley and Mekalonos, 1985; Goshorn & Schlievert, 1988; Weeks & Ferretti, 1986), whilst the SEB gene, although located on the bacterial chromosome, is thought to have originated from either a defective integrated phage or plasmid (Johns & Khan, 1988). In common with SEA, SED, SPEA and SPEB, the *SEC1* gene is reported to be located on mobile elements, in one instance on the same plasmid as the *SEB* gene (Altboum, Hertman & Sarid, 1985).

TSST-1, which is weakly homologous with both the SE and SPE, is not associated with any mobile elements, and it has been suggested that the similar immunological properties of the SE and TSST-1 may have arisen via convergent evolution (Kreisworth *et al.*, 1983).

THE SECONDARY AND TERTIARY STRUCTURES OF BACTERIAL TOXINS

Resolution of the secondary and tertiary structures of the bacterial toxins, and their functional domains, are essential for a complete understanding of their interactions with physiologically important receptors.

A series of early studies attempted to predict the secondary and tertiary structures of SEA, SEB and SEC based on their ultraviolet circular dichroic (CD) and visible optical rotatory dispersion (ORD) spectra (Middlebrook, Spero & Argos, 1980; Munoz, Warren & Noelken, 1976); techniques commonly used to reveal α-helices, β-forms, and the aperiodic conformational content of a given protein. Using these techniques, whilst similarities were observed between SEB and SEC, SEA was noted to be significantly different (Middlebrook *et al.*, 1980). These results are now known to be in accordance with the predicted hierarchical sequence homology of the three toxins (Marrack & Kappler, 1990).

Similarities in the absorption spectra in the 245–290 nm range indicated that the aromatic residues of the three toxins were located in a similar microenvironment, thus suggesting a common tertiary structure for each of the three toxins (Middlebrook *et al.*, 1980). Despite this a more recent, higher resolution analysis of the structure of SEB and SEC1, conducted using a combination of far-UV CD, computer-assisted predictive techniques, and fluorescence spectroscopy, revealed that the microenvironment surrounding each of the tryptophan residues shared by the three toxins was different, thus suggesting fundamental differences in the tertiary structures of SEB and SEC1 (Singh, Evenson & Bergdoll, 1988).

Until the crystal structures of each of the bacterial toxins have been resolved definitively, questions concerning their secondary and tertiary structure will remain unanswered. This work is, however, in progress, and preliminary results concerning the structure of SEB have already been reported (Swaminathan *et al.*, 1988).

THE FUNCTIONAL DOMAINS OF BACTERIAL TOXINS

The majority of studies designed to elucidate the functional domains of bacterial toxins have concentrated on the SEs and TSST-1. These have taken the form of chemically modifying or fragmenting the toxins, then assessing their toxic and immunological activities.

Chemical modifications, such as the alkylation of methionine residues, or the acetylation and guanidation of free amino groups in SEB, have been effective in reducing the emetic effects of the toxin (Bergdoll, 1983). Conclusions concerning the structure–function relationship of these modi-fied toxins are somewhat limited, however, as these treatments also cause profound changes in the secondary and tertiary structure of proteins. Carboxymethylation of SEB has, however, been observed to reduce toxicity whilst retaining the conformational integrity of the toxins and their mito-genic properties (Alber, Hammer & Fleischer, 1990). These experiments, therefore, imply that the toxic properties, and the non-toxic, superantigenic activities of bacterial toxins (described below), are associated with different regions of the toxin molecule.

Both cyanogen bromide and trypsin have been used to fragment the toxins and distinguish which region is responsible for each of the characteristic activities. Experiments in which SEC1 was subjected to limited tryptic hydrolysis concluded that the region conferring mitogenicity was contained in the 6.5 kD N-terminal fragment, whilst the 22 kD C-terminal fragment was associated with emetic activity (Spero & Morlock, 1978; Spero *et al.*, 1976). In contrast, a more recent report found that both the toxic and mitogenic activities of SEC1 were retained by the C-terminal region (Bohach, Handley & Schlievert, 1989). Differences in the solubility of each of the tryptic fragments were suggested to account for the differences observed in these studies.

Using truncated recombinant fusion proteins between SEB and protein A, it has been shown that the immunological activity of SEB resides in the amino terminal region of the molecule within residues 1–138. Interestingly, deletion of the first 30 residues of this molecule stimulated different sub-groups of T-cells *in vitro* (Buelow *et al.*, 1992).

A characteristic disulphide loop present in all SE, has also been found essential for both the superantigenic and emetic functions of SEA and SEB.

Whilst the disulphide loop of SEB can be nicked without any loss of biological function, complete reduction followed by alkylation of the cysteine residues of SEB and SEA abrogates the mitogenic characteristics of the proteins (Grossman *et al.*, 1990). This was not due to a failure of the reduced and alkylated proteins to bind a receptor of superantigenic toxins, major histocompatibility complex (MHC) class II molecules (described in detail below), as the modified molecule successfully competed with native toxin for the MHC binding site (Grossman *et al.*, 1990).

To investigate the role of the disulphide loop further, a synthetic peptide corresponding to the disulphide loop residues 108 to 134 of SEB was synthesized in both a linear and cyclical form. These peptides were found to be incapable of competing for binding to MHC class II molecules or mediating the mitogenic functions characteristic of SEB (J. D. Hayball, unpublished observations). Thus it can be concluded that the disulphide bond is a structural requirement to hold SE in a conformation necessary for their function.

Despite many studies indicating that the active regions of SE may not be resolved as simple peptides, a peptide representing the first 27 amino acids of the mature SEA protein is capable of inhibiting certain mitogenic functions of SEA (Pontzer, Russel & Johnson, 1989). It is not yet clear, however, whether this inhibition operates by competition for binding to MHC class II or the TcR.

The absence of a disulphide loop in TSST-1 argues that the structural requirements determining the activity of TSST-1 are different from those of SE. In accordance with this, it has been reported that a linear peptide of TSST-1 residues 58–78 is fully capable of mimicking the mitogenic activities of the native toxin (Edwin *et al.*, 1991).

The majority of these studies, therefore, indicate that the secondary and tertiary structure of the SE are crucial for their function, and thus fragmentation and the generation of synthetic peptides of toxins are of limited usefulness in the analysis of structure–function relationships. Therefore, to determine the functional domains of SE whilst avoiding disruption of conformation we are currently using subtle techniques such as site directed mutagenesis of critical residues in the SE (J. Hayball, unpublished observations).

BACTERIAL TOXINS POWERFULLY STIMULATE THE IMMUNE SYSTEM

Twenty years have passed since staphylococcal enterotoxins were first shown to stimulate vigorous proliferative responses in lymphocytes (Peavy, Adler & Smith, 1970). At this time, interest in the SE was largely confined to their role in the pathology of staphylococcal food poisoning (Bergdoll, 1970); accordingly, it was suggested that SE mediated activation of the immune system *in vivo* may be of importance in the recovery from infection with staphylococci. (Taranta 1974; Langford, Stanton & Johnson, 1978). More recently, a structural mechanism has been proposed to explain why certain bacterial toxins are mitogenic for some populations of lymphocytes. Based on this, an hypothesis has been constructed suggesting that, far from aiding recovery from infection, toxin mediated mitogenesis actually accounts for their pathogenic effects *in vivo* (Marrack et al., 1990).

PHENOTYPIC SIMILARITIES BETWEEN BACTERIAL TOXINS AND Mls ANTIGENS

In addition to those examining the toxicity of enterotoxins, the discovery that toxins were also mitogenic for T-cells received attention from immunologists investigating the intercellular interactions required for *in vitro* immune responses. In these experiments, SE were found to be more potent inducers of T-cell proliferation than the more commonly used lectin mitogens phytohaemagglutinin (PHA) and concanavalin A (ConA). As such a large proportion of lymphocytes were stimulated in these experiments, SE were classified, along with PHA and ConA, as non-specific mitogens (Kaplan, 1972; Smith & Johnson, 1975; Johnson, Stanton & Baron, 1977; Langford, Stanton & Johnson 1978).

An early indication that the mechanism of enterotoxin mediated mitogenicity differed from that induced by lectin antigens was provided by attempts to identify substances mimicking the T-cell activation induced by products of the Mls loci. (Festenstein, 1973; Acha-Orbea & Palmer, 1991). In this system, vigorous T-cell activation was induced when the T-cells and spleen cells of Mls disparate strains of mice were mixed. Such anti-Mls responses were distinguished from alloreactive responses by their failure to correlate with MHC haplotype. Other characteristics of anti-Mls responses which differed from lectin induced responses, but were similar to enterotoxin induced responses included: a dependence on, but not restriction by, MHC class II antigens and the high clonal frequency of Mls responsive T-cells, estimated to be 2- to 5-fold higher than the frequency of alloreactive T cells (Janeway et al., 1988, 1989).

Comparison of peripheral T-cell responses to lectins and enterotoxins, however, revealed that enterotoxins were more selective in their mode of action than lectin mitogens.

TcRVβ USAGE ACCOUNTS FOR THE SUPERANTIGENIC PROPERTIES OF TOXINS

Although SE were considerably more potent at low concentrations, plateau levels of proliferation were lower than with lectins. This suggested that, whilst acting as polyclonal mitogens, SE only exerted their effects on a subset of T cells (Janeway et al., 1989; White et al., 1989). An explanation for this was first suggested by experiments showing that the capacity of cloned T-cell lines to respond to certain enterotoxins was heterogeneous and clonally expressed (Fleischer & Schrezenmeier, 1988; Janeway et al., 1989). The mechanism of this phenomenon was determined by parallel investigations in the Mls antigen system in which monoclonal antibodies to T-cell antigen receptor (TcR)Vβ regions were used to analyse TcRVβ useage in different strains of mice. These studies demonstrated that, in contrast to TcR recognition of nominal peptide–MHC complexes, in which the Vα/β Jα/β Dβ and N regions of the TcR interact with both the peptide and MHC (Davis & Bjorkman, 1988), TcR Vβ region useage alone was sufficient to accurately predict the specificity of a T-cell for certain Mls antigens (MacDonald et al., 1988). Because of the striking similarities between the mechanisms of enterotoxin and Mls mediated T-cell activation, TcRVβ useage was also examined in SE stimulated murine and human T-cells. As predicted, each SE stimulated a distinct population of T cells in a manner exclusively dependent upon the TcR Vβ region used (Janeway et al., 1989; White et al., 1989; Callahan et al., 1989; Kappler et al., 1989; Choi et al., 1989; Cole et al., 1989a, 1989b; Takimoto et al., 1990; Yagi et al., 1990). Because this unique polyclonal mechanism of activation involved only the Vβ region of the normally antigen specific, clonally distributed TcR, the term 'superantigen' was proposed to describe both the bacterial toxins and Mls antigens (White et al., 1989).

Analysis of the interaction between superantigens and TcRVβ elements has determined that both Mls antigens and enterotoxins interact with parts of the Vβ region distinct from the complementarity determining regions (CDR) involved in binding nominal peptide antigen–MHC complexes (Davis & Bjorkman, 1988). In an initial study, it was noted that, despite the failure of laboratory mouse T cells bearing TcRVβ8.2a to respond to Mls-1[a], natural variants of Vβ8.2, Vβ8.2b and c, isolated from wild mice, responded vigorously to Mls-1[a] spleen cells. By comparing the sequences of Mls-1[a] reactive and non-reactive Vβ8.2 elements, several residues were identified which were strongly associated with reactivity to Mls-1[a] (Pullen et al., 1990a). In a subsequent study, these sites on Vβ8 were mutated, then expressed and their reactivity with Mls-1[a] tested. When the substituted residues affecting reactivity with Mls-1[a] were mapped onto an immunoglobulin based structural model of the TcR (Chothia, Boswell & Lesk, 1988), the Mls-1[a] binding site was found to be located in a region exposed to the aqueous phase on a β-pleated sheet located on the side of the TcR molecule well away from the CDRs (Pullen et al., 1990b). A similar analysis was then conducted using SE and human T cells bearing the naturally variant TcRVβ13.2 and Vβ 13.1 elements which fail to respond, and respond to the enterotoxin SEC2 respectively. Remarkably, this study localized the SEC2 binding site on the TcR to residues 67–77 of the TcRVβ region, the same region as the Mls-1[a] binding site on the murine TcRVβ8.2 (Choi et al., 1990). That this region on the TcRVβ chain functions as the sole superantigen binding site on the TcR is supported by experiments in which an isolated soluble TcRβ chain was sufficient to bind directly to SEA pulsed antigen presenting cells (APC) (Gascoigne & Ames, 1991).

Despite these elegant experiments, the generality of an inflexibly strict relationship between the interaction of particular TcRVβ regions and a certain superantigen has been questioned. In accordance with this, it has been suggested that different TcR bind to superantigens with different affinities and that the specificity of the TcRVβ-superantigen interaction is quantitative rather than qualitative (Fleischer *et al.*, 1991). In support of this, a number of groups have now demonstrated that certain T-cell lines and hybridomas can be activated directly by incubation with superantigens in the complete absence of MHC class II antigen expression (Fleischer and Schrezenmeier, 1988; Matthes *et al.*, 1988; Yagi *et al.*, 1990; Herrmann *et al.*, 1991; Hewitt *et al.*, 1992). Whether the TcR of these cells represent those at the high affinity end of the spectrum of affinities for enterotoxin, or whether other TcR elements such as the α chain are able to influence responsiveness to superantigens (Blackman *et al.*, 1990) has not yet been examined fully.

SUPERANTIGEN PRESENTATION IS MHC DEPENDENT BUT NOT RESTRICTED

Unlike nominal peptide antigens, the T-cell recognition of bacterial super-antigens is not restricted to presentation by a particular allelic MHC antigen. Despite this (noting the exceptions described above), most T-cell–superantigen interactions appear to be dependent upon the presence of MHC class II-bearing accessory cells and can be inhibited by anti-MHC class II antibodies (Fleischer & Schrezenmeier 1988; Carlsson, Fischer & Sjogren, 1988; Janeway *et al.*, 1989). This implies that invariant regions of the MHC class II antigens, outside the polymorphic peptide antigen binding groove, interact with the enterotoxin and present it to T-cells. The region fulfilling this function is so well conserved that allogeneic MHC class II antigens present superantigens to T-cells (Fleischer & Schrezenmeier, 1988; White *et al.*, 1989), and even xenogenic MHC class II antigens suffice (Mollick, Cook & Rich, 1989; Fleischer, Schrezenmeier & Condradt, 1989; Herman *et al.*, 1990).

Despite the prediction of considerable structural homology between MHC class I and class II antigens (Brown *et al.*, 1988), there is only one report suggesting that MHC class I antigens interact with SE (Fraser, 1989). There is no evidence though that these complexes present SE to T-cells, as further illustrated by the absolute requirement for MHC class II antigens in presentation of enterotoxins to the TcRs of CD8[+] cytotoxic T-cells normally restricted to their nominal antigens by MHC class I antigens (Fleischer & Schrezenmeier 1988; Herrmann *et al.*, 1990; Fleischer *et al.*, 1991).

Since MHC class I-enterotoxin complexes have only been observed on cells expressing MHC class II antigens, it has been proposed that MHC class I might interact and thus co-immunoprecipitate with pre-formed MHC class II–enterotoxin complexes (Fraser, 1989).

BACTERIAL SUPERANTIGENS BIND TO MHC CLASS II MOLECULES

The interaction of toxins with MHC class II molecules has been assessed by several approaches, including the immunoprecipitation of toxin/MHC class II complexes with anti-MHC class II or anti-toxin antibodies and the binding of labelled toxins to murine fibroblasts transfected with the genes encoding MHC class II antigens (Fraser, 1989; Mollick *et al.*, 1989; Fischer *et al.*, 1989; Scholl *et al.*, 1990). These studies demonstrate a direct binding interaction between toxins and MHC class II molecules with association and dissociation rates differing considerably from those observed with peptide antigen-MHC binding (Scholl *et al.*, 1989). Although it is disputed whether allelic polymorphism in class II MHC molecules has any effect on toxin binding (Scholl *et al.*, 1990; Herman *et al.*, 1990), most studies agree that individual toxins preferentially bind to selected MHC class II antigen isotypes (Janeway *et al.*, 1989; Yagi *et al.*, 1990; Scholl *et al.*, 1990). SEB and TSST-1, for example, bind to both HLA-DR and HLA-DQ antigens, but fail to bind to HLA-DP antigens (Scholl *et al.*, 1990). Similar studies, in the mouse show SEB to exhibit an isotype preference for I-E over I-A class II molecules (Janeway *et al.*, 1989; Yagi *et al.*, 1990).

Despite the reported interaction of SEA with a synthetic peptide derived from the residues 65–85 of the I-A^b MHC class II β chain, and the possibility that there are differences in SE presentation by different allelic forms of HLA-DR (Herman *et al.*, 1990), there is a large body of evidence supporting the view that MHC class II α chains make the major contribution to toxin binding. Attempts have been made to map the site on MHC class II α chains to which enterotoxins bind using a series of cell lines transfected with mutant MHC class II genes carrying alanine substitutions at each residue of the I-A^k α chain from positions 70–79, the region predicted to form the α helix of the α chain. Mutant α chains were expressed in the presence of wild type I-A^k β chains and the transfectants used to present SEB or nominal peptide antigen to a T-cell hybridoma (Dellabona *et al.*, 1989). Despite the ability of certain substitutions to completely abrogate nominal peptide binding and its presentation to the T-cell hybridoma, the binding of SEB to the mutant MHC molecules and presentation of SEB to the T-cell hybridoma was considerably less sensitive to any of the substitutions. Interestingly, those substitutions which had the greatest negative effect upon the presentation of SEB, (residues 52, 53, 61, 64 and 71), were predicted to lie with their side chains pointing away from the antigen binding groove (Brown *et al.*, 1988) which infers that the outside of the α chain α helix binds SE. These results are consistent with similar experiments which took advantage of the unusual characteristic of TSST-1 to bind with high affinity to HLA-DR and low affinity to HLA-DP (Scholl *et al.*, 1990). Chimaeric α and β chains of HLA-DR and HLA-DP molecules were thus constructed and transfected into murine fibroblasts which were used in a TSST-1 binding assay. The pattern of TSST-1 binding obtained strongly suggested that TSST-1 makes contact with the $\alpha1$ domain of MHC class II antigen (Karp *et al.*, 1990). The remoteness of this site from the peptide binding groove was demonstrated by the inability of TSST-1 to inhibit the binding of an HLA-DR restricted immunogenic peptide (Karp *et al.*, 1990).

Further evidence that SE bind to MHC class II outside the peptide binding groove is derived from similar competition studies. In these experiments, no competition for presentation to T-cells can be demonstrated between lysosyme peptides and SEB (Dellabona *et al.*, 1989). The authors' data demonstrates that an HLA-DR1 restricted peptide antigen is able to interact with a T-cell clone in the presence of SEC2, which binds HLA-DR1, but fails to interact with the TcR of the T cell clone (Hewitt *et al.*, 1992).

SUPERANTIGENS DO NOT REQUIRE ANTIGEN PROCESSING

With the possible exception of MAM (Bauer, Rutenfranz & Kirchner, 1988), there is little data to suggest that enterotoxin superantigens require processing prior to presentation. Much of this evidence is derived from experiments in which paraformaldehyde or glutaraldehyde fixed cells are used as accessory cells for enterotoxin mediated T-cell activation (Fleischer & Schrezenmeier, 1988; Carlsson *et al.*, 1988; Yagi *et al.*, 1990); these data have also been used to corroborate evidence that enterotoxins bind MHC class II antigens outside the peptide binding groove. Others, however, report that SEB mediated stimulation of T-cells is not completely resistant to the fixation of accessory cells, and raise the point that similarities may exist between the lack of processing requirements by enterotoxins and a conventional antigen such as fibrinogen (Dellabona *et al.*, 1989), which is able to stimulate T cells without being processed (Allen, 1987).

Superantigens are, therefore, functionally bivalent molecules with the ability to bind to both MHC class II antigens and the TcR of T-cells bearing certain TcRVβ elements. This results in the induction of a programme of T-cell activation events currently thought to be identical to that induced by nominal peptides presented by appropriate MHC restriction elements.

SUPERANTIGENS MAY MODIFY THE FUNCTIONAL T CELL REPERTOIRE

Investigation of the mitogenicity of bacterial enterotoxins suggested that Mls antigens and enterotoxins may use similar mechanisms to activate T-cells. These similarities extend into the biology of T-cell repertoire selection *in vivo*. Natural mechanisms exist to modify the functional T-cell repertoire *in vivo*, to prevent the expression of self reactive or autoimmune T-cells in the periphery. This is accomplished by either physical deletion or by the functional elimination of self-reactive cells from the mature T-cell repertoire. Clonal deletion is attributed to the neonatal thymus and involves the physical removal of autoreactive thymocytes from the T-cell repertoire as it develops (Kappler, Roehm & Marrack, 1987). Anergy contrasts with deletion in that self-reactive T-cells are functionally, but not physically, removed from the T-cell pool (Lamb *et al.*, 1983). Furthermore, unlike deletion, anergy may be induced in mature T-cells which have already survived thymic deletion, and been released into the periphery. Because of these characteristics, anergy has been proposed to account for tolerance to self neo-antigens and antigens expressed in sites remote from the neonatal thymus.

The ability to induce deletion or anergy in certain populations of T-cells may, therefore, have important applications in the development of novel therapeutic agents designed to modify the functional T-cell repertoire in T-cell-mediated diseases.

SUPERANTIGEN INDUCED CLONAL DELETION

One of the most striking aspects of the Mls system was that the ontogeny of the murine TcR repertoire was critically dependent upon the Mls gene expressed by each strain of mice. In the thymuses of mice expressing a certain Mls gene, thymocytes expressing particular TcRVβ elements were present in the immature (CD4$^+$/CD8$^+$) population but were almost completely deleted from the mature single positive (CD4$^+$ or CD8$^+$) population in both the thymus and periphery (Kappler *et al.*, 1988; MacDonald *et al.*, 1988; Pullen, Marrack & Kappler, 1988; Pullen, Kappler & Marrack, 1989). This finding was of considerable significance as the removal of T-cells recognizing self-antigens by clonal deletion is now established as an important mechanism of self-tolerance.

In common with many other properties of the Mls superantigens, the induction of clonal deletion was also found to be reproducible using bacterial toxins. In one such study, the mature and immature thymocytes of mice, which had been neonatally injected with SEB, were examined for the expression of TcRVβ3 and 8 regions normally associated with reactivity to SEB (White *et al.*, 1989). As in the Mls studies, only a small proportion of mature thymocytes expressing TcRVβ3 or 8 could be identified, thus suggesting that TcRVβ specific deletion had occurred upon neonatal interaction with SEB. In contrast to the initial Mls studies, however, a significant but smaller proportion of immature thymocytes were also deleted (White *et al.*, 1989).

Any useful therapeutic application of superantigen-mediated repertoire control will demand that the mechanism by which unwanted clones of T-cells are deleted will operate on mature, peripheral T-cells. A recent report describes just such a mechanism of peripheral T-cell deletion in mice. This treatment involves a single intravenous injection of SEB (Kawabe & Ochi 1991) and results in the deletion of T-cells bearing $TcRV\beta8.1$ and 8.2. Deletion was induced in both thymic and euthymic mice, thus ruling out a central thymic mechanism and confirming that the deletion observed was the result of interactions occurring in the peripheral T-cell compartment. Interestingly, the mechanism responsible for the death of $TcRV\beta8^+$ T-cells was programmed cell death or apoptosis, a mechanism thought to account for the deletion of immature self-reactive thymocytes in the neonatal thymus (Smith *et al.*, 1989).

SUPERANTIGEN INDUCED T-CELL ANERGY

Unlike clonal deletion of T-cells, in which cells with self-reactive TcR are physically removed from the T-cell repertoire, a second mechanism of repertoire control, antigen specific clonal T-cell anergy, operates by the functional inactivation of self reactive T-cells.

The first indication that superantigens could deliver an anergic signal to T-cells was observed in an *in vivo* study of anergy to antigens of the prototype superantigens of the Mls system. In these experiments, the intravenous immunization of mice possessing cells of one Mls type with cells isolated from mice of a different Mls type, induced a profound state of unresponsiveness to the Mls superantigens expressed by the immunizing cells (Rammensee, Kroschewski & Frangoulis, 1989). In contrast to deletion, functional anergy did not correlate with the disappearance of T-cells bearing the $TcRV\beta$ regions associated with reactivity to the Mls superantigen. Examination of the anergic cells *in vitro* demonstrated that despite the expression of surface IL-2 receptors and normal levels of the TcR and CD4, restimulation with stimulator cells of an appropriate Mls type only induced limited blastogenesis in the absence of IL-2 production. This form of anergy was Mls specific as the response to Mls matched, allogeneic stimulator cells, up to 57 days after immunization, was unaffected. Furthermore, stimulation of the anergic cells with IL-2 failed to overcome anergy, and by mixing anergic cells with untreated cells, no evidence could be found for the involvement of suppressive cells (Rammensee *et al.*, 1989).

Following the success of anergy induction to Mls superantigens, other groups have attempted to induce *in vivo* anergy in mature peripheral T cells using bacterial toxins. In one such study, immunization of mice with SEB induced a state of unresponsiveness in T-cells bearing the appropriate $TcRV\beta$ regions reactive with SEB (Kawabe & Ochi, 1990). Interestingly, unresponsiveness could only be identified in $CD4^+$ T cells, suggesting that the requirements for anergy induction in $CD8^+$ cells might be different to those of $CD4^+$ cells. Unfortunately, mice in this study were re-challenged with SEB at only a single time point 7 days after immunization, and thus no distinction was made between transient T cell refractoriness to re-stimulation and the long-lasting anergy of the type described by (Rammensee *et al.*, 1989).

A similar study in which T-cell unresponsiveness was induced by immunization with SEB demonstrated the induction of a state of anergy in both CD4$^+$ and CD8$^+$ cells which persisted for up to 28 days after immunization (Rellahan *et al.*, 1990). This study also presented evidence that, between 8 and 14 days after immunization, there was a significant decrease in the proportion of T-cells expressing the TcRVβ8 elements reactive with SEB. This contrasts with the failure to identify a relationship between T-cell numbers and anergy in Mls induced anergy (Rammensee *et al.*, 1989), but is in accordance with the concept of SEB induced peripheral deletion (Kawabe & Ochi, 1991).

The differences observed in each of these *in vivo* systems is most likely to be, in part, due to the inherent complexity associated with the analysis of immune function *in vivo*. Therefore, in an attempt to reduce this complexity, the authors have chosen to study the induction of anergy in isolated clones of T-cells with defined specificity and function.

Our interest in clonal T-cell anergy was prompted by a series of *in vitro* experiments in which T-cell clones were stimulated with a wide range of nominal peptide antigen concentrations. In these experiments, T-cells stimulated with high concentrations of peptide failed to proliferate compared to cells stimulated by lower antigen concentrations. Pre-incubation of these clones, with supra-optimal concentrations of peptide antigen, profoundly inhibited the response of the clone to a subsequent challenge with the same antigen (Lamb *et al.*, 1983). This state of unresponsiveness is now known to be due to clonal T-cell anergy.

Since the *in vitro* model of peptide antigen induced anergy involves an initial signal through the TcR, the ability of other TcR ligands, notably the bacterial toxin superantigens, to induce T-cell anergy *in vitro* has been investigated (O'Hehir & Lamb, 1990).

An influenza haemagglutinin peptide specific, TcRVβ3$^+$ T-cell clone HA1.7 was exposed to SEB overnight in the absence of accessory cells. Three days later, despite exhibiting enhanced proliferation in response to exogenous IL-2, HA1.7 was unable to respond to its natural ligand presented in an immunogenic form using accessory cells. In common with peptide-mediated anergy, SEB-induced anergy was also associated with changes in the expression of cell surface CD2, CD3, CD25 and CD28, suggesting that SE can induce a similar state of anergy to that induced by peptide antigens (O'Hehir & Lamb, 1990).

Using this uncomplicated *in vitro* model, the analysis has been extended to determine whether, in the induction of anergy, SEB directly binds to the HA1.7 TcR in the absence of MHC class II (Hewitt *et al.*, 1992). As already mentioned, the direct binding of enterotoxins to the TcR is a controversial issue, but a number of groups have convincing evidence that certain T-cell lines and hybridomas can be directly activated by enterotoxins in the absence of MHC class II antigens (Fleischer & Schrezenmeier, 1988; Matthes *et al.*, 1988; Yagi *et al.*, 1990; Herrmann *et al.*, 1991).

The approach was to use the enterotoxins SEB and SEC2, which cross-compete for a closely related binding site on HLA-DR1 molecules. Of these two enterotoxins, only SEB is able to interact with the TcRVβ3 region of the T-cell clone HA1.7. In competition experiments it was demonstrated that the induction of anergy in HA1.7 by SEB is unaffected by the presence of SEC2 or monoclonal antibodies to MHC class II antigens; thus suggesting that SEB-induced anergy is MHC independent and involves a direct interaction between the TcR and SEB. In order to rule out the possibility that T-cell MHC class II was involved (Hewitt & Feldmann, 1989) and resolve definitively whether SEB bound directly to T-cells in the absence of MHC class II molecules, the cDNAs encoding the HA1.7 TcR were transfected into an MHC class II negative human T-cell line. The addition of SEB to these transfectants resulted in the down-regulation of cell surface TcR expression, an increase in the concentration of intracellular calcium ions, the production of lymphokines and a failure to respond to a subsequent challenge with SEB. It is therefore, concluded that SEB may interact directly with the TcR to induce anergy in HA1.7 in the absence of co-interaction with MHC class II molecules (Hewitt *et al.*, 1992).

The ability to anergize T-cells in the absence of MHC class II antigens using enterotoxins has clinical implications in, for example, autoimmune and allergic diseases (O'Hehir *et al.*, 1990), where there may be restricted TcRVβ usage in the harmful T-cell responses (Paliard *et al.*, 1991) but where the antigen is either not determined or complex, and the involvement of MHC antigens is ill defined. In these cases, enterotoxins modified, as described previously, to separate the emetic effects from the TcR binding activity (Alber *et al.*, 1990) and possibly the TcR from MHC binding activity (Grossman *et al.*, 1990) could be used to anergize T-cells expressing particular TcRVβ regions implicated in disease processes.

THE EVOLUTIONARY SIGNIFICANCE OF SUPERANTIGENS

Interactions between the immune system and bacterial toxin or retroviral Mls superantigens are so similar that it has been speculated that common evolutionary pressures may be responsible for their existence. The lack of sequence homology between superantigens from diverse organisms suggests that convergent evolution, driven by a need to prevent the immune system from generating a response or inhibit infection, may be responsible for their phenotypic similarities (for review see Acha-Orbea & Palmer, 1991). By targeting the TcR and MHC class II antigens, these infectious agents have capitalized on the ability of their superantigens to bind to the invariant regions of two molecules whose function is dependent upon subtle differences in structure, and which the immune system can thus least afford to mutate without loss of function. Bacteria are proposed to benefit from the production of such superantigenic toxins by their ability to immunosuppress the host (Zehavi-Willner, Shenberg & Barnea, 1984; Marrack *et al.*, 1990; Herman *et al.*, 1991). Whilst it seems paradoxical that immunosupression should result from overstimulation of the T-cells, the polyclonal stimulation induced might overwhelm directed responses, or activate cytotoxic T-cells to kill toxin binding MHC class II bearing antigen presenting cells, thus disabling the immune system and immunosuppressing the host (Herman *et al.*, 1991).

If it were advantageous for a pathogen to express superantigens, why then have mice retained endogenous retroviruses encoding the Mls superantigens? It has been suggested that mice retain retroviral self-superantigens, specifically to delete populations of bacterial toxin superantigen reactive T-cells, and thus protect themselves from the pathogenic effects of toxin-secreting bacteria. In support of this, athymic nude mice, and mice genetically constructed to contain only small numbers of T-cells reactive with SEB, were found to be protected from the pathogenic effects of the toxin (Marrack *et al.*, 1990). Further support is derived from an analysis of the T-cell repertoires of wild mice. These mice have also been noted to lack T-cells bearing certain common TcRVβ regions. In some mice, this was due to large deletions in the TcRVβ locus, but in others, evidence was obtained for the involvement of elements with characteristics similar to Mls antigens. It was thus hypothesized that mice with large holes deleted from their TcR repertoires might be at a selective advantage due to protection from the effects of bacterial toxins (Pullen *et al.*, 1990a); the consequent reduction in complexity of their TcR repertoires presumably being balanced by the redundancy of the TcR repertoire. The availability of an Mls-like facility to remove a large part of the TcR repertoire may also maintain flexibility in the T-cell repertoire. This is illustrated by the allelic nature of the Mls-1 locus, only one allele of which, (Mls[a]) is a stimulatory self-superantigen, whilst the other (Mls[b]) is non-stimulatory. Thus, although a pair of mice bearing the Mls-1[a]/Mls-1[b] genotype will have deleted their Mls-1[a] reactive T-cells, at least one in four of their progeny will inherit the Mls-1[b]/Mls-1[b] phenotype and, therefore, retain T-cells bearing the TcRVβ regions deleted by the Mls-1[a+] parents (Herman *et al.*, 1991).

Others suggest that the deletion of certain T-cells by endogenous retroviruses is an immunological accident, and that viral superantigens have persisted in the murine genome simply because they carry little or no selective disadvantage (Acha-Orbea & Palmer, 1991). It is perhaps significant in this respect that, as yet, endogenous retroviral superantigens have not been identified in any other species.

CONCLUSIONS

Bacterial toxins were investigated initially because of their role in various infectious diseases. It is, therefore, appropriate that attempts have been made to link superantigen-mediated stimulation of the immune system with these diseases. Because of their functionally bivalent nature, two routes can be envisaged by which toxin superantigens may exert their pathogenic effects via the immune system. The first route may be through the superantigen presenting cells, in which SE are able to transduce a positive signal through MHC class II antigens resulting in the release of interleukin 1α and tumour necrosis factor α (TNFα) (Jupin *et al.*, 1988; De Azavedo *et al.*, 1988; Fast, Schlievert & Nelson, 1989). The second, and more familiar, route is through T-cells, in which the stimulatory activity of toxin superantigens results in the release of T-cell cytokines such as IFNγ, IL-2 and TNF (Carlsson & Sjogren, 1985; Jupin *et al.*, 1988; Fast *et al.*, 1989). When used therapeutically *in vivo*, activated T-cells and some of the cytokines induced by bacterial toxins are reported to have roles as the mediators of shock, fever, weight loss and osmotic imbalance (Rosenberg & Lotze, 1986; Beutler & Cerami, 1989). It is not inconceivable that cytokines released after the activation of massive numbers of presenting cells and T-cells may be involved in some of the symptoms of diseases in which bacterial toxins are known to have a role (Marrack *et al.*, 1990; Marrack & Kappler, 1990).

Investigation of bacterial toxin superantigens and their effects on the immune system have suggested a potential mechanism, massive T-cell stimulation, to account for their pathogenic properties. Furthermore, by analogy with the endogenous retroviral superantigens, these findings have provided a unique model for dissection of the mechanisms controlling central and peripheral T-cell repertoire selection.

ACKNOWLEDGEMENTS

The authors are supported by The Medical Research Council of Great Britain, and The Wellcome Trust. R. E. O'H holds a Wellcome Senior Research Fellowship in the Clinical Sciences.

REFERENCES

Acha-Orbea, H. & Palmer, E. (1991). Mls–a retrovirus exploits the immune system. *Immunology Today*, **12**, 356–62.

Alber, G., Hammer, D. K. & Fleischer, B. (1990). Relationship between enterotoxic and T lymphocyte-stimulating activity of staphylococcal enterotoxin B. *Journal of Immunology*, **144**, 4501–6.

Allen, P. M. (1987). Antigen processing at the molecular level. *Immunology Today*, **8**, 270–3.

Altboum, Z., Hertman, I. & Sarid, S. (1985). Penicillinase plasmid linked genetic determinants for enterotoxins B and C1. *Infection and Immunity*, **47**, 514–21.

Bartter, T., Dascal, A., Carroll, K. & Curley, F. J. (1988). Toxic streptococcus syndrome. *Archives in Internal Medicine*, **148**, 1421–4.

Bauer, A., Rutenfranz, I. & Kirchner, H. (1988). Processing requirements for T cell activation by mycoplasma arthritidis-derived mitogen. *European Journal of Immunology*, **18**, 2019–22.

Bayles, K. W. & Iandolo, J. J. (1989). Genetic and molecular analyses of the gene encoding Staphylococcal enterotoxin D. *Journal of Bacteriology*, **170**, 2954–60.

Bergdoll, M. S. (1970). Enterotoxins. Microbial Toxins. III. In *Bacterial Protein Toxins*, ed. T. C. Moutie, S. Kadis and S. J. Ajil, pp. 256–326. New York, Academic Press.

Bergdoll, M. S. (1983). Enterotoxins. In *Staphylococci and Staphylococcal Infections*, ed. S. C. F. Easmon and C. Adlam, pp. 559–598. New York, Academic Press.

Bergdoll, M. S., Crass, B. A., Reiser, R. F., Robbins, R. N. & Davis, J. P. (1981). A new staphylococcal enterotoxin, enterotoxin F, associated with toxic shock syndrome isolates. *Lancet*. **i**, 1017–21.

Betley, M. J. & Mekalonos, J. J. (1985). Staphylococcal enterotoxin A is encoded by phage. *Science*, **229**, 185–7.

Beutler, B. & Cerami, A. (1989). The biology of cachectin/TNF-a primary mediator of the host response. *Annual Review in Immunology*, **7**, 625–55.

Blackman, M. A., Gerhard-Burgert, H., Woodland, D. L., Palmer, E., Kappler, J. W. & Marrack, P. (1990). A role for clonal inactivation in T cell tolerance to Mls-1a. *Nature (London)*, **345**, 540–2.

Blomster-Hautamaa, D. A., Kreiswirth, B. N., Kornblum, J. S., Novick, P. P. & Schlievert, P. M. (1986). The nucleotide and partial amino acid sequence of toxic shock syndrome toxin-1. *Journal of Biological Chemistry*, **261**, 15783–6.

Bohach, G. A., Handley, J. P. & Schlievert, P. M. (1989). Biological and immunological properties of the carboxy terminus of Staphylococcal enterotoxin C1. *Infection and Immunity*, **57**, 23–8.

Bohach, G. A. & Schlievert, P. M. (1989). Conservation of the biologically active portions of Staphylococcal enterotoxins C1 and C2. *Infection and Immunity*, **57**, 2249–52.

Brown, J. H., Jardetzky, T., Saper, M. A., Samraoui, B., Bjorkman, P. J. & Wiley, D. C. (1988). A hypothetical model of the foreign antigen binding site of class II histocompatibility molecules. *Nature (London)*, **332**, 845–50.

Buelow, R., O'Hehir, R. E., Schreifels, R., Riley, G., Kummerehl, J. & Lamb, J. R. (1992). Localisation of the immunological activity in the superantigen staphylococcal enterotoxin B using truncated recombinant fusion proteins. *Journal of Immunology*, **148**, 1–6

Callahan, J. E., Herman, A., Kappler, J. W. & Marrack, P. (1989). Stimulation of B10.BR T cells with superantigenic staphylococcal toxins. *Journal of Immunology*, **144**, 2473–9.

Carlsson, R., Fischer, H. & Sjogren, H. O. (1988). Binding of staphylococcal enterotoxin A to accessory cells is a requirement for its ability to activate human T cells. *Journal of Immunology*, **140**, 2484–8.

Carlsson, R. & Sjogren, H. O. (1985). Kinetics of IL-2 and interferon gamma production, expression of IL-2 receptors and cell proliferation in human mononuclear cells exposed to enterotoxin A. *Cellular Immunology*, **96**, 175–83.

Choi, Y., Kotzin, B., Herron, L., Calahan, J., Marrack, P. & Kappler, J. (1989). Interaction of Staphylococcus aureus toxin superantigens with human T cells. *Proceedings of the National Academy of Sciences, USA*, **86**, 8941–5.

Choi, Y. W., Herman, A., DiGiusto, D., Wade, T., Marrack, P. & Kappler, J.

(1990). Residues of the variable region of the T-cell-receptor beta-chain that interact with *S. aureus* toxin superantigens. *Nature (London)* **346**, 471–3.

Chothia, C., Boswell, D. R. & Lesk, A. M. (1988). The outline structure of the T cell $\alpha\beta$ receptor. *EMBO Journal*, **7**, 3745–55.

Cole, B. C. & Atkin, C. L. (1991). The mycoplasma arthritidis T cell mitogen, MAM: a model superantigen. *Immunology Today*, **12**, 271–6.

Cole, B. C., Daynes, R. A. & Ward, J. R. (1981). Stimulation of mouse lymphocytes by a mitogen derived from Mycoplasma arthritidis. *Journal of Immunology*, **127**, 1931–6.

Cole, B. C., Kartchner, D. R. & Wells, D. J. (1989*a*). Stimulation of mouse lymphocytes by a mitogen derived from Mycoplasma arthriditis (MAM) VIII. Selective activation of T cells expressing distinct V beta T cell receptors from various strains of mice by the 'superantigen' MAM. *Journal of Immunology*, **144**, 425–31.

Cole, B. C., Kartchner, D. R. & Wells, D. J. (1989*b*). Stimulation of mouse lymphocytes by a mitogen derived from Mycoplasma arthriditis VII. Responsiveness is associated with expression of a product(s) of the V beta 8 gene family present on the T cell receptor alpha–beta for antigen. *Journal of Immunology*, **142**, 4131–7.

Cole, B. C., Washburn, L. R. & Taylor-Robinson, D. (1985). Mycoplasma induced arthritis. IV. In *The Mycoplasmas*. ed. S. Razin and M. F. Barile, pp. 105–160. New York, Academic Press.

Cone, L. A., Woodard, D. R., Schlievert, P. M. & Tomoroy, G. S. (1987). Clinical and bacteriologic observations of a toxic shock like syndrome due to streptococcus pyrogenes. *New England Journal of Medicine*, **317**, 146–9.

Couch, J. L. & Betley, M. J. (1989). Nucleotide sequence of the type C3 staphylococcal enterotoxin gene suggests that intergenic recombination causes antigenic variation. *Journal of Bacteriology*, **171**, 4507–10.

Couch, J. L., Soltis, M. T. & Betley, M. J. (1988). Cloning and nucleotide sequence of the type E staphylococcal enterotoxin gene. *Journal of Bacteriology*, **170**, 2954–60.

Davis, M. M. & Bjorkman, P. J. (1988). T cell antigen receptor genes and T cell recognition. *Nature (London)*, **334**, 395–402.

De Azavedo, J. C. S., Drumm, A., Jupin, C., Parant, M., Alouf, J. E. & Arbuthnott, J. P. (1988). Induction of tumour necrosis factor by staphylococcal toxic shock toxin 1. FEMS (Fed. Eur. Microbiol. Soc) *Microbiology and Immunology*, **47**, 69.

Dellabona, P., Peccoud, J., Benoist, C. & Mathis, D. (1989). T-cell recognition of superantigens: inside or outside the groove? *Cold Spring Harbor Symposia in Quantitative Biology*, **54**, 375–81.

Edwin, C., Swack, J. A., Williams, K., Bonventre, P. F. & Kass, E. H. (1991). Activation of *in vitro* proliferation of human T cells by a synthetic peptide of toxic shock syndrome toxin-1. *Journal of Infectious Diseases*, **163**, 524–9.

Fast, D. J., Schlievert, P. M. & Nelson, R. D. (1989). Toxic shock syndrome associated staphylococcal and streptococcal pyrogenic toxins are potent inducers of tumour necrosis factor production. *Infection and Immunity*, **57**, 291–4.

Festenstein, H. (1973). Immunogenic and biological aspects of in vitro lymphocyte allotransformation (MLR) in the mouse. *Transplant Reviews*, **15**, 62–88.

Fischer, H., Dohlson, M., Lindvall, M., Sjogren, H. O. & Carlsson, R. (1989). Binding of staphylococcal enterotoxin A to HLA-DR on B cell lines. *Journal of Immunology*, **142**, 3151–7.

Fleischer, B., Gerardy-Schann, R., Metzroth, B., Carrel, S., Gerlach, D. & Kohler,

W. (1991). An evolutionary conserved mechanism of T cell activation by microbial toxins. Evidence for different affinities of T cell receptor–toxin interaction. *Journal of Immunology*, **146**, 11–17.

Fleischer, B. & Schrezenmeier, H. (1988). T cell stimulation by staphylococcal enterotoxins. Clonally variable response and requirement for major histocompatibility complex class II molecules on accessory or target cells. *Journal of Experimental Medicine*, **167**, 1697–707.

Fleischer, B., Schrezenmeier, H. & Condradt, P. (1989). T lymphocyte activation by staphylococcal enterotoxins: role of class II molecules and T cell surface structures. *Cellular Immunology*, **120**, 92–101.

Fraser, J. D. (1989). High affinity binding of staphylococcal enterotoxin A and B to HLA-DR. *Nature (London)*, **339**, 221–3.

Gascoigne, N. R. J. & Ames, K. T. (1991). Direct binding of secreted T cell receptor beta chain to superantigen associated with class II major histocompatibility protein. *Proceedings of the National Academy of Sciences, USA*, **88**, 613–16.

Goshorn, S. C. & Schlievert, P. M. (1988). Nucleotide sequence of streptococcal pyrogenic enterotoxin type C. *Infection and Immunity*, **56**, 2518–20.

Grossman, D., Cook, R. G., Sparrow, J. T., Mollick, J. A. & Rich, R. R. (1990). Dissociation of the stimulatory activities of staphylococcal enterotoxins for T cells and monocytes. *Journal of Experimental Medicine*, **172**, 1831–41.

Herman, A., Croteau, G., Sekaly, R., Kappler, J. W. & Marrack, P. (1990). HLA-DR alleles differ in their ability to present staphylococcal enterotoxins to T cells. *Journal of Experimental Medicine*, **172**, 709–17.

Herman, A., Kappler, J. W., Marrack, P. & Pullen, A. M. (1991). Superantigens: mechanism of T cell stimulation and role in immune responses. *Annual Reviews in Immunology*, **9**, 745–52.

Herrmann, T., Maryanski, J. L., Romero, P., Fleischer, B. & MacDonald, H. R. (1990). Activation of MHC class I restricted CD8+ CTL by microbial T cell antigens. Dependence upon MHC class II expression of the target cells and V beta usage of the responder cells. *Journal of Immunology*, **144**, 1181–6.

Herrmann, T., Romero, P., Sartoris, S., Paiola, F., Accolla, R. S., Maryanski, J. L. & MacDonald, H. R. (1991). Staphylococcal enterotoxin-dependent lysis of MHC class II negative target cells by cytolytic T lymphocytes. *Journal of Immunology*, **146**, 2504–12.

Hewitt, C. R. A. & Feldmann, M. (1989). Human T cell clones present antigen. *Journal of Immunology*, **143**, 762–9.

Hewitt, C. R. A., Lamb, J. R. L., Hayball, J. D., Hill, M., Owen, M. J. & O'Hehir, R. E. (1992). MHC independent clonal T cell anergy by direct interaction of *Staphylococcus aureus* enterotoxin B with the T cell antigen receptor. *Journal of Experimental Medicine*.

Huang, I. Y. & Bergdoll, M. S. (1970). The primary structure of staphylococcal enterotoxin B. III the cyanogen bromide peptides of reduced and aminoethylated enterotoxin B and the complete amino acid sequence. *Journal of Biological Chemistry*, **245**, 3518–25.

Huang, I. Y., Hughes, J. L., Bergdoll, M. S. & Schantz, E. J. (1987). Complete amino acid sequence of staphylococcal enterotoxin A. *Journal of Biological Chemistry*, **262**, 7006–7.

Janeway, C. A., Chalupny, J., Conrad, P. J. & Buxser, S. (1988). An external stimulus that mimics Mls locus responses. *Journal of Immunogenetics*, **15**, 161–8.

Janeway, C. A., Yagi, J., Conrad, P. J., Katz, M. E., Jones, B., Vroegop, S. & Buxser, S. (1989). T cell responses to Mls and to bacterial proteins that mimic its behaviour. *Immunology Reviews*, **107**, 62–88.

Johns, M. B. & Kahn, S. (1988). A Staphylococcal enterotoxin β gene is associated with a discrete genetic element. *Journal of Bacteriology*, **170**, 4033–9.

Johnson, H. M., Stanton, G. J. & Baron, S. (1977). Relative ability of mitogens to stimulate production of interferon by lymphoid cells and to induce supression of the *in-vitro* immune response. *Proceedings of the Society of Experimental Biology and Medicine*, **154**, 138–41.

Jupin, C., Anderson, S., Damais, C., Alouf, J. E. & Parant, M. (1988). Toxic shock syndrome-1 producing *Staphylococcus aureus*. relationship to toxin stimulated production of tumour necrosis factor. *Journal of Experimental Medicine*, **167**, 752–61.

Kaplan, J. (1972). Staphylococcal enterotoxin B induced release of macrophage migration inhibition factor from normal lymphocytes. *Cellular Immunology*, **3**, 245–52.

Kappler, J., Kotzin, B., Herron, L., Gelfand, E. W., Bigler, R. D., Boylston, A., Carrel, S., Posnett, D. N., Choi, Y. & Marrack, P. (1989). V beta-specific stimulation of human T cells by staphylococcal toxins. *Science*, **244**, 811–13.

Kappler, J. W., Roehm, N. & Marrack, P. (1987). T cell tolerance by clonal elimination in the thymus. *Cell*, **49**, 273–80.

Kappler, J. W., Staerz, U., White, J. & Marrack, P. C. (1988). Self-tolerance eliminates T cells specific for Mls-modified products of the major histocompatibility complex. *Nature (London)*, **332**, 35–40.

Karp, D. R., Teletski, C. R., Scholl, P., Geha, R. & Long, E. O. (1990). The alpha 1 domain of the HLA-DR molecule is essential for high affinity binding for the toxic shock syndrome toxin-1. *Nature (London)*, **346**, 474–6.

Kawabe, Y. & Ochi, A. (1990). Selective anergy of V beta 8+,CD4+ T cells in Staphylococcus enterotoxin B-primed mice. *Journal of Experimental Medicine*, **172**, 1065–70.

Kawabe, Y. & Ochi, A. (1991). Programmed cell death and extrathymic reduction of Vbeta8+ CD4+ T cells in mice tolerant to *Staphylococcus aureus* enterotoxin B. *Nature (London)*, **349**, 245–8.

Kreisworth, B. N., Lofdahl, S., Betley, M. J., O'Reilly, M., Schleivert, P. M., Bergdoll, M. S. & Novick, R. P. (1983). The toxic shock syndrome exotoxin structural gene is not detectably transmitted by a prophage. *Nature (London)*, **305**, 709–12.

Lamb, J. R., Skidmore, B. J., Green, N., Chiller, J. M. & Feldmann, M. (1983). Induction of tolerance in influenza virus-immune T lymphocyte clones with synthetic peptides of influenza hemagglutinin. *Journal of Experimental Medicine*, **157**, 1434–47.

Langford, M. P., Stanton, G. J. & Johnson, H. M. (1978). Biological effects of Staphylococcal enterotoxin A on human peripheral lymphocytes. *Infection and Immunity*, **22**, 62–8.

MacDonald, H. R., Schneider, R., Lees, R. K., Howe, R. C., Acha, O. H., Festenstein, H., Zinkernagel, R. M. & Hengartner, H. (1988). T-cell receptor V beta use predicts reactivity and tolerance to Mlsa-encoded antigens. *Nature (London)*, **332**, 40–5.

Marrack, P., Blackman, M., Kushnir, E. & Kappler, J. (1990). The toxicity of staphylococcal enterotoxin B in mice is mediated by T cells. *Journal of Experimental Medicine*, **171**, 455–64.

Marrack, P. & Kappler, J. (1990). The staphylococcal enterotoxins and their relatives. *Science*, **248**, 705–11.

Matthes, M., Schrezenmeier, H., Homfeld, J., Fleischer, S., Malissen, B., Kirchner, H. & Fleischer, B. (1988). Clonal analysis of human T cell activation by the Mycoplasma arthritidis mitogen (MAS). *European Journal of Immunology*, **18**, 1733–7.

Middlebrook, J. L., Spero, L. & Argos, P. (1980). The secondary structure of staphylococcal enterotoxin A, B and C. *Biochemica and Biophysica Acta*, **621**, 233–40.

Mollick, J. A., Cook, R. C. & Rich, R. R. (1989). Class II MHC molecules are specific receptors for staphylococcus enterotoxin A. *Science*, **244**, 817–20.

Munoz, P. A., Warren, J. R. & Noelken, M. E. (1976). Beta structure of aqueous staphylococcal enterotoxin B by spectropolarimetry and sequence based conformational predictions. *Biochemistry*, **15**, 4666–71.

O'Hehir, R. E., Aguilar, B. A., Schmidt, T. J., Gollnick, S. O. & Lamb, J. R. (1990). Functional inactivation of *Dermatophagoides* spp. (house dust mite) reactive T cell clones. *Clinical and Experimental Allergy*, **21**, 209–15.

O'Hehir, R. E. & Lamb, J. R. (1990). Induction of specific clonal anergy in human T lymphocytes by *Staphylococcus aureus* enterotoxins. *Proceedings of the National Academy of Sciences, USA*, **87**, 8884–8.

Paliard, X., West, S. G., Lafferty, J. A., Clements, J. R., Kappler, J. W., Marrack, P. & Kotzin, B. L. (1991). Evidence for the effects of a superantigen in rheumatoid arthritis. *Science*, **253**, 325–9.

Peavy, D. L., Adler, W. H. & Smith, R. T. (1970). The mitogenic effects of endotoxin and Staphylococcal enterotoxin B on mouse spleen cells and human peripheral lymphocytes. *Journal of Immunology*, **105**, 1453–8.

Pontzer, C. H., Russel, J. K. & Johnson, H. M. (1989). Localisation of an immune functional site on staphylococcal enterotoxin A using the synthetic peptide approach. *Journal of Immunology*, **143**, 280–4.

Pullen, A. M., Kappler, J. W. & Marrack, P. (1989). Tolerance to self antigens shapes the T cell repertoire. *Immunology Reviews*, **107**, 125–39.

Pullen, A. M., Marrack, P. & Kappler, J. W. (1988). The T cell repertoire is heavily influenced by tolerance to polymorphic self antigens. *Nature (London)*, **335**, 796–801.

Pullen, A. M., Potts, W., Wakeland, E. K., Kappler, J. W. & Marrack, P. (1990*a*). Surprisingly uneven distribution of the T cell receptor V beta repertoire in wild mice. *Journal of Experimental Medicine*, **171**, 49–62.

Pullen, A. M., Wade, T., Marrack, P. & Kappler, J. W. (1990*b*). Identification of the region of T cell receptor beta chain that interacts with the self-superantigen Mls-1a. *Cell*, **61**, 1365–74.

Rammensee, H. G., Kroschewski, R. & Frangoulis, B. (1989). Clonal anergy induced in mature V beta 6+ T lymphocytes on immunizing Mls-1b mice with Mls-1a expressing cells. *Nature (London)*, **339**, 541–4.

Rellahan, B. L., Jones, L. A., Kruisbeek, A. M., Fry, A. M. & Matis, L. A. (1990). *In vivo* induction of anergy in peripheral V beta 8+ T cells by staphylococcal enterotoxin B. *Journal of Experimental Medicine*, **172**, 1091–100.

Rosenberg, S. A. & Lotze, M. T. (1986). Cancer immunotherapy using interleukin-2 and interleukin-2 activated lymphocytes. *Annual Reviews in Immunology*, **4**, 681–710.

Schmidt, J. J. & Spero, L. (1983). The complete amino acid sequence of enterotoxin C1. *Journal of Biological Chemistry*, **258**, 6300–6.

Scholl, P., Diez, A., Mourand, W., Parsonnet, J., Geha, R. S. & Chatila, T. (1989). Toxic shock syndrome toxin 1 binds to major histocompatibility complex class II molecules. *Proceedings of the National Academy of Sciences, USA*, **86**, 4210–14.

Scholl, P. R., Diez, A., Karr, R., Sekaly, R. P., Trowsdale, J. & Geha, R. S. (1990).

Effect of isotypes and allelic polymorphism on the binding of staphylococcal exotoxins to MHC class II molecules. *Journal of Immunology*, **144**, 226–30.

Singh, B. R., Evenson, M. L. & Bergdoll, M. S. (1988). Structural analysis of staphylococcal enterotoxin B and C1 using circular dichroism and fluorescence spectroscopy. *Biochemistry*, **27**, 8735–41.

Smith, B. G. & Johnson, H. M. (1975). The effect of Staphylococcal enterotoxins on the primary *in-vitro* immune response. *Journal of Immunology*, **115**, 575–8.

Smith, C. A., Williams, G. T., Kingston, R., Jenkinson, E. J. & Owen, J. J. (1989). Antibodies to CD3/T-cell receptor complex induce death by apoptosis in immature T cells in thymic cultures. *Nature (London)*, **337**, 181–4.

Spero, L., Griffin, L. B. Y., Middlebrook, J. L. & Metzger, J. F. (1976). The effect of single and double peptide bound scission by trypsin on the structure and activity of staphylococcal enterotoxin B. *Journal of Biological Chemistry*, **240**, 7279–94.

Spero, L. & Morlock, B. A. (1978). Biological activities of the peptides of staphylococcal enterotoxin C formed by limited tryptic hydrolysis. *Journal of Biological Chemistry*, **253**, 8787–91.

Stevens, D. L., Tanner, M. H. & Winship, J. (1989). Group A streptococcal infections associated with toxic shock like syndrome and scarlet fever toxin. *New England Journal of Medicine*, **321**, 1–7.

Swaminathan, S., Yang, D. S. C., Furey, W., Abrams, L., Pletcher, J. & Sax, M. (1988). Crystallization and preliminary X-ray study of staphylococcal enterotoxin B. *Journal of Molecular Biology*, **199**, 397.

Takimoto, H., Yoshikai, Y., Kishihara, K., Matsuzaki, G., Kuga, H., Otani, T. & Nomoto, K. (1990). Stimulation of all T cells bearing V beta 1, V beta 3, V beta 11, and V beta 12 by staphylococcal enterotoxin A. *European Journal of Immunology*, **20**, 617–21.

Taranta, A. (1974). Lymphocyte mitogens of Staphylococcal origin. *Annals of the New York Academy of Sciences*, **236**, 362–75.

Weeks, C. R. & Ferretti, J. J. (1986). Nucleotide sequence of the type A streptococcal enterotoxin (erythrogenic toxin) gene from streptococcus pyogenes bacteriophage T12. *Infection and Immunity*, **52**, 144–50.

White, J., Herman, A., Pullen, A. M., Kubo, R., Kappler, J. W. & Marrack, P. (1989). The V beta-specific superantigen staphylococcal enterotoxin B: stimulation of mature T cells and clonal deletion in neonatal mice. *Cell*, **56**, 27–35.

Yagi, J., Baron, J., Buxser, S. & Janeway, C. A. (1990). Bacterial proteins that mediate the association of a defined subset of T cell receptor: CD4 complexes with class II MHC. *Journal of Immunology*, **144**, 892–901.

Zehavi-Willner, T., Shenberg, E. & Barnea, A. (1984). *In vivo* effect of staphylococcal enterotoxin A on peripheral blood lymphocytes. *Infection and Immunity*, **44**, 401–5.

THE USE OF MUTANTS FOR DEFINING THE ROLE OF VIRULENCE FACTORS *IN VIVO*

T. J. FOSTER

Microbiology Department, Moyne Institute, Trinity College, Dublin 2, Ireland

INTRODUCTION

Genetic manipulation provides an important experimental approach for defining the relevance of putative virulence factors of pathogenic bacteria in disease processes (Sparling, 1983; see Finlay, this Symposium). Strains differing in the expression of a single factor can be constructed with considerable precision using recombinant DNA technology or transposons.

The principles that should be applied in a rigorous genetic analysis of pathogenicity have been described as a molecular form of Koch's postulates (Falkow, 1988). They can be summarized as follows: (1) The property should always be associated with pathogenic strains; (2) Specific inactivation of the gene that specifies the virulence factor should cause a measurable loss in virulence; and (3) Reversion of the mutated gene or allelic replacement by recombination or by complementation should restore pathogenicity.

It has been pointed out that studying the virulence of a mutant organism which lacks a single virulence factor is analogous to immunizing an animal with purified antigen from a pathogen and determining if infection with the wild-type organism is diminished (Goodwin & Weiss, 1990). A correlation may be obtained between protection induced by the immunogen and a reduction in virulence of the mutant. Alternatively the presence of an antigen could be irrelevant to the outcome of an infection. It is conceivable that a mutant lacking a protective antigen could have the same virulence as the parental strain (e.g. see the section on *Legionella pneumophila* in this article).

The soundness of conclusions that can be drawn from infection experiments with genetically manipulated strains is very much dependent on the relevance of the animal model to the natural disease (Smith, 1989). It can be particularly difficult to prove relevance if the natural host is man. For example, elegant genetic studies have been performed with the gonococcus but there is no ideal animal model to test variants. Where the natural host is a large domestic animal, it may still be desirable to use a small animal as an infection model, despite the fact that the same problems of relevance arise. However, in this case it is possible to check findings with experimental infections in the natural host.

The genetic techniques used to construct strains suitable for testing

Koch's postulates in their molecular form will be described in outline, as well as controls that should be performed, and examples will be given which illustrate the power of genetic manipulation in studying pathogenicity (see also Finlay, this symposium).

SITE-SPECIFIC MUTANTS

Transposons

Transposon mutagenesis has been successfully applied to the inactivation of genes coding for putative virulence factors in both Gram-positive and Gram-negative bacteria. The principles of transposon mutagenesis have been reviewed extensively (for example, see Berg, Berg & Groisman, 1989; Foster, 1984; Kleckner, Roth & Botstein, 1977). To confirm the association between a transposon insertion and a change in virulence phenotype, it is necessary to perform certain control experiments. The first three controls will ensure that the altered phenotype is not caused by a second mutation. (1) Southern hybridization analysis should be performed, using transposon DNA as a probe, to ensure that there is only one copy of the transposon in the chromosome. (2) If possible, linkage between the transposon and the mutant phenotype should be confirmed by transduction or transformation analysis. Selection for the drug resistance marker associated with the transposon in a genetic cross should result in 100% inheritance of the altered virulence phenotype. (3) Reversion of the mutant phenotype should occur when the transposon is lost by precise excision or by allelic replacement with wild-type DNA. (4) Insertion mutations can cause polar effects on the expression of downstream genes in operons. This can be controlled for by performing complementation tests with a recombinant plasmid expressing a functional copy of the gene that has been inactivated by the transposon in the chromosome. Complementation tests are particularly important if the phenotype of the mutant is pleiotropic or if genetic and molecular analysis suggests that the virulence factor gene is not monocistronic. (5) The expression of virulence factors in pathogenic bacteria is often controlled by global regulators (Gross, Aricò & Rappuoli, 1989: Stock, Ninfa & Stock, 1989; see Dorman & Ní Bhriain, this symposium). Inactivation of the regulatory gene would cause pleiotropic changes affecting the expression of several virulence factors simultaneously. It is important to ensure by Southern blotting, using the virulence factor structural gene DNA as a probe, that the coding sequence for the gene under study has indeed been inactivated. This would eliminate the possibility of a regulatory mutation. Derivatives of IS*1* (Joseph-Liauzun, Fellay & Chandler, 1989) and Tn*917* (Camilli, Portnoy & Youngman, 1990) have been constructed which carry an origin of replication which functions in *E. coli* and cloning sites to facilitate direct cloning of flanking chromosomal sequences.

Although transposon mutagenesis can be performed without knowledge of the structure of the gene coding for the putative virulence factor being inactivated, it is essential to have this information before a causal relationship between the transposon-inactivated gene and loss of virulence can be unambiguously established. Bacterial clones carrying transposon DNA (and flanking sequences) can be obtained using labelled transposon DNA as a probe in colony or plaque hybridization. Then DNA flanking the insertion can be isolated and in turn used as a probe to clone a functional copy of the putative virulence factor gene. The gene can then be sequenced and the gene product expressed and analysed.

Allelic replacement

Recombinational allelic replacement is a more precise method for gene inactivation than transposon mutagenesis (Foster, 1991a). However, to design an allelic replacement experiment, it is necessary to have cloned, and preferably to have sequenced, the gene in question. Site-directed insertion, deletion, substitution and point mutations can be constructed by *in vitro* methods (usually using *E. coli* host/vector systems). Similar controls to those described above for transposon mutants should also be performed in order to demonstrate that the mutation constructed *in vitro* has displaced the wild-type allele in the chromosome. *Trans*-complementation tests to eliminate the possibility of polarity are particularly important for larger deletions, although they are probably not necessary for point mutations and in-frame deletions.

It is possible to combine transposon mutagenesis and allelic replacement by inserting a transposon into a cloned virulence-factor gene located on a chimaeric plasmid. Chromosomal recombinants where the wild-type allele has been replaced with the transposon-inactivated gene can be selected in the same way as for *in vitro*-constructed mutations. Care should be taken to ensure that the transposon remains associated with the virulence factor gene. A functional transposon will be able to transpose into other sites in the chromosome and this event will occur along with *bona fide* recombination.

PLASMID-LOCATED VIRULENCE FACTORS

Enteropathogenic E. coli

Before the era of recombinant DNA technology, H. Williams Smith performed pioneering studies which showed that genetic manipulation could be applied to studying bacterial pathogenicity (Williams Smith & Linggood, 1971). In enteropathogenic strains of *E. coli* several important virulence factors, including the K88 fimbrial adhesin, the heat-labile enterotoxin and the haemolysin, are encoded by plasmid-located genes. Williams Smith not only established that these factors were plasmid-specified but also used the transmissibility of the plasmids and plasmid curing to construct strains which lacked the factors or expressed them singly or in combinations. Plasmids were not only cured from enteropathogenic strains but were also transferred into non-pathogenic strains of *E. coli* obtained from the normal gut flora of pigs. It was shown that the two separate plasmids encoding K88 fimbriae and enterotoxin, respectively, were both required to induce diarrhoea in piglets. However, expression of K88 fimbriae alone could induce mild diarrhoea whereas expression of enterotoxin without the fimbrial adhesin did not. In retrospect this approach can be criticized because it did not absolutely establish that the factors studied were responsible for the symptoms. The possibility that other plasmid-specified factors could have played a role was not ruled out.

More recent studies with plasmids have avoided this problem by either comparing a strain carrying a wild-type plasmid with one harbouring a plasmid with a site-specific mutation in the gene in question or by using a recombinant plasmid that is known to carry only the coding sequence for the virulence factor.

Pathogenesis of E. coli *extraintestinal infections*

Some strains of *E. coli* cause peritonitis and urinary tract infections in man. Many of the pathogenic strains express a haemolysin (Hly) which has been strongly implicated as an important virulence factor by genetic studies. The haemolysin is a member of a family of evolutionarily related pore-forming cytotoxins (called RTX; Welch, 1991). A related toxin expressed by *Bordetella pertussis* is discussed in a later section in this review.

The role of haemolysin in peritonitis was studied by introducing the cloned haemolysin gene into a non-haemolytic faecal strain of *E. coli* and measuring virulence in a rat peritonitis model. This strain was compared with one harbouring a derivative of the Hly plasmid with a transposon insertion in the haemolysin gene (Welch *et al.*, 1981). The association of virulence with haemolysin in terms of the number of lethalities produced by intraperitoneal inoculation of organisms was confirmed in a later study in which the level of virulence was shown to correlate with the titre of haemolysin observed in *in vitro*-grown cultures (Welch & Falkow, 1984).

Recombinant plasmids have also been used to study virulence factors of pyelonephritis strains of *E. coli*. Genes expressing adhesins, serum resistance and haemolysin are chromosomally located in these strains. A spontaneous mutant lacking these factors was avirulent in a rat pyelonephritis model (Marre *et al.*, 1986) which is considered by several criteria to reflect accurately the human disease. These putative virulence attributes were reintroduced singly and in combinations into the avirulent variant on recombinant plasmids. Expression of fimbriae or haemolysin alone caused moderate increases in virulence. The reconstituted strain expressing all three attributes had a high level of virulence but did not quite achieve that of the wild-type. This could have been due to another virulence factor that had not been restored or to the fact that the replacement haemolysin differed from that of the parental strain. This study confirmed that the pathogenesis of *E. coli* pyelonephritis is multifactorial and unequivocally demonstrated an important role for the haemolysin.

Staphylococcus saprophyticus *urease*

S. saprophyticus is a frequent cause of urinary tract infection in young females. The urease expressed by this organism has been implicated as a virulence factor by genetic studies involving transfer of a chimaeric plasmid bearing the urease gene to complement a mutant lacking urease. A urease-negative mutant was isolated by chemical mutagenesis (Gatermann, John & Marre, 1989). Normally this method of generating mutants is fraught with the danger of multiple changes. However, the reduced virulence of the mutant in an ascending unobstructed rat pyelonephritis model was restored by introduction of the cloned urease gene on a chimaeric plasmid (Gatermann & Marre, 1989).

Staphylococcus aureus *TSST-1*

Genetic manipulation has played an important part in determining that Toxic Shock Syndrome Toxin number 1 (TSST-1) is the major toxin involved in Toxic Shock Syndrome caused by *S. aureus*. However, it is now clear that staphylococcal enterotoxins can cause similar symptoms but these are not normally involved in the tampon-associated disease (Bohach *et al.*, 1990).

The cloned TSST-1 gene was introduced into a non-TSST-1-producing vaginal isolate of *S. aureus* (de Azevedo *et al.*, 1985). The bacteria were inoculated into a diffusion chamber which was then surgically implanted into the uterus of an adult rabbit. The TSST-1-expressing strains were virulent in this model. The animals suffered many of the symptoms associated with staphylococcal toxic shock syndrome in humans. Strains carrying a plasmid with a deletion in the *tst* gene lacked virulence showing that only the cloned *tst* gene was involved. This conclusion was confirmed recently with a TSST-1-deficient mutant of a toxic shock strain constructed by allelic replacement (Sloane *et al.*, 1991).

GENETIC STUDIES WITH MULTIFACTORIAL PATHOGENS: TRANSPOSON AND ALLELIC REPLACEMENT MUTANTS STUDIED *IN VIVO*

Bordetella pertussis

Among the earliest reports of site-specific mutants defective in virulence factors were studies by Weiss *et al.* (1983, 1984) describing Tn5 mutants of

Bordetella pertussis defective in pertussis toxin (Ptx), adenylate cyclase (Adc), haemolysin (Hly) and filamentous haemagglutinin (Fha). Adc$^-$ mutants were also Hly$^-$, whereas some Hly$^-$ mutants were Adc$^+$. These phenotypes were initially attributed to polar effects of the Tn5 insertion in the *adc* gene which was lying 5' to the *hly* gene in the same transcription unit. However, it is now known that both activities are specified by the same polypeptide. The calmodulin-activated adenylate cyclase activity was localized to the N-terminal 400 residues by subcloning from the *adc* gene and by constructing a gene fusion with the *lacZ* gene for β-galactosidase (Glaser *et al.*, 1989). Recently site-directed mutagenesis has identified key residues in the ATP-binding and calmodulin-binding domains (Glaser *et al.*, 1991).

The first infection experiments with *B. pertussis* mutants measured the LD$_{50}$ in mice by intravenous injection (Weiss *et al.*, 1983, 1984). In this model Ptx$^-$, Hly$^-$ and Adc$^-$ Hly$^-$ double mutants were less virulent than the wild-type strain. Also the Adc$^-$ Hly$^-$ double mutant was less virulent than the Hly$^-$ Adc$^+$ strain which suggested that both Hly and Adc were required for full virulence. More recently the virulence of mutants has been evaluated in mouse infection models by aerosol and intranasal inoculation to reflect more accurately the disease in humans and to determine both colonization and toxic activity (Goodwin & Weiss, 1990; Kimura *et al.*, 1990; Roberts *et al.*, 1991; Weiss & Goodwin, 1989).

Infection experiments in mice with Ptx$^-$ mutants have consistently shown that pertussis toxin is an important virulence factor. Infection with Ptx$^-$ mutants of *B. pertussis*, as well as with the normally non-toxigenic *B. parapertussis* expressing recombinant Ptx, attributed leukocytosis, anaphylaxis and histamine sensitivity to expression of Ptx. Furthermore, Goodwin & Weiss (1990) showed that a Ptx$^-$ mutant did not cause lethal infection in the infant mouse model by intranasal inoculation. When low doses of bacteria were used (sublethal infection), the Ptx$^-$ mutant was no less virulent than the wild-type strain but was much less associated with subsequent development of lung infection.

Goodwin & Weiss (1990) also showed that an Adc$^-$ mutant was less virulent because it was defective in initial colonization and was also impaired in lethal activity. Conversely at high challenge doses the Adc$^-$ mutant persisted longer than the wild-type strain but less bacterial growth occurred. This was attributed to the possibility that the mutant invaded host cells better than the wild-type strain and consequently escaped the host's immune response. However, the conclusion drawn by Goodwin & Weiss (1990) that Adc is critical for colonization must be tempered by the fact that the mutant they used was also defective in haemolysin. Studies with an in-frame deletion mutant that remains Hly$^+$ are required.

There have been contradictory reports about the requirement for Fha in murine infections. Fha has been shown to promote adhesion to cultured human cells *in vitro* via an RGD amino-acid sequence that promotes interaction with integrin proteins on the mammalian cell surface (Relman *et al.*, 1990). Several studies with Fha$^-$ mutants have failed to demonstrate that the haemagglutinin is a virulence factor (Weiss *et al.*, 1983, 1984; Roberts *et al.*, 1991). However, one report (Kimura *et al.*, 1990) using non-lethal aerosol infection in adult mice showed that the Fha$^-$ mutants had reduced ability to colonize the trachea early in infection and that Fha was not required for later infection of the lung. The reason for this discrepancy is not clear.

Similar findings were obtained with a mutant defective in the P.69 outer membrane protein adhesin. Infection experiments in mice by aerosol administration with a P.69$^-$ mutant and also with a Fha$^-$ P.69$^-$ double mutant did not demonstrate reduced virulence (Roberts *et al.*, 1991). These results could be due to lack of receptors for Fha and P.69 on murine cells, to the presence of additional bacterial adhesins that can function in murine infections or to the possibility that adhesins do not play a major role in the murine disease. This illustrates some of the problems that can be encountered when using animal infection models to characterize bacterial virulence factors.

Staphylococcus aureus

S. aureus is a Gram-positive pathogen of great complexity. It can cause a variety of different infections in man and animals and can express an array of potential virulence factors. Many of these exoproteins have been purified and their modes of action *in vitro* have been well characterized. These include cytolytic toxins (α-, β-, γ-, δ-toxins and leucocidin), the mitogenic enterotoxins, coagulase and protein A. Genetic manipulation has been used to construct strains of *S. aureus* that can be studied in suitable animal infection models to determine the roles of individual factors in pathogenesis.

Mutants lacking α-toxin (Hla$^-$) displayed reduced ability to kill mice after intraperitoneal or intramammary inoculation, confirming that α-toxin is the major lethal toxin of *S. aureus* (O'Reilly *et al.*, 1986; Patel *et al.*, 1987; Bramley *et al.*, 1989). Also, when injected subcutaneously in mice, smaller abscesses were formed by the Hla$^-$ mutant, suggesting that α-toxin interferes with phagocytic cells and promotes necrosis of host tissue (Patel *et al.*, 1987).

The mouse mastitis model has been used to examine pathological effects of α- and β-toxins (Bramley *et al.*, 1989). In the mouse mammary gland bacteria rapidly became associated with macrophages, but they were not eliminated. Indeed, an Hla$^+$ strain of *S. aureus* appeared to proliferate inside macrophages and eventually to lyse them. After 24 h no macrophages were seen in the area of bacterial growth. These studies implied that α-toxin is responsible for preventing macrophage chemotaxis and function. It is attractive to think that intracellular expression of α-toxin promotes escape from the phagosome.

This notion is supported by studies with cultured endothelial cells which *S. aureus* cells can specifically attach to and subsequently enter. Hla$^+$ organisms killed endothelial cells but these cells survived infection with an Hla$^-$ mutant because the bacteria could not lyse the phagosome (Vann & Proctor, 1988). Thus, the expression of α-toxin might play an important role in the pathogenesis of invasive endocarditis. It would be very interesting to compare the virulence of Hla$^+$ and Hla$^-$ bacteria in animal infection models for staphylococcal endocarditis.

Prior to the study by Bramley *et al.* (1989) there was little evidence that β-toxin (Hlb) might be an important virulence factor of *S. aureus*. However, studies with Hla$^-$ and Hlb$^-$ mutants in the mouse mastitis model suggested that β-toxin promotes growth in the infected mammary gland. An Hla$^-$ Hlb$^-$ mutant grew poorly in the gland while an Hla$^-$ Hlb$^+$ mutant was recovered at much higher numbers. Indeed, the Hla$^-$ Hlb$^+$ and Hla$^+$ Hlb$^-$ mutants grew more profusely than the parental strain. The reduced growth of the wild-type strain compared to the mutants was attributed to antagonism between the two toxins.

Site-specific mutants affected in Hla and Hlb have been constructed from a wild-type bovine strain of *S. aureus* and the mutants are currently being tested in experimental infections of lactating cows (K. Kenny, F. D. Bastida-Corcuera, T. J. Foster and N. L. Norcross, unpublished data). An Hla$^-$ Hlb$^-$ double mutant caused much less severe infection that the parental Hla$^+$ Hlb$^+$ strain. The mutant bacteria were completely cleared from nearly 50% of the infected animals. However, all first lactation animals became infected by the double mutant. The somatic cell count was high, which suggests that factors other than α- or β-toxin can attract polymorphonuclear leucocytes into the infected gland. It will be very interesting to study infections caused by mutants defective in either α- or β-toxin alone.

Epidemiological evidence favours the view that β-toxin is important in the bovine disease. The majority of bovine isolates are β-toxigenic whereas human isolates rarely express β-toxin. The *hlb* gene is inactivated by lysogenization of *S. aureus* by a converting bacteriophage, which often leads simultaneously to expression of staphylokinase (Coleman *et al.*, 1989, 1991). The converting phage inserts at an attachment site located within the structural gene for β-toxin. Occasionally bovines can become infected with typical human strains. However, such organisms isolated from mastitic milk have all the characteristics of human strains of *S. aureus* except that they express β-toxin and fail to express staphylokinase (Hummel, Witte & Kemmer, 1978). This could be explained if strains that were cured of the bacteriophage and consequently became β-toxigenic had a growth advantage *in vivo*. This notion could be tested by experimental infection of lactating cows with a non-β-toxigenic lysogen and testing isolates for expression of β-toxin.

Infection experiments in mice with protein A-deficient (Spa$^-$) mutants have provided evidence that this surface protein is a virulence factor. A Spa$^-$ mutant of *S. aureus* strain 8235-4 was slightly less virulent in peritonitis and subcutaneous infections (Patel *et al.*, 1987) but no differences were observed in the mouse mastitis model. Surprisingly a mutant defective in Agr, the global regulator of virulence factor expression in *S. aureus* (Recsei *et al.*, 1986), was quite virulent in the mastitis model (Foster, O'Reilly & Bramley, 1990). This contrasts with *B. pertussis* where mutants in the regulatory locus *vir/bvg* are completely avirulent (Weiss *et al.*, 1983). The *agr* mutation is known to increase expression of protein A and coagulase, while reducing expression of exoproteins such as the α- and β-toxins (Recsei *et al.*, 1986). Derivatives of an Agr$^-$ mutant which lacked protein A or β-toxin or both factors were less virulent than the Agr$^-$ Spa$^+$ Hlb$^+$ parental strain. It was concluded that the mouse virulence of the Agr$^-$ mutant was due to expression of β-toxin and to a high level of protein A.

Infection experiments in several mouse models with an allele replacement mutant which eliminated expression of coagulase have consistently failed to identify this protein as a virulence factor (Phonimdaeng *et al.*, 1990). This contradicts earlier reports using chemically induced mutants (Jonsson *et al.*, 1985), where the reduced virulence reported must have been due to lesions in other genes. It is possible that coagulase is a virulence factor in animals other than mice but this remains to be tested.

The role of the capsule as a virulence factor of *S. aureus* has been investigated by generating capsule-defective mutants with transposons. Strains that produce serotype 1 macrocapsules are highly virulent because of impaired opsonophagocytosis. Two transposon Tn*551* mutants of a macrocapsular strain were isolated; one was devoid of detectable capsular polysaccharide and the other expressed a microcapsule similar in size to those of most clinical isolates (Lee *et al.*, 1987). The microcapsular mutant had the same virulence for mice as the non-capsulated mutant suggesting that the microcapsule does not enhance virulence. Clinically relevant strains produce serotype 5 or serotype 8 capsular polysaccharide in the form of a microcapsule (reviewed by Foster, 1991*b*). The macrocapsule is never seen with these serotypes. Mutants defective in expression of type 5 capsular polysaccharide also exhibited the same virulence as the wild-type strain (Albus, Arbeit & Lee, 1991). This result is not in agreement with *in vitro* phagocytosis experiments which suggested that both serotype 5 and serotype 8 microcapsules inhibited phagocytosis (Karakawa *et al.*, 1988; Fattom *et al.*, 1990). Despite the uncertainty about the role of capsular polysaccharide in virulence, there is mounting evidence that anti-capsular polysaccharide antibodies promote phagocytosis and killing and that anti-capsular polysaccharide immunity is protective.

GENETIC STUDIES WITH INTRACELLULAR PATHOGENS

Genetic manipulation is playing an important role in dissecting the pathogenesis of infections which involve bacterial invasion of host cells. This is exemplified by the detailed genetic analysis of chromosomal and plasmid-encoded factors of *Shigella* spp. where mutants blocked at different stages of invasion and intracellular movement have been isolated (reviewed by Hale, 1991; see also Sansonetti, this symposium). Much of the analysis of mutants of *Shigella* has been conducted using *in vitro* assays with cultured cells, with less emphasis on *in vivo* tests. This is understandable because of the doubtful relevance to human shigellosis of *in vivo* models such as oral infection of opiate-treated, starved guinea pigs or the Séreny (keratoconjunctivitis) test. Most animals are innately resistant to oral infection with *Shigella*.

However, macaque monkeys are susceptible to oral infection and suffer an illness that is very similar to human shigellosis, although this model has obvious limitations. Nevertheless infection studies with monkeys inoculated intragastrically with an allelic replacement mutant of *S. dysenteriae* type 1 defective in shiga toxin showed that shiga toxin was not required for invasiveness but was responsible for destroying blood capillaries in connective tissue and for an influx of inflammatory cells into the lumen of the intestine (Fontaine, Arondel & Sansonetti, 1988). Both the wild-type and mutant strains caused lethal dysentery; the amount of pus and mucus in the stools was indistinguishable and the numbers and severities of abscesses were similar. The major difference was the absence of blood in the stools of the animals infected with the shiga toxin mutant and the lack of haemorrhage in the colon. Thus, shiga toxin is not a diarrhoeagenic toxin. It seems to exacerbate the infection by damaging endothelial cells in blood vessels of the large intestine (see Wren, this symposium, for further discussion of shiga toxin).

In contrast to *Shigella*, tissue-culture-cell invasion studies with mutants of *Listeria monocytogenes* have been supported by infection studies in mice. The mouse infection model is a realistic model for the human disease. In view of this and the fact that the genetics of *Shigella* virulence have been reviewed recently (Hale 1991; see Sansonetti, this symposium), recent studies with *Listeria* and to a lesser extent on *Legionella pneumophila* will be highlighted.

The genetics of *Salmonella typhimurium* is second only to that of *E. coli* K-12, yet until recently there was little genetic input into studying the pathogenesis of salmonellosis. This is even more surprising given the availability of well characterized mouse infection models with which to test mutants (e.g. see Brown *et al.*, 1987; Maskell *et al.*, 1987). Consequently there is no clear picture of the number or function of genes involved in invasion and intracellular survival. Undoubtedly, genetics will play an important part in unravelling the pathogenic processes of members of this genus (see also Finlay & Chatfield *et al.*, this volume).

Legionella pneumophila

L. pneumophila is a Gram-negative bacterium that occurs in the environment. It is also a facultative intracellular pathogen and can cause a serious form of pneumonia in man if encountered in aerosols. Guinea-pig lung infection models have been established which have allowed mutants defective in potential virulence factors to be evaluated.

The Major Secretory Protein (Msp) of *L. pneumophila* is a metalloprotease with *in vitro* cytotoxic activity. It has been shown to act as a protective immunogen for guinea-pigs yet paradoxically the protein is not required for virulence. Thus, guinea-pigs infected with sublethal doses of Msp$^+$ *L. pneumophila* developed strong cell-mediated immunity against Msp and vaccination with purified Msp protected against the lethal infection (Blander & Horwitz, 1989). A mutation generated by Tn9 insertion in the cloned *msp* gene in *E. coli* was transferred into *L. pneumophila* by allelic exchange (Szeto & Shuman, 1990). The Msp$^-$ mutant exhibited the same level of virulence as the wild-type strain in the guinea-pig infection model, with animals infected with the mutant showing similar clinical signs to those infected with the wild-type strain (Blander *et al.*, 1990). In addition the Msp$^+$ and Msp$^-$ strains had the same LD$_{50}$, multiplied at the same rate in lung tissue and produced similar pathological lesions. The *in vivo* results correlated well with the ability of the bacteria to grow within and to kill human macrophages. This clearly demonstrates that a protective role as an immunogen need not necessarily correlate with a requirement for virulence.

In contrast genetic manipulation has demonstrated conclusively that a 24 kD basic surface protein, the Macrophage Infectivity Potentiation (Mip) protein, is a major virulence factor (Cianciotto *et al.*, 1989, 1990). A site-specific Mip$^-$ mutant had impaired ability to initiate infection in a human macrophage cell line as well as explanted human alveolar macrophages. This correlated with reduced virulence after intratracheal infection of guinea pigs. The Mip$^-$ mutant produced fewer illnesses and fewer lethal infections than the wild-type strain. Infectivity both *in vitro* and *in vivo* was restored by reintroduction of the cloned *mip* gene into the Mip$^-$ mutant.

Listeria monocytogenes

L. monocytogenes is a ubiquitous soil organism that can cause serious infections in man. The key to its pathogenesis is the ability of the bacteria to survive within host macrophages. Like *Shigella*, *L. monocytogenes* can escape from the phagosome, a process which has been shown by genetic manipulation to require the haemolysin (listeriolysin). This advance in

understanding intracellular survival of *L. monocytogenes* was made possible
by the development of transposon mutagenesis and allelic replacement
techniques and by the use of a mouse infection model to support studies with
cultured cells *in vitro*.

Initial genetic analysis of the virulence of *L. monocytogenes* virulence
concentrated on listeriolysin, a member of the family of thiol-activated
cytolysins related to streptolysin O (Smyth & Duncan, 1978). The first
mutants were isolated with the conjugative transposons Tn*916* and Tn*1545*
and were shown to be avirulent in the mouse model (Kathariou *et al.*, 1987;
Gaillard, Berche & Sansonetti, 1986). That this reduction in virulence was
due exclusively to the loss of listeriolysin was confirmed by complementation tests with a recombinant plasmid that carried only the listeriolysin
gene (Cossart *et al.*, 1989). Also discrete mutations in the listeriolysin gene
generated by *in vitro* oligonucleotide mutagenesis, which led to expression
of gene products that differed only from the wild-type listeriolysin by single
amino-acid changes, were introduced into the chromosome by allelic
exchange with an antibiotic resistance-conferring Tn*916* mutation (Michel *et
al.*, 1990). The level of virulence of the mutants was proportional to residual
haemolytic activity expressed *in vitro*.

In an alternative genetic approach, expression of listeriolysin by *Bacillus
subtilis* allowed these heterologous bacteria to escape from the phagosome
after being internalized by macrophages (Bielecki *et al.*, 1990). This directly
demonstrated the role of listeriolysin in escape of bacteria from the phagosome but many other factors are required for virulence.

More recent genetic studies by French and American research groups
have suggested that the ability to escape from the phagocytic vacuole and to
survive in and to multiply in host cells also requires a phosphatidylinositol-
specific phospholipase C (Plc; Mengaud, Braun-Breton & Cossart, 1991*a*;
Camilli, Goldfine & Portnoy, 1991). This was discovered because some
mutants defective in plaque formation in monolayers (a reflection of the
ability of bacteria to enter and kill host cells *in vitro*) remained haemolytic
(Camilli *et al.*, 1991). These Hly$^+$ plaque-deficient mutants were shown to
be defective in Plc. Mengaud *et al.* (1991*a*) identified the *plcA* gene by
sequencing DNA adjacent to *hlyA*, the gene encoding listeriolysin (Fig. 1).
They subsequently inactivated the *plcA* gene by allelic exchange. The *plcA*
mutants isolated by both groups were less virulent in the mouse model. The
100-fold difference in virulence is probably due to different polar effects on
prfA, a gene encoding a positive transcriptional regulator of several virulence factors which is cotranscribed with *plcA* (Mengaud *et al.*, 1991*b*).
Complementation tests to demonstrate that only the *plcA* gene was affected
are in order. The precise role of Plc in pathogenicity is as yet unclear. There
is the fascinating possibility that Plc could break the glycosyl phosphatidy-
linositol links that anchor proteins to the mammalian cell surface.

Genetic engineering is continuing to play a key role in dissecting the

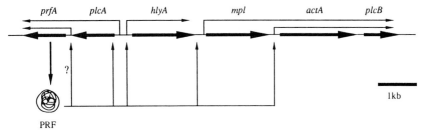

Fig. 1. A virulence locus of *Listeria monocytogenes*. The horizontal bold arrows indicate the position of, the approximate size of, and direction of, transcription of genes linked to *hlyA*, the gene encoding listeriolysin. The other lines show the positions of promoters regulated by the positive regulator PrfA. The other genes are abbreviated as follows: *plcA*, phosphatidylinositol-specific phospholipase C; *plcB*, phosphatidylcholine-specific phospholipase C; *mpl*, metalloprotease; *actA*, actin-binding surface protein.

virulence mechanisms of *L. monocytogenes*. Lying directly adjacent to *prfA*, *plcA* and *hlyA* is a 5.7 kb transcription unit which encodes at least three proteins with possible roles in virulence (Fig. 1). The first gene *mpl* encodes a protein having homology with metalloproteases (Mengaud, Geoffrey & Cossart, 1991c; Domann *et al.*, 1991). However, proteolytic activity has not yet been associated with Mpl. A *mpl* mutant generated by transposon insertion had a lecithinase-negative phenotype due to a polarity effect on *plcB*, a 3′ gene in the same transcription unit (Fig. 1). *plcB* encodes a protein having homology with phosphatidylcholine-specific phospholipases C (Vasquez-Boland *et al.*, 1992). This protein may be required along with listeriolysin for the bacteria to escape the phagolysosome. Located between the *plcB* and *mpl* genes is *actA*, a gene which encodes a surface-located actin-binding protein involved in intracellular bacterial movement (Vasquez-Boland *et al.*, 1992; Kocks *et al.*, 1992). The roles of these proteins in the pathogenicity of *L. monocytogenes* needs to be evaluated by testing non-polar mutants defective in each gene using the mouse infection model.

Gaillard *et al.* (1991) identified a bacterial surface protein called internalin, which appears to be required by *L. monocytogenes* to invade cultured epithelial cells, by selecting for transposon mutants defective in killing cultured mammalian cells. There is evidence that the natural disease involves invasion of non-professional phagocytes as well as survival within macrophages. This protein was shown to be different to the protein which is missing in spontaneous rough mutants that are also invasion deficient (Kuhn & Goebel, 1989). It will be interesting to determine if the internalin mutants lack virulence in the mouse infection model.

CONCLUDING REMARKS

Genetic manipulation can now be applied to almost any bacterial pathogen with the development of methods for transferring chimaeric plasmids into diverse species (Chassy, Mercenier & Flickinger, 1988; Luchansky, Muriana & Klaenhammer, 1988) and broad host-range suicide systems for transposon mutagenesis and allelic replacement (Simon, Priefer & Puhler, 1983; de Lorenzo *et al.*, 1990; Schafer *et al.*, 1990). Pathogens not previously subjected to genetic analysis can now be studied and, indeed, more virulence factors can be scrutinized in those organisms where genetic systems are well developed. As mentioned (see Introduction) the major limitations for the use of mutants for defining roles for virulence factors *in vivo* are animal infection models. These models must be relevant to the natural disease and have sufficient sensitivity to reflect differences in virulence factor expression.

ACKNOWLEDGEMENTS

Work in the author's laboratory is supported by the Wellcome Trust and by the Health Research Board of Ireland. The assistance of Teresa Hogan is gratefully acknowledged and thanks are due to Catherine O'Connell, Damien McDevitt, Brian Sheehan and Niamh Kinsella for comments on the manuscript. Brian Sheehan also helped by composing Fig. 1.

REFERENCES

Albus, A., Arbeit, R. D. & Lee, J. C. (1991). Virulence of *Staphylococcus aureus* mutants altered in type 5 capsule production. *Infection and Immunity*, **59**, 1008–14.

Bielecki, J., Youngman, P., Connelly, P. & Portnoy, D. A. (1990). *Bacillus subtilis* expressing a haemolysin gene from *Listeria monocytogenes* can grow in mammalian cells. *Nature*, **345**, 175–6.

Berg, C. M., Berg, D. E. & Groisman, F. A. (1989). Transposable elements and genetic engineering of bacteria. In *Mobile DNA*, eds D. E. Berg & M. M. Howe, pp. 879–925. Washington D.C.: American Society for Microbiology.

Blander, S. J. & Horwitz, M. A. (1989). Vaccination with the major secretory protein of *Legionella pneumophila* induces cell-mediated and protective immunity in a guinea pig model of Legionnaire's disease. *Journal of Experimental Medicine*, **169**, 691–705.

Blander, S. J., Szeto, L., Shuman, H. A. & Horwitz, M. A. (1990). An immunoprotective molecule, the major secretory protein of *Legionella pneumophila*, is not a virulence factor in a guinea pig model for Legionnaire's disease. *Journal of Clinical Investigation*, **86**, 817–24.

Bohach, G. A., Fast, D. J., Nelson, R. D. & Schlievert, P. M. (1990). Staphylococcal and streptococcal pyrogenic toxins involved in toxic shock syndrome and related illnesses. *CRC Critical Reviews in Microbiology*, **17**, 251–72.

Bramley, A. J., Patel, A. H., O'Reilly, M., Foster, R. & Foster, T. J. (1989). Roles for alpha-toxin and beta-toxin in virulence of *Staphylococcus aureus* for the mouse mammary gland. *Infection and Immunity*, **57**, 2489–94.

Brown, A., Hormaeche, C. E., Demarco de Hormaeche, R., Winther, M., Dougan, G., Maskell, D. J. & Stocker, B. A. D. (1987). An attenuated *aroA Salmonella typhimurium* vaccine elicits humoral and cellular immunity to cloned β-galactosidase in mice. *Journal of Infectious Diseases*, **155**, 86–92.

Camilli, A., Goldfine, H. & Portnoy, D. A. (1991). *Listeria monocytogenes* mutants lacking phosphatidylinositol-specific phospholipase C are avirulent. *Journal of Experimental Medicine*, **173**, 751–4.

Camilli, A., Portnoy, D. A. & Youngman, P. (1990). Insertional mutagenesis of *Listeria monocytogenes* with a novel Tn*917* derivative that allows direct cloning of DNA flanking transposon insertions. *Journal of Bacteriology*, **172**, 3738-44.

Chassy, B. M., Mercenier, A. & Flickinger, J. (1988). Transformation of bacteria by electroporation. *Trends in Biotechnology*, **6**, 303–9.

Cianciotto, N. P., Eisenstein, B. I., Mody, C. H., Toews, G. B. & Engleberg, N. C. (1989). A *Legionella pneumophila* gene encoding a species-specific surface protein potentiates initiation of intracellular infection. *Infection and Immunity*, **57**, 1255–62.

Cianciotto, N. P., Eisenstein, B. I., Mody, C. H. & Engleberg, N. C. (1990). A mutation in the *mip* gene results in an attenuation of *Legionella pneumophila* virulence. *Journal of Infectious Diseases*, **162**, 121–6.

Coleman, D. C., Sullivan, D. J., Russell, R. J., Arbuthnott, J. P., Carey, B. F. & Pomeroy, H. M. (1989). *Staphylococcus aureus* bacteriophages mediating the simultaneous lysogenic conversion of β-lysin, staphylokinase and enterotoxin A: molecular mechanisms of triple-conversion. *Journal of General Microbiology*, **135**, 1679–97.

Coleman, D., Knights, J., Russell, R., Shanley, D., Birkbeck, T. H., Dougan, G. & Charles, I. (1991). Insertional inactivation of the *Staphylococcus aureus* β-toxin by bacteriophage φ13 genome. *Molecular Microbiology*, **5**, 933–9.

Cossart, P., Vincente, M. F., Mengaud, J., Baquero, F., Perez-Diaz, J. C. & Berche, P. (1989). Listeriolysin O is essential for virulence of *Listeria monocytogenes*: direct evidence obtained by gene complementation. *Infection and Immunity*, **57**, 3629–36.

de Azevedo, J. C. S., Foster, T. J., Hartigan, P. J., Arbuthnott, J. P., O'Reilly, M., Kreiswirth, B. N. & Novick, R. P. (1985). Expression of the cloned toxic shock syndrome toxin 1 gene (*tst*) in vivo with a rabbit uterine model. *Infection and Immunity*, **50**, 304–9.

de Lorenzo, V., Herrero, M., Jakubzik, U. & Timmis, K. N. (1990). Mini-Tn*5* transposon derivatives for insertion mutagenesis, promoter probing, and chromosomal insertion of cloned DNA in Gram-negative bacteria. *Journal of Bacteriology*, **172**, 6568–72.

Domann, E., Leimeister-Wachter, M., Goebel, W. & Chakraborty, T. (1991). Molecular cloning, sequencing, and identification of a metalloprotease gene from *Listeria monocytogenes* that is species specific and physically linked to the listeriolysin gene. *Infection and Immunity*, **59**, 65–72.

Falkow, S. (1988). Molecular Koch's postulates applied to microbial pathogenicity. *Reviews of Infectious Diseases*, **10**, S274–6.

Fattom, A., Schneerson, R., Szu, S. C., Vann, W. F., Shiloach, J., Karakawa, W. F. & Robbins, J. B. (1990). Synthesis and immunologic properties in mice of vaccines composed of *Staphylococcus aureus* type 5 and type 8 capsular polysaccharides conjugated to *Pseudomonas aeruginosa* exotoxin A. *Infection and Immunity*, **58**, 2367–74.

Fontaine, A., Arondel, J. & Sansonetti, P. J. (1988). Role for shiga toxin in the pathogenesis of bacillary dysentery, studied by using a Tox⁻ mutant of *Shigella dysenteriae* 1. *Infection and Immunity*, **56**, 3099–109.

Foster, T. J. (1984). Analysis of plasmids with transposons. In *Methods in Microbiology*, Vol. 17, eds P. M. Bennett & J. Grinsted, pp. 197–226. London: Academic Press.

Foster, T. J. (1991a). Genetic analysis of the *in vivo* role of bacterial toxins. In *Sourcebook of Bacterial Protein Toxins*, eds J. E. Alouf & J. H. Freer, pp. 445–59. London: Academic Press.

Foster, T. J. (1991b). Potential for vaccination against infections caused by *Staphylococcus aureus*. *Vaccine*, **9**, 221–7.

Foster, T. J., O'Reilly, M. & Bramley, A. J. (1990). Genetic studies of *Staphylococcus aureus* virulence factors. In *Pathogenesis of Wound and Biomaterial-Associated Infections*, eds T. Wadström, I. Eliasson, I. Holder & Å. Ljungh, pp. 35–46. London: Springer-Verlag.

Gaillard, J.-L., Berche, P. & Sansonetti, P. (1986). Transposon mutagenesis as a tool to study the role of hemolysin in the virulence of *Listeria monocytogenes*. *Infection and Immunity*, **52**, 50–5.

Gaillard, J.-L., Berche, P., Frehel, C., Gouin, E. & Cossart, P. (1991). Entry of *L. monocytogenes* into cells is mediated by internalin, a repeat protein reminiscent of surface antigens from Gram-positive cocci. *Cell*, **65**, 1127–41.

Gatermann, S., John, J. & Marre, R. (1989). *Staphylococcus saprophyticus* urease: characterization and contribution to uropathogenicity in unobstructed urinary tract infection of rats. *Infection and Immunity*, **57**, 110–16.

Gatermann, S. & Marre, R. (1989). Cloning and expression of *Staphylococcus saprophyticus* urease gene sequences in *Staphylococcus carnosus* and contribution of the enzyme to virulence. *Infection and Immunity*, **57**, 2998–3002.

Glaser, P., Elmaoglou-Lazaridou, A., Krin, E., Ladant, D., Barzu, O. & Danchin, A. (1989). Identification of residues essential for catalysis and binding of calmodulin in *Bordetella pertussis* adenylate cyclase by site-directed mutagenesis. *EMBO Journal*, **8**, 967–72.

Glaser, P., Munier, H., Gilles, A., Krin, E., Porumb, T., Barzu, O., Sarfati, R., Pellecuer, C. & Danchin, A. (1991). Functional consequences of single amino acid substitutions in calmodulin-activated adenylate cyclase of *Bordetella pertussis*. *EMBO Journal*, **10**, 1683–8.

Goodwin, M. S. & Weiss, A. A. (1990). Adenylate cyclase toxin is critical for colonization and pertussis toxin is critical for lethal infection by *Bordetella pertussis* on infant mice. *Infection and Immunity*, **58**, 3445–47.

Gross, R., Aricò, B. & Rappuoli, R. (1989). Families of signal-transducing proteins. *Molecular Microbiology*, **3**, 1661–7.

Hale, L. T. (1991). Genetic basis of virulence in *Shigella* species. *Microbiological Reviews*, **55**, 206–24.

Hummel, R., Witte, W. & Kemmer, G. (1978). Zur frage der wechselseitigen übertragung von *Staphylococcus aureus* zwischen mensch und rind und der milieuadaptation der hämolysin- und fibrinolysinbildung. *Archiv für Experimentelle Veterinärmedizin*, **32**, 287–98.

Jonsson P., Lindberg, M., Haraldsson, I. & Wadström, T. (1985). Virulence of *Staphylococcus aureus* in a mouse mastitis model: studies of alpha hemolysin, coagulase and protein A as possible virulence determinants with protoplast fusion and gene cloning. *Infection and Immunity*, **49**, 765–9.

Joseph-Liauzun, E., Fellay, R. & Chandler, M. (1989). Transposable elements for efficient manipulation of a wide range of Gram-negative bacteria: promoter probes and vectors for foreign genes. *Gene*, **85**, 83–9.

Karakawa, W. W., Sutton, A., Schneerson, R., Karpas, A. & Vann, W. F. (1988). Capsular antibodies induce type-specific phagocytosis of capsulated *Staphylococcus aureus* by human polymorphonuclear leukocytes. *Infection and Immunity*, **56**, 1090–5.

Kathariou, S., Metz, P., Hof, P. & Goebel, W. (1987). Tn916-induced mutations in the hemolysin determinant affecting virulence of Listeria monocytogenes. Journal of Bacteriology, 169, 1291–7.

Kimura, A., Mountzouros, K. T., Relman, D. A., Falkow, S. & Cowell, J. L. (1990). Bordetella pertussis filamentous hemagglutinin: evaluation as a protective antigen and colonization factor in a mouse respiratory infection model. Infection and Immunity, 58, 7–16.

Kleckner, N., Roth, J. & Botstein, D. (1977). Genetic engineering in vivo using translocatable drug-resistance elements—new methods in bacterial genetics. Journal of Molecular Biology, 116, 125–59.

Kocks, C., Gouin, E., Tabouret, M., Ohayon, H. & Cossart, P. (1992). L. monocytogenes-induced actin assembly requires the actA gene product, a surface protein. Cell, 68, 521–31.

Kuhn, M. & Goebel, W. (1989). Identification of an extracellular protein of Listeria monocytogenes possibly involved in intracellular uptake by mammalian cells. Infection and Immunity, 57, 55–61.

Lee, J. C., Betley, M. J., Hopkins, C. A., Perez, N. E. & Pier, G. B. (1987). Virulence studies, in mice, of transposon-induced mutants of Staphylococcus aureus differing in capsule size. Journal of Infectious Diseases, 156, 741–50.

Luchansky, J. B., Muriana, P. M. & Klaenhammer, T. R. (1988). Application of electroporation for transfer of plasmid DNA to Lactobacillus, Lactococcus, Leuconostoc, Listeria, Pediococcus, Bacillus, Staphylococcus, Enterococcus and Propionibacterium. Molecular Microbiology, 2, 637–46.

Marre, R., Hacker, J., Henkel, W. & Goebel, W. (1986). Contribution of cloned virulence factors from uropathogenic Escherichia coli strains to nephropathogenicity in an experimental rat pyelonephritis model. Infection and Immunity, 54, 761–7.

Maskell, D. J., Sweeney, K. J., O'Callaghan, D., Hormaeche, C. E., Liew, F. Y. & Dougan, G. (1987). Salmonella typhimurium aroA mutants as carriers of the Escherichia coli heat-labile enterotoxin B subunit to the murine secretory and systemic immune systems. Microbial Pathogenesis, 2, 211–21.

Mengaud, J., Braun-Breton, C. & Cossart, P. (1991a). Identification of phosphatidylinositol-specific phospholipase C activity in Listeria monocytogenes: a novel type of virulence factor? Molecular Microbiology, 5, 367–72.

Mengaud, J., Dramsi, S., Gouin, E., Vazquez-Boland, J. A., Milon, G. & Cossart, P. (1991b). Pleiotropic control of Listeria monocytogenes virulence factors by a gene that is autoregulated. Molecular Microbiology, 5, 2273–83.

Mengaud, J., Geoffroy, C. & Cossart, P. (1991c). Identification of a new operon involved in Listeria monocytogenes virulence: its first gene encodes a protein homologous to bacterial metalloproteases. Infection and Immunity, 59, 1043–9.

Michel, E., Reich, K. A., Favier, R., Berche, P. & Cossart, P. (1990). Attenuated mutants of the intracellular bacterium Listeria monocytogenes obtained by single amino acid substitutions in listeriolysin O. Molecular Microbiology, 4, 2167–78.

O'Reilly, M., de Azevedo, J. C. S., Kennedy, S. & Foster, T. J. (1986). Inactivation of the alpha-haemolysin gene of Staphylococcus aureus 8325-4 by site-directed mutagenesis and studies on expression of its haemolysins. Microbial Pathogenesis, 1, 125–38.

Patel, A. H., Nowlan, P., Weavers, E. D. & Foster, T. J. (1987). Virulence of protein A-deficient mutants of Staphylococcus aureus isolated by allele-replacement. Infection and Immunity, 55, 3103–10.

Phonimdaeng, P., O'Reilly, M., Nowlan, P., Bramley, A. J. & Foster, T. J. (1990).

The coagulase of *Staphylococcus aureus* 8325-4. Sequence analysis and virulence of site-specific coagulase-deficient mutants. *Molecular Microbiology*, **4**, 393–404.

Recsei, P., Keiswirth, B., O'Reilly, M., Schlievert, P., Gruss, A. & Novick, R. P. (1986). Regulation of exoprotein gene expression in *Staphylococcus aureus* by *agr*. *Molecular and General Genetics*, **202**, 58–61.

Relman, D., Tuomanen, E., Falkow, S., Golenbock, D. T., Saukkonen, K. & Wright, S. D. (1990). Recognition of a bacterial adhesin by an integrin: macrophage CR3 ($a_M\beta_2$, CD11b/CD18) binds filamentous hemagglutinin of *Bordetella pertussis*. *Cell*, **61**, 1375–82.

Roberts, M., Fairweather, N. F., Leininger, E., Pickard, D., Hewlett, E. L., Robinson, A., Hayward, C., Dougan, G. & Charles, I. G. (1991). Construction and characterization of *Bordetella pertussis* mutants lacking the *vir*-regulated P.69 outer membrane protein. *Molecular Microbiology*, **5**, 1393–404.

Schafer, A., Kalinowski, J., Simon, R., Seep-Feldhaus, A.-H. & Puhler, A. (1990). High-frequency conjugal plasmid transfer from Gram-negative *Escherichia coli* to various Gram-positive coryneform bacteria. *Journal of Bacteriology*, **172**, 1663–6.

Simon, R., Priefer, U. & Puhler, A. (1983). A broad host-range mobilization system for *in vivo* genetic engineering: transposon mutagenesis in Gram-negative bacteria. *Biotechnology*, **1**, 784–91.

Sloane, R., de Azevedo, J. C. S., Arbuthnott, J. P., Hartigan, P. J., Kreiswirth, B., Novick, R. & Foster, T. J. (1991). A toxic shock syndrome toxin mutant of *Staphylococcus aureus* isolated by allelic replacement lacks virulence in a rabbit uterine model. *FEMS Microbiology Letters*, **78**, 239–44.

Smith, H. (1989). The mounting interest in bacterial and viral pathogenicity. *Annual Review of Microbiology*, **43**, 1–22.

Smyth, C. J. & Duncan, J. L. (1978). Thiol-activated (oxygen-labile) cytolysins. In *Bacterial Toxins and Cell Membranes*, eds J. Jeljaszewicz and T. Wadström, pp. 129–83. New York: Academic Press.

Sparling, P. F. (1983). Applications of genetics to studies of bacterial virulence. *Philosophical Transactions of the Royal Society*, **B303**, 199–207.

Stock, J. B., Ninfa, A. J. & Stock, M. (1989). Protein phosphorylation and regulation of adaptive responses in bacteria. *Microbiological Reviews*, **53**, 450–90.

Szeto, L. & Schuman, H. A. (1990). The *Legionella pneumophila* major secretory protein, a protease, is not required for intracellular growth or cell killing. *Infection and Immunity*, **58**, 2585–92.

Vann, J. M. & Proctor, R. A. (1988). Cytotoxic effects of ingested *Staphylococcus aureus* on bovine endothelial cells: role of *S. aureus* α-hemolysin. *Microbial Pathogenesis*, **4**, 443–53.

Vasquez-Boland, J. A., Kocks, C., Dramsi, S., Ohayon, H., Geoffroy, C., Mengaud, J. & Cossart, P. (1992). Nucleotide sequence of the lecithinase operon of *Listeria monocytogenes* and possible role of lecithinase in cell-to-cell spread. *Infection and Immunity*, **60**, 219–30.

Weiss, A. A. & Goodwin, M. S. (1989). Lethal infection by *Bordetella pertussis* mutants in the infant mouse model. *Infection and Immunity*, **57**, 3757–64.

Weiss, A. A., Hewlett, E. L., Myers, G. A. & Falkow, S. (1983). Tn5-induced mutations affecting virulence factors of *Bordetella pertussis*. *Infection and Immunity*, **42**, 33–41.

Weiss, A. A., Hewlett, E. L., Myers, G. A. & Falkow, S. (1984). Pertussis toxin and extracytoplasmic adenylate cyclase as virulence factors of *Bordetella pertussis*. *Journal of Infectious Diseases*, **150**, 219–22.

Welch, R. A. (1991). Pore-forming cytolysins of Gram-negative bacteria. *Molecular Microbiology*, **5**, 521–28.

Welch, R. A., Dellinger, E. P., Minshew, B. & Falkow, S. (1981). Haemolysin contributes to virulence of extraintestinal *E. coli* infections. *Nature* (London), **294**, 665–7.

Welch, R. A. & Falkow, S. (1984). Characterization of *Escherichia coli* hemolysins conferring quantitative differences in virulence. *Infection and Immunity*, **43**, 156–60.

Williams Smith, H. & Linggood, M. A. (1971). Observations on the pathogenic properties of the K88, Hly and Ent plasmids of *Escherichia coli* with particular reference to porcine diarrhoea. *Journal of Medical Microbiology*, **4**, 467–85.

GLOBAL REGULATION OF GENE EXPRESSION DURING ENVIRONMENTAL ADAPTATION: IMPLICATIONS FOR BACTERIAL PATHOGENS

CHARLES J. DORMAN AND NIAMH NÍ BHRIAIN

Molecular Genetics Laboratory
Department of Biochemistry
University of Dundee
Dundee DD1 4HN
Scotland

INTRODUCTION

The pathogenic bacterium seeks to colonize the surface or interior of its host for its own benefit. The infection route may be direct or vector mediated and the source of the infection may be a host of the same species, or an intermediate, heterologous host, or infection may follow a period of free-living by the bacterium in an external environment. To survive at each stage of the process, the pathogenic bacterium must adapt to a series of ecological niches or mini-environments. Those niches which are host associated may be superficial or internal, and the latter may be extracellular or intracellular. Successful pathogenesis requires the invader to avoid or evade the host's defences and to possess a strategy for infecting additional hosts. All of this makes considerable demands on a unicellular organism which typically possesses just a few thousand genes; without the ability to devote specialized tissues to solve the problems associated with its complex lifestyle, the bacterium must adapt and readapt its cellular composition to ensure fitness at each stage. Since bacteria are manifestly capable of such adaptation, it is implicit that they possess the means to gather and analyse environmental information and to transduce this information to the cellular apparatus concerned with mounting a response. Frequently, this is the apparatus which is concerned with regulating gene expression. Multidisciplinary studies into mechanisms of signal transduction and response regulation in bacteria have begun to reveal something of the molecular nature of these aspects of prokaryotic biology.

ENVIRONMENTAL CONSIDERATIONS IN BACTERIAL PATHOGENESIS

Any environment encountered by any bacterium during any type of infection may be defined in terms of parameters such as temperature, osmotic pressure, oxygen status, pH and nutrient availability. Of these, only temperature and the availability of one nutrient (iron) have been shown to be crucial *in vivo* (Smith, 1990). This low level of *in vitro/in vivo* correlation reflects a lack of experimentation rather than doubt about the *in vivo* relevance of the other factors listed. *In vitro* experiments have been conducted in many laboratories to detect bacterial genes whose expression is modulated by changes in these parameters and these surveys have yielded a rich harvest of environmentally controlled genes, some of which contribute to virulence. The philosophy behind such experiments includes an expectation that virulence genes are regulated rather than constitutive and that the regulatory signal is provided by the environment. By fusing bacterial genes to readily assayable reporter genes and then screening for genes induced or repressed by changes in environmental parameters believed to be significant *in vivo*, it is hoped that virulence determinants will be discovered. Ultimately, the gene detected can be assessed for importance in pathogenicity by conducting a suitable virulence assay.

This approach is likely to reveal both dedicated virulence genes (such as toxins) with obvious roles in pathogenesis and accessory genes whose precise contribution to infectivity may be much more difficult to elucidate. It is unlikely to yield a great deal of information about how the different genes relate to one another in the infection process. To understand these relationships, a search for regulators must be conducted. This is known as the 'top-down' approach and has the advantage of allowing several virulence (and other) genes which share a common regulator to be identified and grouped as a regulatory unit or 'regulon'.

REGULONS

These are defined as groups of genes under the control of a common regulator. The regulator is usually a DNA binding protein which recognizes and binds to a specific DNA sequence in the control regions of its subservient genes, thereby altering their expression. The ability of the regulator to carry out its task is under environmental control, through mechanisms such as covalent modification and/or cofactor binding. Examples of regulons in the bacteria are legion and most of them are concerned primarily with the control of functions essential in commensal life (Neidhardt & Van Bogelen, 1987). They include systems to cope with carbon, nitrogen or phosphate starvation, starvation for amino acids (the stringent response system), survival of DNA damage (the SOS response), survival of oxidative and osmotic stress and the switch to anaerobic growth. Virulence regulons have been described in several bacterial species and, confusingly, some housekeeping regulons have been found to contribute to pathogenicity. Furthermore, the strict segregation between regulons which is implied by the specific nature of the interaction of the regulator with its cognate control sequences has been found to be breached in several instances. This points to a subtle networking of gene control circuits which, while making life difficult for investigators, greatly assists the bacteria by increasing the flexibility of their environmental response systems.

For historical reasons, the maltose system of enteric bacteria, which is concerned with transport and utilization of maltose and maltodextrins, can be regarded as an archetypal regulon which has played a founding role in the development of the concept of coordinated control of gene expression (Schwartz, 1987). In *E. coli*, three unlinked loci (two of which contain divergently transcribed operons) are coordinately regulated by MalT, a transcription activator which is itself allosterically activated by binding maltose. However, MalT is not the only regulator of the regulon. The cAMP receptor protein (CRP) is also required for *mal* expression, making transcription of the maltose operons carbon-source regulated and linking them to the other cAMP–CRP controlled operons in the cell, a total of at least 200 genes (Busby, 1986) (Fig. 1). Interestingly, the possession of an active cAMP–CRP system is crucial to virulence, at least in the Gram-negative pathogen, *Salmonella typhimurium* (Curtiss & Kelly, 1987). Although the ability to transport and utilize maltose is probably of minor importance in bacterial pathogenesis, it appears that the commensal-associated *mal* regulon shares its cAMP–CRP requirement with genes whose products make a direct contribution to bacterial virulence. This is an example of regulon networking within a regulatory hierarchy. The more pleiotropic the regulator, the more senior its position in the hierarchy. Thus the highly pleiotropic cAMP–CRP system ranks above MalT as a global regulator of gene expression. However, there is no reason to believe that bacterial control hierarchies consist of ranked regulators ultimately governed by an omnipotent controller. Instead, a range of largely equally ranked, higher order regulators control gene expression in overlapping networks. This theme is developed further below (see the section on 'Networking of Regulons').

REGULONS WITH 'TWO-COMPONENT' REGULATORS AND ROLES IN BOTH COMMENSAL AND PATHOGENIC LIFE

The enteric EnvZ/OmpR system is an example of a regulon which is networked to several other regulons and plays a role in both commensal and pathogenic processes and in which the regulator is governed by covalent modification (Fig. 2). The EnvZ and OmpR proteins are encoded by the *ompB* operon which maps to 74' in *E. coli* and *S. typhimurium* (Bachmann, 1990; Sanderson & Hurley, 1987). They are involved in regulating the expression of outer membrane proteins OmpC and OmpF and also regulate the *tppB* tripeptide permease operon and some other uncharacterized genes in *S. typhimurium* (Gibson *et al.*, 1987).

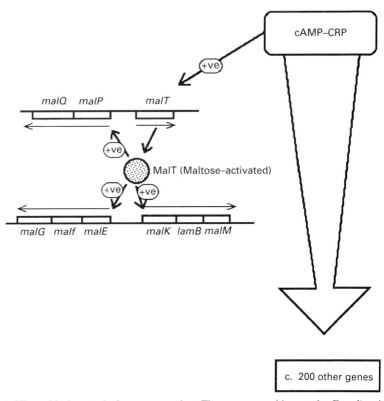

Fig. 1. Hierarchical control of gene expression. The operons making up the *E. coli* maltose regulon are shown, in which *mal* gene expression is under the control of the MalT transcriptional activator. However, *mal* is also positively regulated by the much more pleiotropic cAMP–CRP system.

EnvZ is a 50 kD, 450 amino acid inner membrane protein concerned with transducing osmotic signals to OmpR, a 27 kD, 239 amino acid cytoplasmic DNA-binding protein (Comeau *et al.*, 1985; Forst *et al.*, 1987; Jo *et al.*, 1986; Liljeström, 1986; Norioka *et al.*, 1986). These proteins belong to a family of control elements called the 'two-component' family or, more accurately, the histidine protein kinase / response regulator family. These systems are central to the regulation of cellular processes required for both commensal and pathogenic life (see below) and some members play a key role in regulating development in the Gram-positive bacterium *Bacillus* (for a review, see Stock, Ninfa & Stock, 1989). In these two-component partnerships, one protein is a sensor of changes in an environmental parameter and transmits information to the regulator protein by phosphorylating it. In the *ompB*-encoded system (Fig. 2), EnvZ is the sensor and OmpR the response regulator. High osmolarity causes OmpR to activate transcription of *ompC* and to repress transcription of *ompF*; low osmolarity has the opposite effect. EnvZ becomes autophosphorylated on His-243 within its C-terminal domain in the presence of ATP and rapidly transfers the phosphate group to the N-terminal domain of OmpR, probably to an Asp residue that is conserved among the response regulators (Aiba, Mizuno & Mizushima, 1989; Forst, Delgada & Inouye, 1989; Stock *et al.*, 1989). Phosphorylation makes OmpR proficient for DNA binding via the DNA binding domain within its C-terminus. The differential control of *ompC* and *ompF* expression is thought to be achieved through modulation of the levels of phospho-OmpR in the cell. At low osmolarity, EnvZ does not autophosphorylate and the level of phospho-OmpR in the cell is low. The available phospho-OmpR is bound by a high affinity site in the regulatory region of *ompF*, activating its transcription. The *ompC* gene possesses a low-affinity binding site for phospho-OmpR and is not preferentially activated under these conditions. At high osmolarity, the rate of EnvZ autophosphorylation increases and the cellular concentration of phospho-OmpR is elevated. Binding to the low affinity site in the *ompC* control region activates this gene while binding of phospho-OmpR to a low affinity site in the *ompF* control region inhibits that gene, perhaps via a mechanism that includes DNA looping (Slauch & Silhavy, 1991; Stock *et al.*, 1989). There is evidence that phospho-OmpR interacts directly with the alpha subunit of RNA polymerase while both molecules are bound to the control region of the repressed *ompF* gene (Garrett & Silhavy, 1987; Matsuyama & Mizushima, 1987).

The physiological significance of the differential regulation of *ompC* and *ompF* transcription is thought to be due to the types of trimeric porin which their gene products form in the outer membrane. These are of different sizes, with the pore size of the OmpF type (1.16 nm) being significantly greater than that of the OmpC type (1.08 nm) in terms of its ability to admit to the periplasm harmful solutes such as bile salts found during life in the mammalian gut (Nikaido & Vaara, 1987). Thus, expression of OmpF in the gut, a high osmolarity environment, could be detrimental to the cell, whereas expression of OmpF in an external, aquatic, low osmolarity environment may assist in scavenging for scarce nutrients.

The role of porin regulation in environmental adaptation has prompted investigations of the possibility that this regulon could contribute to the pathogenicity of enteric bacteria. When tested in a mouse model system, mutations in *ompC* or in *ompF* did not affect the virulence of *S. typhimurium*. When combined in the same strain, *ompC* and *ompF* mutations dramatically attenuated the virulence of *S. typhimurium* (Chatfield *et al.*, 1991). Significantly, a mutation in *ompB* which deprives the cell of both EnvZ and OmpR attenuates *S. typhimurium* mouse virulence to an even greater extent than that seen in the *ompC ompF* double mutant, indicating that additional, unknown *ompB*-dependent genes also contribute to virulence in this pathogen (Dorman *et al.*, 1989). The contribution of *ompB* to

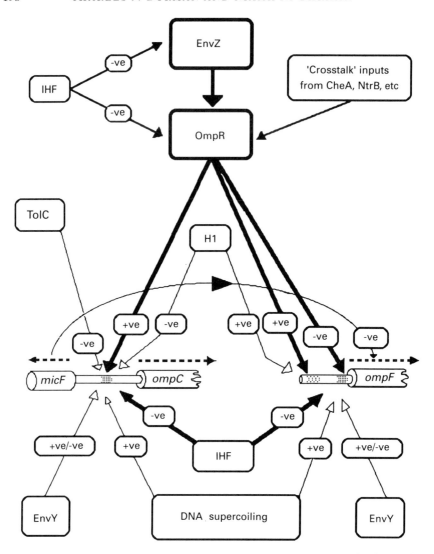

Fig. 2. Multiple and overlapping regulators of *ompC* and *ompF* gene expression in enteric bacteria. The *ompC* and *ompF* structural genes and the *micF* antisense RNA regulatory gene are represented by large horizontal cylinders. Dotted horizontal lines represent the transcripts. Upstream regulatory regions are represented by the narrow horizontal cylinders while stippling represents binding sites within these upstream regions for the OmpR regulatory protein; (....) = low affinity binding sites and (. . .) = high affinity site. Regulatory inputs are shown by the arrows. These are drawn as thick lines where the molecular detail is well understood and thin lines where the situation is less well defined or controversial. The primary osmotic regulators EnvZ and OmpR are indicated (also see Fig. 3), as are phosphorylation inputs from related histidine protein kinases (for review see Stock *et al.*, 1989; Fig. 3). Regulatory contributions from the envelope proteins EnvY (Lundrigan & Earhardt, 1984) and TolC are poorly defined at the molecular level, although the TolC input at *ompF* is believed to be achieved via *micF* (Misra

Salmonella virulence may be dependent upon the route of infection, with *ompB* mutants being attenuated when infecting by the natural, oral route but not when introduced directly to the peritoneal cavity of the host (Dorman *et al.*, 1989; Miller, Kukral & Mekalanos, 1989; Groisman & Saier, 1990). The *ompB* locus has also been found to be required for virulence in *Shigella flexneri*, another enteroinvasive pathogen (Bernardini, Fontaine & Sansonetti, 1990).

Genetic experiments have shown that the *phoP phoQ* operon is required for *S. typhimurium* virulence in mice (Miller *et al.*, 1989*a*). [These genes should not be confused with *phoB phoR* which code for the sequence-specific DNA binding protein PhoB and the histidine protein kinase PhoR, respectively (Fig. 3). The PhoR protein is membrane located and in response to phosphate starvation it phosphorylates the cytoplasmic PhoB protein, thus permitting it to activate transcription of the component genes of the phosphate starvation regulon (*phoA, phoE, phoS* etc) (Makino *et al.*, 1989; Wanner, 1987).] Based on the predicted sequences of the gene products deduced from the nucleotide sequence, the PhoP and PhoQ proteins also belong to the histidine protein kinase / response regulator family (Fig. 3). The PhoQ amino acid sequence predicts that this is a transmembrane protein with a cytoplasmic site for autophosphorylation. The PhoP protein shares significant sequence homology throughout its length with the DNA binding protein, OmpR (Miller *et al.*, 1989*a*). The *phoP* locus was originally identified as a regulator of *phoN*, the gene for acid phosphatase (Kier, Weppelman & Ames, 1979). It has now been shown to regulate additional genes, including *pagC*, which is required for full virulence in *S. typhimurium*. Evidence from protein fusion experiments has shown that PagC is a cell envelope protein (Miller *et al.*, 1989*a*). Mutations in *phoP* render *S. typhimurium* sensitive to antimicrobial cationic peptides called defensins, which are produced by granules within macrophages and neutrophils (Fields, Groisman & Heffron, 1989; Groisman & Saier, 1990). This suggests that the PhoP/PhoQ system positively regulates the expression of unidentified genes concerned with defensin resistance. By analogy with the role of EnvZ/OmpR in transducing osmotic stress signals to the *ompC* and *ompF* promoters, the environmental signals which activate the PhoP/PhoQ-dependent gene *phoN* should also activate the defensin genes and give clues to the environmental signals which activate the defensin resistance pathway *in vivo*. The key factors in *phoN* activation are starvation for

---

Caption for Fig. 2 (*cont.*)

& Reeves, 1987). Effects on *ompC* and *ompF* gene expression due to the nucleoid-associated proteins IHF (Huang *et al.*, 1990; Tsui *et al.*, 1988) and H1 (Graeme-Cook *et al.*, 1989) have been described, as has a role for alterations in DNA supercoiling levels brought about by DNA gyrase-inactivating antibiotic treatments or mutations in genes coding for topoisomerases (Graeme-Cook *et al.*, 1989).

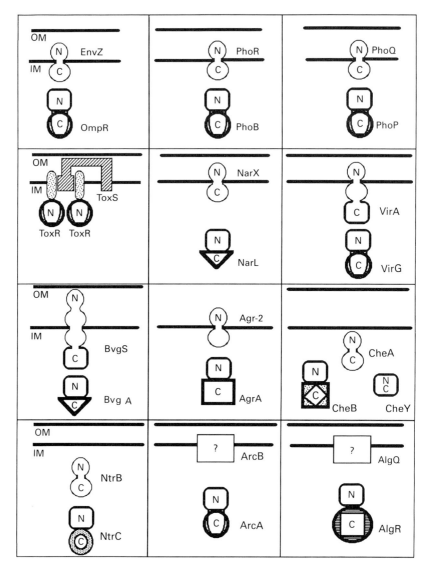

Fig. 3. Bacterial two-component signal transduction systems. The 12 signal transduction systems discussed in the text are illustrated to show their modular nature. Detailed descriptions of each may be found in Stock *et al*. (1989). The Che and Ntr systems are exceptional in that the histidine protein kinase sensors are cytoplasmic rather than membrane associated (the cellular locations of ArcB and AlgQ have not been determined). The Che proteins are also unusual in that DNA interaction does not form part of their control circuit. Of the 12 systems illustrated, Agr is the only one from a Gram-positive organism. The ToxR/ToxS system is included only on the basis of the ToxR N-terminal homology to the conserved C-termini of the OmpR/PhoB/PhoR/VirG/ArcA subgroup of DNA-binding response regulators. The DNA-binding C-terminal domains of NarL and BvgA share homology; those of AgrA, NtrC and AlgR are unique. The most highly conserved region in all of these systems is the N-terminus of the

carbon, nitrogen and phosphorus and growth at low pH (Kier *et al.*, 1979; Miller *et al.*, 1989*a*). By analogy with the role of high osmolarity as a cue for induction of *ompC* and repression of *ompF* transcription as part of adaptation to life in the mammalian gut, these conditions may signal to the cell the need to adapt to the intracellular environment of the macrophage.

Such ecological considerations have prompted assessments of the importance of other regulon control elements in bacterial virulence. The *oxyR* gene codes for a regulator of the oxidative stress response (Christman *et al.*, 1985) and *S. typhimurium* mutants deficient in this function display reduced virulence *in vivo* in some studies (Fields *et al.*, 1986) but not in others (Miller *et al.*, 1989*a*). Adaptation to anaerobic conditions requires a switch to anaerobic respiration (Iuchi & Lin, 1991) and a pleiotropic positive regulator of genes required for this pathway is encoded by *fnr* in *E. coli* and *S. typhimurium*, whose product, Fnr, is homologous to the CRP DNA-binding protein discussed above (Shaw, Rice & Guest, 1983). However, transposon insertion mutations in *fnr* have no effect on mouse virulence in *S. typhimurium* (C. J. Dorman, S. N. Chatfield & G. Dougan, unpublished results). Fnr and OxyR, are not members of the two-component family. However, ArcA/ArcB and NarX/NarL, two other regulators of the aerobic/anaerobic switch, are (Drury & Buxton, 1985; Iuchi, Cameron & Lin, 1989; Nohno, Noji & Saito, 1989; Stewart, Parales & Merkel, 1989; Fig. 3). The possible contributions of these two-component regulators to bacterial virulence has not yet been assessed, although given the highly pleiotropic nature of mutations in *arcA* (Stock *et al.*, 1989), a study of its role in virulence may prove to be rewarding.

DEDICATED VIRULENCE REGULONS

The ToxR-dependent virulence genes of *Vibrio cholerae* constitute a regulon concerned primarily with the pathogenic processes of this bacterium (Miller, Mekalanos & Falkow, 1987). ToxR-dependent genes include the *ctxAB* operon which codes for the A and B subunits of cholera toxin, as well as genes coding for adhesins and outer membrane proteins. Transcription of these genes is modulated by growth phase and by changes in several environmental factors, including aeration, osmolarity, pH, temperature and

---

Caption for Fig. 3 (*cont.*)

response regulators; this is conserved in all except ToxR/ToxS, but this is not a true two-component system (DiRita & Mekalanos, 1991). Among the histidine protein kinase sensors, VirG and BvgA stand out by virtue of possessing additional modules. Each has at its C-terminus a copy of the sequence normally found in the N-termini of response regulators. This sequence may serve to attenuate the message being transmitted to the regulator by mopping up excess signal.

Abbreviations: OM, outer membrane; IM, inner membrane.

the availability of some amino acids (Betley, Miller & Mekalanos, 1986; Miller & Mekalanos, 1988; Parsot & Mekalanos, 1990; Taylor, 1989).

ToxR is encoded by the *toxR* gene which is divergently transcribed from the *V. cholerae htpG* heat shock gene. Expression of *toxR* is thermoregulated and it has been proposed that this is mediated by the *V. cholerae* analogue of $E\sigma^{32}$ RNA polymerase (Parsot & Mekalanos, 1990). ToxR is a 32.5 kD protein with a number of unusual properties. It is a cytoplasmic membrane-located DNA binding protein which recognises and binds to the nucleotide sequence TTTTGAT which is found tandemly repeated upstream of the ToxR-dependent *ctxAB* operon. ToxR may not activate transcription of all of the genes in the ToxR regulon directly but may act via the ToxR-dependent *toxT* gene in some cases. For example, the ToxR-dependent *tcpA*, *tcpI*, *aldA* and *tagA* genes have been shown to be activated in *E. coli* by *toxT* but not by ToxR (DiRita *et al.*, 1991). Thus, the ToxR regulon contains a regulatory hierarchy in which some genes, such as *ctxAB* are directly activated by ToxR while others require an intermediate regulator, ToxT. The cascade may be even more extensive for some genes; for example, the ToxT-dependent *tcpI* gene is itself a regulator of *tcp* gene expression (Taylor, 1989).

Despite speculation, based on amino acid sequence homologies, that ToxR represents a one-component 'two-component' signal transduction system, it is clearly something rather different. The active form of ToxR is thought to be a dimer and another membrane protein, ToxS, appears to be required to ensure efficient dimerization when cellular concentrations of ToxR are at wild-type levels; overproduction of ToxR removes the need for ToxS (Miller, DiRita & Mekalanos, 1989*b*). The *toxS* gene is located downstream of, and forms an operon with, *toxR* (DiRita & Mekalanos, 1991). The N-terminus of ToxR displays significant homology to the DNA-binding region within the C-terminus of OmpR (Miller *et al.*, 1987) and it is this that has prompted speculation that it may be appropriate to classify ToxR as a member of the two-component family of signal transducers (Gross, Aricò & Rappuoli, 1989; Ronson, Nixon & Ausubel, 1987). However, it now appears that, in terms of transmembrane signalling, ToxR is mechanistically closer to eukaryotic membrane receptor protein tyrosine kinases (DiRita & Mekalanos, 1991) with the caveat that the biological activity on the cytoplasmic side is DNA binding rather than phosphotransfer (Fig. 3).

Bordetella pertussis is the causative agent of whooping cough in humans and possesses at least one dedicated virulence gene regulon. This is composed of the genes coding for pertussis toxin, adenylate cyclase, filamentous haemagglutinin, dermonecrotic toxin and haemolysin (Weiss & Hewlett, 1986). Expression of the genes coding for these virulence determinants is under the control of a regulatory locus formerly designated *vir* but now referred to as *bvg* (for *Bordetella* Virulence Gene). The *bvg* locus is required for modulation of virulence gene expression in response to changes in environmental stimuli such as temperature and the concentration of nicotinic acid and $MgSO_4$ (McPheat, Wardlaw & Novotny, 1983; Melton & Weiss, 1989). [Nucleotide sequence analysis of *bvg* originally suggested that it was composed of three regulatory proteins which contained homologies to the conserved domains of classical 'two-component' regulators and led to models of Bvg protein function which proposed that this was a three component 'two component' system (Aricò *et al.*, 1989; Gross *et al.*, 1989). Genetic data from complementation analyses proved to be inconsistent with the three gene model (Stibitz, 1990) and the revised nucleotide sequence of *bvg* shows it to code for just two proteins, albeit with significant homologies to the histidine protein kinase / response regulator family (the nucleotide sequencing data of Stibitz which led to the revision of the Bvg model are discussed in Scarlato *et al.*, 1990).] The nucleotide sequence of *bvg* shows it to code for two proteins with significant homologies to the histidine protein kinase / response regulator family (Fig. 3). One protein, BvgS, is thought to be a transmembrane signal transducer while the other, BvgA, is a cytoplasmic DNA-binding protein which activates transcription of the Bvg regulon genes in response to inputs (presumed to be in the form of phosphotransfer) from BvgS (Scarlato *et al.*, 1990). The regulatory circuitry of the Bvg regulon includes an element of autoregulation; transcription of *bvgA* is enhanced by activated BvgA protein (Scarlato *et al.*, 1990). Apart from self-activation, BvgA has only been shown directly to activate transcription of the closely linked *fha* gene, which codes for filamentous haemagglutinin (Roy, Miller & Falkow, 1989). The complexities of the Bvg regulon are compounded by the phenomenon of phase variation, which is now known to occur via a frameshift mutation which inactivates *bvg*, although whether or not this is a programmed event remains unclear (Stibitz *et al.*, 1989).

The production and export of alginate is an important virulence characteristic of *Pseudomonas aeruginosa* strains which cause opportunistic infections in the lungs of cystic fibrosis sufferers (DeVault *et al.*, 1989). A chromosomal gene, *algD*, appears to be central to the development of the virulent, mucoid phenotype. This gene encodes guanine diphosphate–mannose dehydrogenase and is subject to environmental control through a set of overlapping regulatory circuits. The key environmental factors involved appear to be osmolarity, oxygen tension and nitrogen starvation, with at least osmolarity and nitrogen starvation acting at the level of transcription, the latter via σ^{54}-programmed RNA polymerase (Bayer *et al.*, 1990; Berry, DeVault & Chakrabarty, 1989; DeVault *et al.*, 1989; Kimbara & Chakrabarty, 1989). Genetic analyses have identified a number of regulatory loci which are required for normal control of *algD* transcription. These include the *algR* gene, whose product is a protein with significant homology to the response-regulator members of the two-component system family (Fig. 3). Furthermore, it has been found that *algR* can be replaced functionally by *ompR* in *E. coli* (Deretic *et al.*, 1989). The candidate for the gene coding for the kinase required to modify the *algR* gene product in *Ps. aeruginosa* is currently believed to be *algQ* (Deretic & Konyecsni, 1989; Fig. 3).

The two-component signal transduction model is also in use in bacteria which are pathogenic for plants. The induction of virulence gene expression in *Agrobacterium tumefaciens* in response to plant exudates requires a membrane-located histidine protein kinase, VirA, and a sequence-specific DNA-binding response-regulator protein, VirG, with the complication that VirA also has C-terminal domain homologies normally associated with response regulators (Hooykaas, 1989; Jin *et al.*, 1990*a,b,c*; Winans *et al.*, 1986; Fig. 3). Such an apparent mixing of functions within one polypeptide is also seen in the BvgS membrane protein of *Bordetella pertussis* (Gross *et al.*, 1989; Scarlato *et al.*, 1990). The *A. tumefaciens* genes coding for this signal transduction system, *virA* and *virG*, are located on the Ti (tumour-inducing) virulence plasmid where they are subject to autoregulation in response to plant exudates and in response to changes in pH or phosphate levels through a chromosomal gene, *chvD* (Winans, Kerstettler & Nester, 1988).

Nitrogen starvation is an important trigger for transcriptional activation of several virulence genes in bacterial pathogens (see above). In enteric bacteria, an important part of the response to low nitrogen levels is controlled by the NtrB (NR_{II}) and NtrC (NR_I) proteins. Both are cytoplas-mic proteins and NtrB is a histidine protein kinase which phosphorylates the DNA-binding protein, NtrC, making it proficient for activation of the genes of the Ntr regulon, some of which also require $E\sigma^{54}$ RNA polymerase (for review see Stock *et al.*, 1989; Fig. 3). Together with EnvZ and the chemotactic regulator CheA, NtrB represents one of the prototypic examples of a bacterial histidine protein kinase and was one of the first for which biochemical evidence for phosphotransfer became available (Ninfa & Magasanik, 1986). The extent of the contribution of the NtrB/NtrC system to bacterial virulence is currently unknown. However, the gene coding for σ^{54} (variously known as *ntrA*, *glnF* or *rpoN*) does not contribute to mouse virulence in *S. typhimurium* (Miller *et al.*, 1989).

NETWORKING OF REGULONS

Apparently independent regulons have been found to be coordinately controlled at several levels. The overarching regulatory role of cAMP–CRP has already been alluded to in terms of its ability to link together the control of expression of up to 200 genes (Busby, 1986). This is a pleiotropic effect which involves protein–DNA interactions. Other examples involve protein–protein cross-talking, in which sensor molecules from one signal transduc-tion pathway modify response regulators from another. Biochemical evi-dence is now available for interactions of this type between members of the two-component family, in particular EnvZ/OmpR, NtrB/NtrC, CheA/CheY and CheA/CheB (Ninfa *et al.*, 1988; Stock *et al.*, 1989). Other regulatory factors which cross regulon boundaries include alternative sigma factors which re-programme RNA polymerase to recognize non-standard promoters which are used to express genes in response to environmental changes (Hoopes & McClure, 1987) and small molecules such as guanosine 3'-diphosphate-5'-diphosphate which activate some genes required for survival in difficult circumstances, i.e. members of the stringent response regulon (Cashel & Rudd, 1987).

Proteases can serve pleiotropic functions which make important contributions to the control of virulence or to survival in stressful environments (these two generally go together to some extent). For example, the Lon protease plays an important part in the control of the life cycle of a virus (bacteriophage lambda) (Gottesman *et al.*, 1981), ensures cellular survival by degrading the cell division inhibitor SulA following an SOS response (Mizusawa & Gottesman, 1983) and is also needed for control of expression of colanic acid (also called M antigen) (Torres-Cabassa & Gottesman, 1987) (Fig. 4). Lon-dependent extracellular polysaccharide is an important virulence factor in human infections caused by *Klebsiella pneumoniae* and in plant infections by species of *Erwinia* (Nassif *et al.*, 1989; Roberts & Coleman, 1991). Lon also contributes to the regulation of part of the phosphate starvation response in *E. coli* (Wanner, 1983). Significantly, the *lon* gene is a member of the stringent response regulon and is regulated by guanosine 3'-diphosphate-5'-diphosphate levels (Voellmy & Goldberg, 1980). Lon is also a heat-shock protein, with transcription of the *lon* gene being under the control of the heat-shock sigma factor, RpoH (σ^{32}) (Gayda *et al.*, 1985) (Fig. 4).

The heat shock regulon includes several proteins with important roles in cellular survival in hostile environments and in virulence. Prototypic members include the GroES and GroEL proteins of *E. coli*, detected originally because of their contributions to the life-cycle of bacteriophage lambda, which are now known to be involved in assisting in lambda head assembly (Friedman *et al.*, 1984). These proteins are essential for cell viability in *E. coli* where they fulfil a crucial role as molecular chaperonins (Hemmingsen *et al.*, 1988). The sequences and functions of the GroEL protein have been highly conserved among both prokaryotes and eukaryotes and GroES is well conserved among prokaryotes, with the *Coxiella burnetti* heat shock-inducible HtpA and HtpB Q-fever antigens being homologues of GroES and GroEL, respectively (Hemmingsen *et al.*, 1988; Vodkin & Williams, 1988). The 12 kD BCGa antigen of *Mycobacterium tuberculosis* and *M. bovis* is 45% identical to GroES (cited in Young *et al.*, 1988). The *E. coli* GrpE, DnaJ and DnaK heat-shock proteins were also detected initially because they contribute to lambda replication, in which they assist in the organization of the complicated nucleoprotein initiation

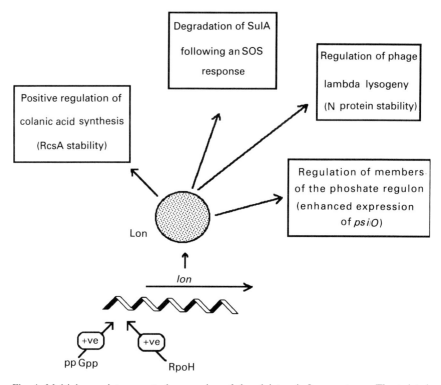

Fig. 4. Multiple regulators control expression of the pleiotropic Lon protease. The twisted ribbon represents *lon* DNA with *lon* mRNA and Lon protein being represented by the horizontal arrow and stippled disc, respectively. Cellular functions subject to Lon-mediated control are boxed. Inputs into the control of *lon* expression from the stringent response (ppGpp) and the heat shock response (RpoH) are also shown.

complex (Georgopulos *et al.*, 1990). The DnaK protein is highly homologous to the Hsp70 family of heat-shock proteins from eukaryotes (Bradwell & Craig, 1987). The 71 kD antigen of *M. tuberculosis*, which is a target for both cellular and humoral immune responses during infection, and the 70 kD antigen of *M. leprae* cross-react with antiserum against the DnaK protein and induced by heat shock (Mehlert & Young, 1989; Young *et al.*, 1988) Homologues of DnaJ have been found in non-enteric prokaryotes, including species of *Mycobacterium* (Lathigra *et al.*, 1988).

A second heat shock regulon exists in enteric bacteria. Its best characterized member, the HtrA (DegP) protein of *E. coli* is a serine protease which degrades abnormally folded periplasmic proteins (Lipinska, Zylicz & Georgopoulos, 1990; Strauch & Beckwith, 1988). The *E. coli htrA* gene is transcribed exclusively by σ^{24}-directed RNA polymerase, a sigma factor with a role in regulating expression of *rpoH*, the gene coding for σ^{34}, the major heat-shock sigma factor (Erickson & Gross 1989; Georgopoulos *et al.*,

Stringent response

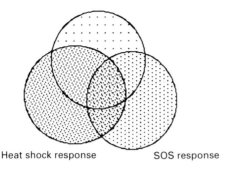

Heat shock response   SOS response

Fig. 5. Overlapping regulons. This simple diagram illustrates the observation that genes may be members of more than one regulon. A detailed treatment of this subject may be found in Van Bogelen *et al.* (1987).

1990). The *htrA* homologue in *S. typhimurium*, has been shown to be required for mouse virulence (Johnson *et al.*, 1991).

The heat-shock protein family may serve a dual purpose in the cell, helping to fold proteins into the correct conformations and assisting in the degradation of proteins which become incorrectly folded. Furthermore, it has been recognized for some time that the heat shock proteins are really stress proteins and that their synthesis may be induced by many factors other than heat shock. Apart from a rapid increase in temperature, a heat shock response in *E. coli* may be elicited by exposure to cadmium chloride, alkylating and methylating agents, ethanol, hydrogen peroxide, nalidixic acid, by starvation for amino acids and by bacteriophage lambda infection (for review see Neidhardt & Van Bogelen, 1987). Some heat shock proteins are also induced by acid treatment (Heyde & Portalier, 1990). Interestingly, several of the treatments listed here (ethanol, phage infection and starvation for amino acids) produce an increase in guanosine 3'-diphosphate-5'-diphosphate levels, a classical feature of the stringent response (Cashel & Rudd, 1987). The cellular target for nalidixic acid is the A subunit of DNA gyrase, indicating a link between the heat shock response, the topoisomerase required to negatively supercoil bacterial DNA (see below) and the SOS response (Van Bogelen, Kelley & Neidhardt, 1987). Phage lambda infection causes a drop in cAMP levels in the cell, implying a link with the cAMP-CRP regulon (Court, Gottesman & Gallo, 1980). The issue of overlaps between stress responses in *E. coli* has been formally addressed by Van Bogelen *et al.* (1987) who found it possible to demonstrate cross induction of some of the proteins of the heat shock (*rpoH*-dependent), oxidative stress (*oxyR*-dependent) and SOS (*lexA*-dependent) responses (Fig. 5).

Taken to its ultimate conclusion, this search for genetic networks can lead to groupings of genes dependent upon factors so pleiotropic that meaningful studies on relationships between the genes become almost impossible. For example, all σ^{70}-dependent promoters in *E. coli* may be regarded as networked since they rely on the same form of RNA polymerase for transcription initiation. However, as this group includes the majority of genes in the cell, the complexity of the group defies study at all but the most general level. At this low level of resolution, it may be more profitable to conduct studies on the genome itself, i.e. the entire complement of cellular genes considered at the level of chromatin structure. This type of study will be reviewed in the next section where it will be seen that such general studies of genes frequently lead to new information about gene regulation in particular cases.

THE BACTERIAL GENOME

Most of the available information about the nature of the bacterial genome comes from studies of one organism, *E. coli*, but the principal features found there appear to obtain in most other prokaryotes. Bacterial genomes consist of the macromolecules carrying the genetic information, i.e. the chromosome (in *E. coli* K-12 this is a covalently closed circular DNA molecule of approximately 4750 kb) together with resident mobile genetic elements and any integrated or autonomous plasmids and bacteriophage (Kohara, Akiyama & Isono, 1987; Krawiec & Riley, 1990). The chromosome is organized structurally at a number of levels, typically being folded into 43 ± 10 topologically independent loops (Sinden & Pettijohn, 1981) with the DNA in the loops being complexed with proteins to form bacterial chromatin (Griffith, 1976; for reviews see Drlica & Riley, 1990).

This nucleoprotein complex is called the nucleoid (Drlica, 1987) and an understanding of its structure and organization is fundamental to any overview of global programmes for the control of gene expression in pathogenic or commensal bacteria. It poses some interesting problems in terms of storage (the 1 mm long chromosome and its associated macromolecules must be fitted into a cell measuring 2 μm long by 1 μm wide), replication (the topological perturbations caused by replication fork passage must be reconciled with the other DNA processes taking place simultaneously) and gene expression (which occurs exclusively at the nucleoid surface). The storage problem is solved by the folding of the chromosome and its organization into chromatin (for review see Kellenberger, 1990). Supercoiling of the DNA also helps to compact it (see below). Disturbances caused by replication may be reconciled with the demands of transcription by avoiding collisions between the fast-moving DNA polymerase and the slower RNA polymerase by aligning the most heavily transcribed genes with the direction taken by DNA polymerase during replication (Brewer, 1988). Since transcription seems to be confined to the lobed and cleft surface of the nucleoid (Drlica, 1987; Hobot *et al.*, 1985), the cell would appear to require a mechanism for retrieving genes stored in the bulk nucleoid, bringing them to the surface for transcription and then returning them.

DNA SUPERCOILING AND ITS BIOLOGICAL CONSEQUENCES

The key topological parameters of supercoiled DNA may be illustrated by a simple example. In Fig. 6 a notional linear B-form DNA molecule of 105 bp is shown. This molecule has ten helical turns with each turn containing 10.5 bp. The ends of this DNA fragment may be ligated to form a circular molecule, also with ten helical turns. This circular DNA experiences no torsional stress (i.e. it is relaxed) and so lies in a flat plane. It has an inherent topological property known as its linking number (Lk) which is determined by the number of times each strand of the duplex crosses the other. Providing that the strands of a DNA molecule are intact, the linking number of the molecule is invariant. Changes to the linking number can only be brought about by strand breakage. Circular molecules which have identical nucleotide sequences but different linking numbers are topological isomers (topoisomers). The enzymes which interconvert topoisomers by transient strand breakage are called topoisomerases (see below). Circularity is not essential for the linking number to be fixed in value; linear molecules arranged as contrained loops (perhaps closed by proteins) may also be described by specific linking values.

The linking number value assigned to the relaxed or ground state of the molecule is Lk^{o}. If two complete turns are removed from the closed circular molecule (for example, due to the action of a topoisomerase), the value of Lk is reduced by 2. The change in Lk is given by:

$$\Delta Lk = Lk - Lk^{o} \tag{1}$$

In this case, $\triangle Lk = -2$.

This change in topological state may be described in another way by considering a further topological parameter of the plasmid, the twist (Tw). Tw describes the intertwining of the two strands of the duplex about their common axis. If the Tw value of the molecule in the figure has been reduced by 2 and if this closed circular molecule is forced to lie in a plane, it is seen that the linking number has changed by an integral value equal to the change in twist. Underwinding the molecule by the equivalent of two turns introduces considerable strain (torsional stress) and so this closed circle is no longer relaxed. As can be seen in Fig. 6, this strain favours strand separation and will assist those reactions of DNA which depend on strand separation in order to proceed.

If the plasmid is not forced to lie in a plane but allowed to find its equilibrium conformation, it will writhe about its helical axis revealing a third topological property, the writhe (Wr). The underwinding of the molecule in Fig. 6 by two turns introduces two compensatory writhing turns

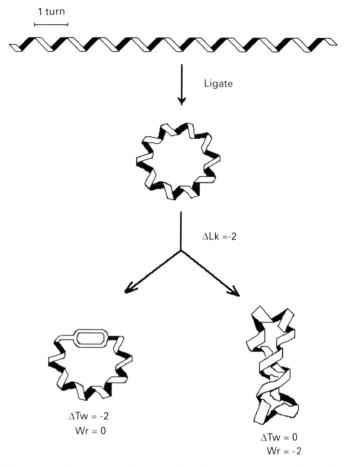

Fig. 6. DNA topological parameters. The topological changes undergone in the simple example described in the text are illustrated. DNA is represented in the diagram as a twisted ribbon and each turn of the ribbon equals one turn of the B-DNA helix. Since each turn is composed of 10.5 bp of DNA, this sequence is 105 bp in length. The DNA is also shown as a ligated closed circle in a relaxed state and as a negatively supercoiled topoisomer differing from its relaxed form by a linking difference (ΔLk) value of -2 (see eqn. 1 in the text). When this underwound species is constrained to lie in a plane as at lower left, it loses two helical turns. When the constraint is removed, the topoisomer converts the melted region to two negative writhing turns (see eqns. 2 and 3 in the text). The parameters Lk (linking number), Tw (twist) and Wr (writhe) are explained in the text.

of negative sign. The relationship of the three topological characters to each other is given by:

$$Lk = Tw + Wr \tag{2}$$

with changes in linking number being given by:

$$\Delta Lk = \Delta Tw + Wr \tag{3}$$

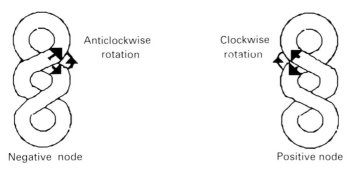

Fig. 7. Rules for node sign assignments. To determine the sign of any node (i.e. crossover point) in DNA, track along the duplex placing arrows pointing away from the exits of nodes. Then determine the shortest distance needed to rotate the arrow on the underpassing section in order to align it with the arrow on the overpassing section. If this rotation involves an anticlockwise turn (as at left in the figure) then the node has a negative sign. A clockwise rotation is indicative of a positive node. The diagram at left shows a duplex with negative nodes. Coincidently, this duplex is also negatively supercoiled. The molecule at right has positive nodes and is positively supercoiled. Note that the sign of supercoiling is a function of the ΔLk value of the topoisomer with respect to its relaxed form (see Fig. 6) and the fact that the node sign convention assigns the same sign to it is coincidental.

In practice, changes in the linking number of closed circular molecules are usually partitioned between twist and writhe. Writhing approximates to the intuitive meaning of supercoiling. Supercoiling is negative when the change in linking number is negative with respect to the relaxed state (i.e. the molecule has been underwound) and positive when the change in linking number is positive with respect to the relaxed state (i.e. when the molecule is overwound) as described in eqn. 1 above.

It should be noted that when one part of the duplex crosses another, signs are given to the resulting nodes according to the following convention: Arrows are drawn at the exits of the node as one tracks along the DNA in a particular direction; then the arrow on the under-crossing segment is aligned with that of the overhead segment by the shortest route; if that route involves a clockwise rotation, then the sign of the node is positive; an anticlockwise rotation results in a negative node (Fig. 7). By comparing Fig. 6 and Fig. 7, it will be seen for negatively supercoiled DNA, the signs on the duplex nodes have negative signs. It must be stressed that the sign describing the supercoiling is a function of the ΔLk; it is simply a useful coincidence that the arbitrary assignments arising from the application of this convention match the description of supercoiling. An introductory account of the mathematical parameters used to describe DNA topology may be found in Cozzarelli, Boles & White (1990).

Bacteria maintain their DNA in an underwound, negatively supercoiled state and the duplex experiences torsional stress when isolated from the cell (Drlica, 1990). The existence of negatively supercoiled DNA was first described by Vinograd *et al.* (1965) and it is now recognized that DNA from most biological sources is in this state. In bacteria, there is now convincing evidence that the DNA experiences torsional strain while still inside the cell, with the tendency to melt which characterizes negatively supercoiled DNA favouring reactions such as transcription, recombination, replication and transposition (Fisher, 1984; Lilley, 1986*a,b*; Smith, 1981). Positively super-coiled DNA is uncommon in nature but is an important by-product of reactions in which protein complexes move between the strands of the duplex, as happens during transcription and DNA replication. It may also be the normal topological state of DNA in some archaebacteria isolated from extreme environments, particularly those characterized by very high tem-peratures (Kikuchi & Asai, 1984). Perhaps the overwinding of the positively supercoiled DNA in these prokaryotes stabilizes it against thermally induced melting.

In bacteria, negative supercoils are put into DNA by an ATP-dependent topoisomerase called DNA gyrase (Gellert *et al.*, 1976). This enzyme introduces supercoils by making transient double-stranded breaks in the DNA duplex reducing the linking number in steps of 2 (Fig. 8). This is refered to as the type II reaction and DNA gyrase is classed as a Type II topoisomerase. DNA gyrase also removes positive supercoils from DNA by a process which is mechanistically identical to that used to introduce negative ones.

Negative supercoils are removed from DNA by topoisomerase I, a type I enzyme (Wang, 1971). The mechanism does not require ATP and changes the linking number of the substrate in steps of 1 by making a transient single-stranded break in DNA and using a 'swivelase' action to rotate the broken strand about the axis provided by its intact complementary strand (Fig. 9). (Cozzarelli, 1980*a,b*; Drlica, 1990; Wang, 1985, 1987). *E. coli* possesses at least two further topoisomerases, a type I enzyme, topoisom-erase III, coded for by the *topB* gene and a Type II enzyme, topoisomerase IV, coded for by the *parC* and *parE* genes. These enzymes relax super-coiled DNA and are thought to be primarily concerned with decatenating the daughter chromosomes at the end of each chromosome replication cycle (DiGate & Marians, 1988, 1989; Kato *et al.*, 1990*b*; Srivenugopal, Lockshon & Morris, 1984).

Supercoiling levels are believed to be homeostatically balanced through the countervailing actions of DNA gyrase and DNA topoisomerase I. In support of this model, mutations in *topA*, the gene coding for topoisomerase I, are compensated by mutations in *gyrA* or *gyrB*, the genes coding for the A and B subunits of DNA gyrase, respectively, or in *toc* (*t*opoisomerase *o*ne *c*ompensation), a genetic locus now known to include the *parC* and *parE* genes which each code for one of the subunits of topoisomerase IV (DiNardo *et al.*, 1982; Dorman *et al.*, 1989; Kato *et al.*, 1990*b*; Pruss, Manes & Drlica, 1982; Raji *et al.*, 1985). Further evidence in favour of homeostatic

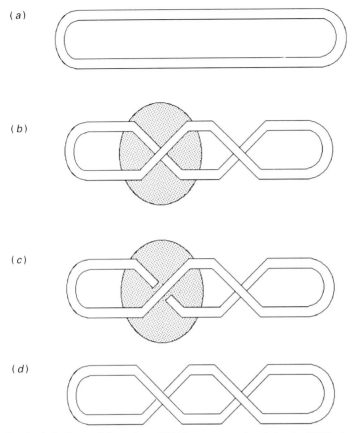

Fig. 8. Topological effects of the action of DNA gyrase on closed circular DNA. A relaxed, closed circular B-DNA molecule is shown (step a). The stippled disc represents DNA gyrase. This introduces a positive node and a compensatory negative node into the DNA (step b). Compare with the molecules illustrated in Fig. 7. The topological event catalyzed by gyrase is mechanistically equivalent to introducing a double-stranded cleavage in the underpassing section of the positive node (step c) with passage of the intact overpassing section through the gap to the back followed by sealing of the double stranded break. DNA gyrase requires ATP to carry out this reaction. This reaction results in a sign reversal at this node, converting what had been a positive node to a negative node (step d). The negative node introduced at step b to compensate for the positive node is retained; thus the linking number of this molecule has been reduced by 2 (i.e. a ΔLk of -2). Linking number changes in steps of 2 are characteristic of Type II topoisomerases (see text).

control of DNA supercoiling comes from regulatory studies on the topo-isomerase genes. The *topA* promoter is induced by increased negative supercoiling (Tse-Dinh & Beran, 1988) while the *gyrA* and *gyrB* promoters are induced by a relaxation of DNA (Menzel & Gellert, 1987).

Measurements of the degree to which DNA is supercoiled in bacterial cells have produced results inconsistent with a situation in which all of the

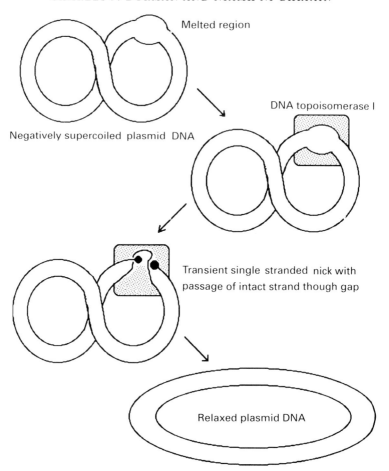

Fig. 9. Topological effects of the action of DNA topoisomerase I on closed circular plasmid DNA. A negatively supercoiled molecule of B-DNA is illustrated. The torsional stress due to negative supercoiling has produced a region of melting in the duplex. Such regions may serve as recognition sites for DNA topoisomerase I. The topoisomerase is represented by the stippled box. It breaks one strand of the duplex and passes the intact strand through the gap. This can be achieved by swivelling one broken end around the axis of the intact strand and then resealing the nick. No external energy is required for any step in this reaction. The product is a relaxed topoisomer with a linking number increased by one with respect to the substrate DNA (i.e. a $\triangle$Lk of + 1). Linking number changes in steps of one are characteristic of Type I topoisomerases (see text).

DNA experiences torsional stress. While it is generally accepted that bacterial DNA is supercoiled *in vivo*, evidence from independent experimental approaches indicates that about half of the supercoils are constrained, probably through interactions with proteins in nucleoprotein complexes analogous to eukaryotic chromatin (Bliska & Cozzarelli, 1987;

Lilley, 1986*a*; Pettijohn, 1982; Pettijohn & Pfenninger, 1980). Much research has been conducted into the nature of bacterial chromatin and the molecular biology of nucleoid-associated proteins. Recently, the roles of these proteins in controlling transcription and other DNA reactions crucial to commensal and pathogenic bacterial life have become apparent.

NUCLEOID-ASSOCIATED PROTEINS AND THE REGULATION OF GENE EXPRESSION

The most abundant member of this class of histone-like proteins is HU, present in about 20 000 copies per genome equivalent in exponentially growing cells. This is a basic heterodimeric protein of 9.5 kDa subunits which binds to and wraps DNA non-specifically. Its subunits are encoded by the unlinked genes *hupA* and *hupB* (Bachmann, 1990). HU homologues are found throughout the eubacterial kingdom, in both Gram-positive and Gram-negative organisms (Drlica & Rouvière-Yaniv, 1987). It is believed that 8 to 10 HU dimers interact with between 275 and 290 bp of DNA to form a nucleosome (Broyles & Pettijohn, 1986; Rouvière-Yaniv, Germond & Yaniv, 1979). Given the level of HU in the cell and the size of the genome, much of the DNA is probably HU-free most of the time. HU does not appear to be confined to the transcriptionally active surface of the nucleoid, as was originally believed (Dürrenberger *et al.*, 1988). Instead, it is evenly distributed throughout the nucleoid (Shellman & Pettijohn, 1991). Genetic studies have shown that HU is required for a normal cell cycle and *in vitro* studies show that it is involved in chromosome replication, transposition and site-specific recombination events (for review see Rouvière-Yaniv *et al.*, 1990).

The integration host factor (IHF) is a close relative of HU which was originally detected as an *E. coli* protein required for Int-promoted bacteriophage lambda integration and excision (Friedman, 1988). IHF is a heterodimeric, sequence-specific DNA-binding protein which introduces bends of up to 140° into the B-DNA helix (Kosturko, Daub & Murialdo, 1989). Its subunits are encoded by the unlinked genes *himA* and *himD* (Bachmann, 1990). IHF levels fluctuate with the growth cycle, peaking at the end of exponential growth. Thus, this protein serves as a link between IHF-dependent reactions and cellular physiology (Bushman *et al.*, 1985). Its regulatory role in some transcription, transposition and recombination reactions is thought to be due to its ability to organize local DNA topology in a manner which lowers the activation energy for these processes (Friedman, 1988). Recently, attention has been drawn to the frequent cooperativity between IHF and promoters transcribed by the σ^{54} RNA polymerase (Collado-Vides, Magasanik & Gralla, 1991; Gralla, 1991).

IHF-dependent processes may be regarded tentatively as members of a regulon. The EnvZ/OmpR-dependent *ompC* and *ompF* genes discussed above require IHF for normal regulation and IHF is a negative effector of *ompB* transcription in growing cells (Huang, Tsui & Freundlich, 1990; Tsui, Helu & Freundlich, 1988; Tsui, Huang & Freundlich, 1991; Fig. 2). Similarly, the site-specific recombination system controlling expression of Type I fimbriae in *E. coli* requires IHF, as does the promoter of the fimbrial subunit gene (Dorman & Higgins, 1987; Eisenstein *et al.*, 1987). Circumstances may be envisaged in which the cell needs to coordinate the expression of its major outer membrane proteins and Type I fimbriae. This is an example of an ability to network expression of two functions which at first appear to be completely unrelated.

The H1 protein (also called HNS) is an important component of the nucleoid in *E. coli* and *S. typhimurium* (reviewed in Higgins, Dorman & Ní Bhriain, 1990). This basic protein occurs in the cell in three isoforms (a, b and c) with the H1a form predominating in stationary phase cells and having the ability to strongly compact DNA into structures resembling nucleosomes (Spassky *et al.*, 1984). The gene coding for H1a is part of the *osmZ* locus and mutations in *osmZ* produce pleiotropic effects. These include derepression of transcription of the osmotically inducible *proU* operon, derepression of the cryptic *bgl* operon, and deregulation of the *ompC* and *ompF* genes (Fig. 2) and of the Type I fimbrial site-specific recombination system (Higgins *et al.*, 1988). Thus, the cell might be regarded as possessing an *osmZ* regulon, with some members (*ompC*, *ompF* and the Type I fimbrial operon) also belonging to the IHF-dependent group. Mutations in the *virR* locus of *Shigella flexneri*, which is the counterpart of the *E. coli osmZ* locus, deregulate expression of the thermally inducible invasion regulon on the *Sh. flexneri* virulence plasmid (Dorman *et al.*, 1990). The *E. coli osmZ* locus is also involved in regulating expression of the thermally controlled Pap pilus operon (Göransson *et al.*, 1990). Thus, this genetic locus plays important roles in controlling functions with parts to play in environmental response in both pathogenic and commensal life.

Other low molecular weight nucleoid-associated proteins include FIS and HLP-I. FIS was originally discovered because of its role in promoting site-specific DNA inversions, hence the name *Factor for Inversion Stimulation*. It cooperates with HU to promote flagellar phase variation in *Salmonella*, binding to an enhancer sequence within the *hin* gene (reviewed in Glasgow, Hughes & Simon, 1989). FIS also assists lambda excision under conditions in which the Xis excisionase function is limiting and it enhances transcription of genes coding for stable RNA by binding to upstream enhancer sequences (Nilsson *et al.*, 1990). Like IHF, FIS is sequence-specific for binding and it bends DNA, a feature which is probably crucial to its biological activity (Thompson & Landy, 1988). HLPI is a 17 kDa histone-like protein and is encoded by the *firA* gene. It is membrane associated and might be important during the period of DNA replication when the chromosome is in contact with the cell envelope (Hirvas *et al.*, 1990; Kusano *et al.*, 1984).

DNA TOPOLOGY AND THE REGULATION OF GENE EXPRESSION: IMPLICATIONS FOR BACTERIAL PATHOGENICITY

Since the negatively supercoiled DNA of bacteria is under torsional stress, this could provide the cell with a regulatory motif for the global control of gene expression. In this model, any fluctuation in DNA supercoiling levels would be sensed instantly by all of the promoters and their proficiency for transcription initiation modulated. This model would be particularly attractive if DNA supercoiling levels were environmentally responsive.

It is now recognized that the responses of bacterial promoters to changes in DNA supercoiling levels are very variable and that each promoter must be studied in its own right. Some are stimulated by elevated supercoiling (e.g. the *topA* and *lac* promoters, Sanzey, 1979; Tse-Dinh & Beran, 1988), others prefer relaxed DNA (e.g. *gyrA*, *gyrB* and *tonB*, Dorman *et al.*, 1988; Menzel & Gellert, 1987) while still others appear to be indifferent to supercoiling fluctuations, at least within the ranges achievable *in vivo* (e.g. the *trp* promoter, Sanzey, 1979). Some promoters display supercoiling sensitivity when on the chromosome but not when carried on plasmids (e.g. the *Salmonella leu500* promoter, Richardson, Higgins & Lilley, 1988); others are sensitive to DNA supercoiling on plasmids but not in their natural context on the bacterial chromosome (e.g. the *tyrT* gene of *E. coli*; Lamond, 1985). Yet others are sensitive both in the natural context and when carried on cloning vectors (e.g. the *S. typhimurium tonB* promoter, Dorman *et al.*, 1988).

A survey of over 60 random Mu d1–8 Lac$^+$ fusions in *S. typhimurium* demonstrated that most showed some fluctuation in expression in response to the DNA gyrase-inhibiting antibiotic coumermycin (Jovanovich & Lebowitz, 1987). While most of the changes in expression detected were mild, the possibilities for resetting the transcriptional profile of the cell in response to supercoiling fluctuations are self-evident. Recent data have shown that changes in environmental parameters, many of which are encountered by bacteria during infections, can alter bacterial DNA supercoiling dramatically. These include changes in osmolarity (Higgins *et al.*, 1988), temperature (Dorman, Ní Bhriain & Higgins, 1990; Goldstein & Drlica, 1984) and the anaerobic switch (Dorman *et al.*, 1988; Yamamoto & Droffner, 1985). DNA supercoiling levels also vary with growth phase (Dorman *et al.*, 1988) and with changes in carbon sources (Balke & Gralla, 1987). Explanations of how changes in the external environment produce changes in the supercoiling of intracellular DNA include the possibility that fluctuations in the [ATP]/[ADP] ratio in response to osmotic or anaerobic stress might modulate the activity of ATP-dependent DNA gyrase (Hsieh, Burger & Drlica, 1991*a*; Hsieh, Rouvière-Yaniv & Drlica, 1991*b*) and the possibility that changes in intracellular levels of monovalent cations might alter DNA supercoiling following osmotic stress (Higgins *et al.*, 1990).

It would be surprising if bacteria did not exploit changing DNA super-coiling levels as part of their regulatory repertoire for environmental adaptation. A corollary to this is the possibility that this control motif may be utilized by pathogens during infections (Dorman, 1991). Mutations affect-ing the expression of virulence genes have been identified in several organisms which also alter DNA-supercoiling levels. These include mu-tations in *virR* in *Sh. flexneri* which alters the thermoregulation of invasion gene expression (Dorman *et al.*, 1990; Maurelli & Sansonetti, 1988), mutations in *topA* in *S. typhimurium* which alter invasion gene expression (Galan & Curtiss, 1990), mutations in the *osmZ* gene of *S. typhimurium* which attenuate the organism *in vivo* (J. Harrison, C. J. Dorman, C. Hormacche & G. Dougan, unpublished data) and mutations in the *ymo* gene of *Yersinia enterocolitica* which alter expression of the *yop* virulence regulon (Cornelis *et al.*, 1991; see Cornelis, this symposium). The *osmZ*, *virR* and *ymoA* loci encode proteins with properties associated with histone-like, DNA-compacting proteins and this probably explains why strains which harbour mutations in these genes display aberrant levels of DNA supercoiling. In all of these bacteria, virulence factor expression is affected by environmental parameters known to alter DNA supercoiling. *Salmonella* invasiveness is modulated by anaerobiosis, growth phase and osmolarity (Galan & Curtiss, 1990; Lee & Falkow, 1990). Invasiveness in *Sh. flexneri* is thermoregulated (Hale, 1991) as is *yop* expression in *Y. enterocolitica* (Cornelis *et al.*, 1989a).

Examples of bacterial genes which respond to changes in DNA supercoil-ing alone are rare. In most cases, this type of regulation appears to form the basal layer of a hierarchy of control. For example, the *ompC* and *ompF* genes of *S. typhimurium* contribute to virulence in that organism and are regulated in response to changes in anaerobiosis, osmolarity, pH and temperature by a combination of the two-component signal transduction system EnvZ/OmpR, the TolC and EnvY envelope proteins, the histone-like DNA-bending protein IHF, antisense RNA (the *micF* transcript), the H1 nucleoid-associated protein and changes in DNA supercoiling (Ander-sen *et al.*, 1989; Chatfield *et al.*, 1991; Graeme-Cook *et al.*, 1989; Heyde & Portalier, 1987; Huang *et al.*, 1990; Lundrigran & Earhart, 1984; Misra & Reeves, 1987; Ní Bhriain, Dorman & Higgins, 1989; Tsui *et al.*, 1988, 1991; Fig. 2). The thermoregulated invasion genes of *Sh. flexneri* are also under the control of a pleiotropic transcriptional activator, VirF (Hale, 1991) which is homologous to members of the AraC family of DNA-binding proteins (Dorman, 1992). Similarly, an AraC homologue (also called VirF) is required to activate the *Y. enterocolitica* Yop regulon in response to increased temperature (Cornelis *et al.*, 1989b; see Cornelis, this sym-posium). Environmental regulation of the *algD* gene of *Ps. aeruginosa* which is required for alginate biosynthesis is also achieved via a two-component regulator (probably AlgQ/AlgR), a histone-like protein

(AlgR3) and changes in DNA supercoiling (DeVault, Kimbara & Chakrabarty, 1990; Kato, Misra & Chakrabarty, 1990*a*).

In the Gram-positive pathogen, *Staphylococcus aureus*, the promoter of the *eta* gene, which codes for epidermolytic toxin A, is regulated by changes in osmolarity and growth phase via the Agr two-component signal transduction system. Mutations in *agr* are highly pleiotropic and define a regulon of genes coding for exoproteins with important roles in virulence (Rescei *et al.*, 1986). There is a strong correlation between *eta* expression and the *in vivo* level of DNA supercoiling (Sheehan *et al.*, 1992). Given the many genera in which this pattern of regulation has now been described, it is likely that the most usual contribution of DNA supercoiling to bacterial gene regulation is in a cooperative role involving more conventional modes of transcriptional control.

PERSPECTIVE

The reductionist philosophy which has served biology so well in the past is now giving way to an integrationist approach in which it is possible to link apparently discrete pieces of information in order to provide a unified picture of the system of interest. This development is best illustrated in studies of bacteria such as *E. coli* and *S. typhimurium* where the wealth of molecular detail available make these the ideal biological systems in which to conduct research into global control mechanisms. The strategies which these and other prokaryotes employ to acquire and transmit environmental information are beginning to become apparent. The discovery of overlaps between different regulons and of the pleiotropic contributions of DNA supercoiling and the bacterial nucleoid-associated proteins has revealed the existence of gene control networks with a great diversity of membership and responsiveness. As the interactions between different parts of the bacterium are explained, it will increasingly become possible to consider these as functioning unicellular organisms rather than as collections of discrete molecular components participating in separate biochemical events. The advantages of this advance in understanding are manifold, extending from the considerable satisfaction to be gained from a much more profound insight into the nature of at least one manifestation of life to an enhanced ability to combat bacterial pathogens through a more complete knowledge of how they work.

ACKNOWLEDGEMENTS

CJD is a Royal Society 1983 University Research Fellow and is also supported by the Agricultural and Food Research Council, the Medical Research Council and the Wellcome Trust.

REFERENCES

Aiba, H., Mizuno, T. & Mizushima, S. (1989). Transfer of phosphoryl group between two regulatory proteins involved in osmoregulatory expression of the *ompF* and *ompC* genes in *Escherichia coli. Journal of Biological Chemistry*, **264**, 8563–7.

Andersen, J., Forst, S. A., Zhao, K., Inouye, M. & Delihas, N. (1989). The function of *micF* RNA. *Journal of Biological Chemistry*, **264**, 17961–70.

Aricò, B., Miller, J. F., Roy, C., Stibitz, S., Monack, D., Falkow, S., Gross, R. & Rappuoli, R. (1989). Sequences required for expression of *Bordetella pertussis* virulence factors share homology with prokaryotic signal transduction proteins. *Proceedings of the National Academy of Sciences, USA*, **86**, 6671–5.

Bachmann, B. J. (1990). Linkage map of *Escherichia coli* K–12, edition 8. *Microbiological Reviews*, **54**, 130–97.

Balke, V. L. & Gralla, J. D. (1987). Changes in linking number of supercoiled DNA accompany growth transitions in *Escherichia coli. Journal of Bacteriology*, **169**, 4499–506.

Bayer, A. S., Eftekhar, F., Tu, J., Nast, C. & Speert, D. P. (1990). Oxygen-dependent up-regulation of mucoid exopolysaccharide (alginate) production in *Pseudomonas aeruginosa. Journal of Bacteriology*, **58**, 1344–9.

Bernardini, M. L., Fontaine, A. & Sansonetti, P. J. (1990). The two-component regulatory system OmpR-EnvZ controls the virulence of *Shigella flexneri. Journal of Bacteriology*, **172**, 6274–81.

Berry, A., DeVault, J. D. & Chakrabarty, A. M. (1989). High osmolarity is a signal for enhanced *algD* transcription in mucoid and nonmucoid *Pseudomonas aeruginosa* strains. *Journal of Bacteriology*, **171**, 2312–17.

Betley, M. J., Miller, V. L. & Mekalanos, J. J. (1986). Genetics of bacterial endotoxins. *Annual Review of Microbiology*, **40**, 577–605.

Bliska, J. B. & Cozzarelli, N. R. (1987). Use of site-specific recombination as a probe of DNA structure and metabolism *in vivo. Journal of Molecular Biology*, **194**, 205–18.

Bradwell, J. C. A. & Craig, E. A. (1987). Eukaryotic M_r 83,000 heat shock protein has a homologue in *Escherichia coli. Proceedings of the National Academy of Sciences, USA*, **84**, 5177–81.

Brewer, B. J. (1988). When polymerases collide: replication and the transcriptional organisation of the *E. coli* chromosome. *Cell*, **53**, 679–686.

Broyles, S. S. & Pettijohn, D. E. (1986). Interaction of the *Escherichia coli* HU protein with DNA: evidence for formation of nucleosome-like structure with circular double-stranded DNA. *Journal of Molecular Biology*, **187**, 47–60.

Busby, S. J. W. (1986). Positive regulation in gene expression. In *Regulation of Gene Expression—25 Years On*, Booth, I. R. & Higgins, C. F., eds, pp. 52–77. Cambridge University Press.

Bushman, W., Thompson, J. F., Vargas, L. & Landy, A. (1985). Control of directionality in lambda site-specific recombination. *Science*, **230**, 906–11.

Cashel, M. & Rudd, K. E. (1987). The stringent response. In Escherichia coli *and* Salmonella typhimurium: *Cellular and Molecular Biology*, Neidhardt, F. C., Ingraham, J. L., Low, K. B., Magasanik, B., Schaechter, M. & Umbarger, H. E., eds, pp. 1410–1438. Washington, DC: American Society for Microbiology.

Chatfield, S. N., Dorman, C. J., Hayward, C. & Dougan, G. (1991). Role of *ompR*-dependent genes in *Salmonella typhimurium* virulence: mutants deficient in both OmpC and OmpF are attenuated *in vivo. Infection and Immunity*, **59**, 449–52.

Christman, M. F., Morgan, R. W., Jacobson, F. S. & Ames, B. N. (1985). Positive control of a regulon for defenses against oxidative stress and some heat shock proteins in *Salmonella typhimurium*. *Cell*, **41**, 753–62.

Collado-Vides, J., Magasanik, B. & Gralla, J. D. (1991). Control site location and transcriptional regulation in *Escherichia coli*. *Microbiological Reviews*, **55**, 371–94.

Comeau, D. E., Ikenaka, K., Tsung, K. & Inouye, M. (1985). Primary characterisation of the protein products of the *Escherichia coli ompB* locus: structure and regulation of synthesis of the OmpR and EnvZ proteins. *Journal of Bacteriology*, **164**, 578–84.

Cornelis, G. R., Biot, T., Lambert de Rouvroit, C., Michiels, T., Mulder, B., Sluiters, C., Sory, M.-P., Van Bouchaute, M. & Vanooteghem, J.-C. (1989*a*). The *Yersinia yop* regulon. *Molecular Microbiology*, **3**, 1455–9.

Cornelis, G. R., Sluiters, C., Delor, I., Geib, D., Kaniga, K., Lambert de Rouvroit, C., Sory, M.-P., Vanooteghem, J.-C. & Michelis, T. (1991). *ymoA*, a *Yersinia enterocolitica* chromosomal gene modulating the expression of virulence functions. *Molecular Microbiology*, **5**, 1023–34.

Cornelis, G., Sluiters, C., Lambert de Rouvroit, C. & Michiels, T. (1989*b*). Homology between VirF, the transcriptional activator of the *Yersinia* virulence regulon, and AraC, the *Escherichia coli* arabinose operon regulator. *Journal of Bacteriology*, **171**, 254–62.

Court, D., Gottesman, M. & Gallo, M. (1980). Bacteriophage lambda *hin* function. I. Pleiotropic alteration in host physiology. *Journal of Molecular Biology*, **138**, 715–29.

Cozzarelli, N. R. (1980*a*). DNA gyrase and the supercoiling of DNA. *Science*, **207**, 953–60.

Cozzarelli, N. R. (1980*b*). DNA topoisomerases. *Cell*, **22**, 327–8.

Cozzarelli, N. R., Boles, T. C. & White, J. H. (1990). Primer on the topology and geometry of DNA supercoiling. In *DNA Topology and its Biological Effects*, Cozzarelli, N. R. & Wang, J. C., eds, pp. 139–184. Cold Spring Harbor Laboratory Press, Cold Spring Harbor, New York.

Curtiss, R., III & Kelly, S. M. (1987). *Salmonella typhimurium* deletion mutants lacking adenylate cyclase and cyclic AMP receptor protein are avirulent and immunogenic. *Infection and Immunity*, **55**, 3035–43.

Deretic, V., Dikshit, R., Konecsni, W. M., Chakrabarty, A. M. & Misra, T. K. (1989). The *algR* gene, which regulates mucoidy in *Pseudomonas aeruginosa*, belongs to a class of environmentally responsive genes. *Journal of Bacteriology*, **171**, 1278–83.

Deretic, V. & Konyecsni, W. M. (1989). Control of mucoidy in *Pseudomonas aeruginosa*: transcriptional regulation of *algR* and identification of the second regulatory gene, *algQ*. *Journal of Bacteriology*, **171**, 3680–8.

DeVault, J. D., Berry, A., Misra, T. K., Darzins, A. & Chakrabarty, A. M. (1989). Environmental sensory signals in microbial pathogenesis: *Pseudomonas aeruginosa* infection in cystic fibrosis. *Biotechnology*, **7**, 352–7.

DeVault, J. D., Kimbara, K. & Chakrabarty, A. M. (1990). Pulmonary dehydration and infection in cystic fibrosis: evidence that ethanol activates gene expression and induction of mucoidy in *Pseudomonas aeruginosa*. *Molecular Microbiology*, **4**, 737–45.

DiGate, R. J. & Marians, K. J. (1988). Identification of a potent decatenating enzyme from *Escherichia coli*. *Journal of Biological Chemistry*, **263**, 13366–73.

DiGate, R. J. & Marians, K. J. (1989). Molecular cloning and DNA sequence analysis of *Escherichia coli topB*, the gene encoding topoisomerase III. *Journal of Biological Chemistry*, **264**, 17924–30.

DiNardo, S., Voekel, K. A., Sternglanz, R., Reynolds, A. E. & Wright, A. (1982). *Escherichia coli* DNA topoisomerase I mutants have compensatory mutations in DNA gyrase genes. *Cell*, **31**, 43–51.

DiRita, V. J. & Mekalanos, J. J. (1991). Periplasmic interaction between two membrane regulatory proteins, ToxR and ToxS, results in signal transduction and transcriptional activation. *Cell*, **64**, 29–37.

DiRita, V. J., Parsot, C., Jander, G. & Mekalanos, J. J. (1991). Regulatory cascade controls virulence in *Vibrio cholerae*. *Proceedings of the National Academy of Sciences, USA*, **88**, 5403–7.

Dorman, C. J. (1991). DNA supercoiling and environmental regulation of gene expression in pathogenic bacteria. *Infection and Immunity*, **59**, 745–9.

Dorman, C. J. (1992). The VirF protein from *Shigella flexneri* is a member of the AraC transcription factor superfamily and is highly homologous to Rns, a positive regulator of virulence genes in enterotoxigenic *Escherichia coli*. *Molecular Microbiology*, **6**, 1575.

Dorman, C. J., Barr, G. C., Ní Bhriain, N. & Higgins, C. F. (1988). DNA supercoiling and the anaerobic and growth phase regulation of *tonB* gene expression. *Journal of Bacteriology*, **170**, 2816–26.

Dorman, C. J., Chatfield, S., Higgins, C. F., Hayward, C. & Dougan, G. (1989). Characterisation of porin and *ompR* mutants of a virulent strain of *Salmonella typhimurium*: *ompR* mutants are attenuated *in vivo*. *Infection and Immunity*, **57**, 2136–40.

Dorman, C. J. & Higgins, C. F. (1987). Fimbrial phase variation in *Escherichia coli*: dependence on integration host factor and homologies with other site-specific recombinases. *Journal of Bacteriology*, **169**, 3840–3.

Dorman, C. J., Lynch, A. S., Ní Bhriain, N. & Higgins C. F. (1989). DNA supercoiling in *Escherichia coli*: *topA* mutations can be suppressed by amplifications involving the *tolC* locus. *Molecular Microbiology*, **3**, 531–40.

Dorman, C. J., Ní Bhriain, N. & Higgins, C. F. (1990). DNA supercoiling and environmental regulation of virulence gene expression in *Shigella flexneri*. *Nature*, **344**, 789–92.

Drlica, K. (1987). The nucleoid. In Escherichia coli *and* Salmonella typhimurium: *Cellular and Molecular Biology*, Neidhardt, F. C., Ingraham, J. L., Low, K. B., Magasanik, B., Schaecter, M. & Umbarger, H. E., eds, pp. 91–103. Washington, DC: American Society for Microbiology.

Drlica, K. (1990). Bacterial topoisomerases and the control of DNA supercoiling. *Trends in Genetics*, **6**, 433–7.

Drlica, K. & Riley, M. (eds) (1990). *The Bacterial Chromosome*, American Society for Microbiology, Washington, DC.

Drlica, K. & Rouvière-Yaniv, J. (1987). Histonelike proteins of bacteria. *Microbiological Reviews*, **51**, 303–19.

Drury, L. S. & Buxton, R. S. (1985). DNA sequence of the *dye* gene of *Escherichia coli* reveals amino acid homology between the *dye* and *ompR* proteins. *Journal of Biological Chemistry*, **260**, 4236–42.

Dürrenberger, M., Bjornst, M.-A., Uetz, T., Hobot, J. & Kellenberger, E. (1988). Intracellular location of the histonelike protein HU in *Escherichia coli*. *Journal of Bacteriology*, **170**, 4757–68.

Eisenstein, B. I., Sweet, D., Vaughn, V. & Friedman, D. I. (1987). Integration host factor is required for the DNA inversion event that controls phase variation in *Escherichia coli*. *Proceedings of the National Academy of Sciences, USA*, **84**, 6506–10.

Erickson, J. W. & Gross, C. A. (1989). Identification of the σ^E subunit of

Escherichia coli RNA polymerase: a second alternative σ factor involved in high-temperature gene expression. *Genes and Development*, **3**, 1462–71.

Fields, P. I., Groisman, E. A. & Heffron, F. (1989). A *Salmonella* locus that controls resistance to microbiocidal proteins from phagocytic cells. *Science*, **243**, 1059–62.

Fields, P. I., Swanson, R. V., Haidaris, C. G. & Heffron, F. (1986). Mutants of *Salmonella typhimurium* that cannot survive within the macrophage are avirulent. *Proceedings of the National Academy of Sciences, USA*, **83**, 5189–93.

Fisher, L. M. (1984). DNA supercoiling and gene expression. *Nature*, **307**, 686–7.

Forst, S., Comeau, D. E., Norioka, S. & Inouye, M. (1987). Localisation and membrane topology of EnvZ, a protein involved in osmoregulation of OmpF and OmpC in *Escherichia coli*. *Journal of Biological Chemistry*, **262**, 16433–8.

Forst, S., Delgado, J. & Inouye, M. (1989). Phosphorylation of OmpR by the osmosensor EnvZ modulates the expression of the *ompF* and *ompC* genes in *Escherichia coli*. *Proceedings of the National Academy of Sciences, USA*, **86**, 6052–6.

Friedman, E. I., Olson, E. R., Georgopoulos, C., Tilly, K., Herskowitz, I. & Banuett, F. (1984). Interactions of bacteriophage and host macromolecules in the growth of bacteriophage λ. *Microbiological Reviews*, **48**, 299–325.

Friedman, D. I. (1988). Integration host factor: a protein for all reasons. *Cell*, **55**, 545–54.

Galan, J. E. & Curtiss III, R. (1990). Expression of *Salmonella typhimurium* genes required for invasion is regulated by changes in DNA supercoiling. *Infection and Immunity*, **58**, 1879–85.

Garrett, S. & Silhavy, T. J. (1987). Isolation of mutations in the alpha operon of *Escherichia coli* that suppress the transcription defect conferred by a mutation in the porin regulatory gene *envZ*. *Journal of Bacteriology*, **169**, 1379–85.

Gayda, R. C., Stephens, P. E., Henwick, R., Schoemaker, J. M., Dreyer, W. J. & Markovitz, A. (1985). Regulatory region of the heat shock-inducible *capR (lon)* gene: DNA and protein sequences. *Journal of Bacteriology*, **162**, 271–5.

Gellert, M., Mizuuchi, M., O'Dea, M. H. & H. Nash. (1976). DNA gyrase: an enzyme that introduces superhelical turns into DNA. *Proceedings of the National Academy of Sciences, USA*, **73**, 3872–6.

Georgopoulos, C., Ang, D., Maddock, A., Raina, S., Lipinska, B. & Zylicz, M. (1990). Heat shock response of *Escherichia coli*. In *The Bacterial Chromosome*, Drlica, K. & Riley, M., eds. pp. 405–419. Washington, D.C.: American Society for Microbiology.

Gibson, M. M., Ellis, E. M., Graeme-Cook, K. A. & Higgins, C. F. (1987). OmpR and EnvZ are pleiotropic regulatory proteins: positive regulation of the tripeptide permease (*tppB*) of *Salmonella typhimurium*. *Molecular and General Genetics*, **207**, 120–9.

Glasgow, A. C., Hughes, K. T. & Simon, M. I. (1989). Bacterial DNA inversion systems. In *Mobile DNA*, Berg, D. E. & Howe, M. M., eds. pp. 637–659. American Society for Microbiology, Washington, DC.

Goldstein, E. & Drlica, K. (1984). Regulation of bacterial DNA supercoiling: plasmid linking numbers vary with growth temperature. *Proceedings of the National Academy of Sciences, USA*, **81**, 4046–50.

Göransson, M., Sonden, B., Nilsson, P., Dagberg, B., Forsman, K., Emanuelsson, K. & Uhlin, B. E. (1990). Transcriptional silencing and thermoregulation of gene expression in *Escherichia coli*. *Nature*, **344**, 682–5.

Gottesman, S., Gottesman, M., Shaw, J. E. & Pearson, M. I. (1981). Protein degradation in *E. coli*: the *lon* mutation and bacteriophage lambda N and CII protein stability. *Cell*, **24**, 225–33.

Graeme-Cook, K. A., May, G., Bremer, E. & Higgins, C. F. (1989). Osmotic regulation of porin expression: a role for DNA supercoiling. *Molecular Microbiology*, **3**, 1287–94.

Gralla, J. D. (1991). Transcriptional control: lessons from an *E. coli* data base. *Cell*, **66**, 415–418.

Griffith, J. D. (1976). Visualisation of prokaryotic DNA in a regularly condensed chromatin-like fibre. *Proceedings of the National Academy of Sciences, USA*, **73**, 563–7.

Groisman, E. A. & Saier, M. H. (1990). *Salmonella* virulence: new clues to intramacrophage survival. *Trends in Biochemistry*, **15**, 30–3.

Gross, R., Aricò, B. & Rappuoli, R. (1989). Families of bacterial signal transducing proteins. *Molecular Microbiology*, **3**, 1661–7.

Hale, L. T. (1991). Genetic basis of virulence in *Shigella* species. *Microbiological Reviews*, **55**, 206–24.

Hemmingsen, S. M., Woolford, C., van der Vies, S. M., Tilly, K., Dennis, D. T., Georgopoulos, C. P., Hendrix, R. W. & Ellis, R. J. (1988). Homologous plant and bacterial proteins chaperone oligomeric protein assembly. *Nature*, **333**, 330–4.

Heyde, M. & Portalier, R. (1987). Regulation of major outer membrane porin proteins of *Escherichia coli* K–12 by pH. *Molecular and General Genetics*, **208**, 511–17.

Heyde, M. & Portalier, R. (1990). Acid shock proteins of *Escherichia coli*. *FEMS Microbiology Letters*, **69**, 19–26.

Higgins, C. F., Dorman, C. J. & Ní Bhriain, N. (1990). Environmental influences on DNA supercoiling: a novel mechanism for the regulation of gene expression. In *The Bacterial Chromosome*, Drlica, K. & Riley, M., eds, pp. 421–432. American Society for Microbiology: Washington, DC.

Higgins, C. F., Dorman, C. J., Stirling, D. A., Waddell, L., Booth, I. R., May, G. & Bremer, E. (1988). A physiological role for DNA supercoiling in the osmotic regulation of gene expression in *S. typhimurium* and *E. coli*. *Cell*, **52**, 569–84.

Higgins, C. F., Hinton, J. C. D., Hulton, C. S. J., Owen-Hughes, T., Pavitt, G. D. & Seirafi, A. (1990). Protein H1: a role for chromatin in the regulation of bacterial gene expression and virulence? *Molecular Microbiology*, **4**, 2007–12.

Hirvas, L., Coleman, J., Koski, P. & Vaara, M. (1990). Bacterial 'histone-like protein I' (HLP-I) is an outer membrane constituent? *FEMS Microbiology Letters*, **262**, 123–6.

Hobot, J. A., Villiger, W., Escaig, J., Maeder, M., Ryter, A. & Kellenberger, E. (1985). Shape and fine structure of nucleoids observed on sections of ultrarapidly frozen and cryosubstituted bacteria. *Journal of Bacteriology*, **162**, 960–71.

Hoopes, B. C. & McClure, W. R. (1987). Strategies in regulation of transcription initiation. In Escherichia coli *and* Salmonella typhimurium: *Cellular and Molecular Biology*, Neidhardt, F. C., Ingraham, J. L., Low, K. B., Magasanik, B., Schaechter, M. & Umbarger, H. E., eds, pp. 1231–1240. American Society for Microbiology, Washington, DC.

Hooykaas, P. J. J. (1989). Tumorigenicity of *Agrobacterium* on plants. In *Genetics of Bacterial Diversity*, Hopwood, D. A. & Chater, K. F. eds, pp. 373–391. Academic Press, London.

Hsieh, L.-S., Burger, R. M. & Drlica, K. (1991*a*). Bacterial DNA supercoiling and [ATP]/[ADP] changes associated with a transition to anaerobic growth. *Journal of Molecular Biology*, **219**, 443–50.

Hsieh, L.-S., Rouvière-Yaniv, J. & Drlica, K. (1991*b*). Bacterial DNA supercoiling and [ATP]/[ADP] ratio: changes associated with salt shock. *Journal of Bacteriology*, **173**, 3914–17.

Huang, L., Tsui, P. & Freundlich, M. (1990). Integration host factor is a negative effector of *in vivo* and *in vitro* expression of *ompC* in *Escherichia coli*. *Journal of Bacteriology*, **172**, 5293–98.

Luchi, S., Cameron, D. C. & Lin, E. C. C. (1989). A second global regulator gene (*arcB*) mediating repression of enzymes in aerobic pathways of *Escherichia coli*. *Journal of Bacteriology*, **171**, 868–73.

Iuchi, S. & Lin, E. C. C. (1991). Adaptation of *Escherichia coli* to respiratory conditions: regulation of gene expression. *Cell*, **66**, 5–7.

Jin, S., Prusti, R. K., Roitsch, T., Ankenbauer, R. G. & Nester, E. W. (1990*a*). Phosphorylation of the VirG protein of *Agrobacterium tumefaciens* by the auto-phosphorylated VirA protein: essential role in biological activity of VirG. *Journal of Bacteriology*, **172**, 4945–50.

Jin, S., Roitsch, T., Ankenbauer, R., Gordon, M. & Nester, E. (1990*b*). The VirA protein of *Agrobacterium tumefaciens* is autophosphorylated and is essential for *vir* gene regulation. *Journal of Bacteriology*, **172**, 525–30.

Jin, S., Roitsch, T., Christie, P. & Nester, E. (1990*c*). The regulatory protein specifically binds to a *cis*-acting regulatory sequence involved in transcriptional activation of *Agrobacterium tumefaciens* virulence genes. *Journal of Bacteriology*, **172**, 531–7.

Yo, Y.-L., Nara, F., Ichikara, S., Mizuno, T. & Mizushima, S. (1986). Purification and characterisation of the OmpR protein, a positive regulator involved in osmoregulatory expression of the *ompF* and *ompC* genes in *Escherichia coli*. *Journal of Biological Chemistry*, **261**, 15252–6.

Johnson, K., Charles, I., Dougan, G., Pickard, D., O'Gaora, P., Costa, G., Ali, T., Miller, I. & Hormaeche, C. (1991). The role of a stress-response protein in *Salmonella typhimurium* virulence. *Molecular Microbiology*, **5**, 401–7.

Jovanovich, S. B. & Lebowitz, J. (1987). Estimation of the effect of coumermycin A_1 on *Salmonella typhimurium* promoters by using random operon fusions. *Journal of Bacteriology*, **169**, 4431–5.

Kato, J., Misra, T. K. & Chakrabarty, A. M. (1990*a*). AlgR3, a protein resembling eukaryotic histone H1, regulates alginate synthesis in *Pseudomonas aeruginosa*. *Proceedings of the National Academy of Sciences, USA*, **87**, 2887–91.

Kato, J.-I., Nishimura, Y., Imamura, R., Niki, H., Hiraga, S. & Suzuki, H. (1990*b*). New topoisomerase essential for chromosome segregation in *E. coli*. *Cell*, **63**, 393–404.

Kellenberger, E. (1990). Intracellular organisation of the bacterial genome. In *The Bacterial Chromosome*, Drlica, K. & Riley, M., eds, pp. 173–186. American Society for Microbiology: Washington, DC.

Kikuchi, A. & Asai, K. (1984). Reverse gyrase—a topoisomerase which introduces positive superhelical turns into DNA. *Nature*, **309**, 677–81.

Kimbara, K. & Chakrabarty, A. M. (1989). Control of alginate synthesis in *Pseudomonas aeruginosa*: regulation of the *algR1* gene. *Biochemical and Biophysical Research Communications*, **164**, 601–8.

Kier, L. D., Weppelman, R. M. & Ames, B. N. (1979). Regulation of nonspecific acid phosphatase in *Salmonella: phoN* and *phoP* genes. *Journal of Bacteriology*, **138**, 155–61.

Kohara, Y., K. Akiyama & K. Isono. (1987). The physical map of the whole *E. coli* chromosome: application of a new strategy for rapid analysis and sorting of a large genomic library. *Cell*, **50**, 495–508.

Kosturko, L. D., Daub, E. & Murialdo, H. (1989). The interaction of *E. coli* integration host factor and λ *cos* DNA: multiple complex formation and protein-induced bending. *Nucleic Acids Research*, **17**, 317–34.

Krawiec, S. & Riley, M. (1990). Organisation of the bacterial chromosome. *Microbiological Reviews*, **54**, 502–39.

Kusano, T., Steinmetz, D., Hendrickson, W. G., Murchie, J., King, M., Benson, A. & Schaechter, M. (1984). Direct evidence for specific binding of the replication origin of the *Escherichia coli* chromosome to the membrane. *Journal of Bacteriology*, **158**, 313–16.

Lamond, A. I. (1985). Supercoiling response of a bacterial tRNA gene. *EMBO Journal*, **4**, 501–7.

Lathigra, R. B., Young, D. B., Sweetser, D. & Young R. A. (1988). A gene from *Mycobacterium tuberculosis* which is homologous to the DnaJ heat shock protein of *E. coli*. *Nucleic Acids Research*, **16**, 1636.

Lee, C. A. & Falkow, S. (1990). The ability of *Salmonella* to enter mammalian cells is affected by bacterial growth state. *Proceedings of the National Academy of Sciences, USA*, **87**, 4304–8.

Liljeström, P. (1986). The EnvZ protein of *Salmonella typhimurium* LT-2 and *Escherichia coli* K-12 is located in the cytoplasmic membrane. *FEMS Microbiology Letters*, **36**, 145–50.

Lilley, D. M. J. (1986*a*). Bacterial chromatin: a new twist to an old story. *Nature*, **320**, 14–15.

Lilley, D. M. J. (1986*b*). The genetic control of DNA supercoiling, and vice versa. In *Regulation of Gene Expression—25 Years On*, Booth, I. R. & Higgins, C. F., eds, pp. 105–126. Cambridge University Press.

Lipinska, B., Zylicz, M. & Georgopoulos, C. (1990). The HtrA (DegP) protein, essential for *Escherichia coli* survival at high temperatures is an endopeptidase. *Journal of Bacteriology*, **172**, 1791–7.

Lundrigran, M. D. & Earhart, C. F. (1984). Gene *envY* of *Escherichia coli* K-12 affects thermoregulation of major porin expression. *Journal of Bacteriology*, **157**, 262–8.

Makino, K., Shinagawa, H., Amemura, M., Kawamoto, T., Yamada, M. & Nakata, A. (1989). Signal transduction in the phosphate regulon of *Escherichia coli* involves phosphotransfer between PhoR and PhoB proteins. *Journal of Molecular Biology*, **210**, 551–9.

Matsuyama, S. & Mizushima, S. (1987). Novel *rpoA* mutation that interferes with the function of OmpR and EnvZ, positive regulators of the *ompF* and *ompC* genes that code for outer membrane proteins in *Escherichia coli* K-12. *Journal of Molecular Biology*, **195**, 847–53.

Maurelli, A. T. & Sansonetti, P. J. (1988). Identification of a chromosomal gene controlling temperature-regulated expression of *Shigella* virulence. *Proceedings of the National Academy of Sciences, USA*, **85**, 2820–4.

McPheat, W. L., Wardlaw, A. C. & Novotny, P. (1983). Modulation of *Bordetella pertussis* by nicotinic acid. *Infection and Immunity*, **41**, 516–22.

Mehlert, A. & Young, D. B. (1989). Biochemical and antigenic characterisation of the *Mycobacterium tuberculosis* 71 kD antigen, a member of the 70 kD heat shock protein family. *Molecular Microbiology*, **3**, 125–30.

Melton, A. R. & Weiss, A. A. (1989). Environmental regulation of expression of virulence determinants in *Bordetella pertussis*. *Journal of Bacteriology*, **171**, 6206–12.

Menzel, R. & Gellert, M. (1987). Fusions of the *Escherichia coli gyrA* and *gyrB* control regions to the galactokinase gene are inducible by coumermycin treatment. *Journal of Bacteriology*, **169**, 1272–8.

Miller, J. F., Mekalanos, J. J. & Falkow, S. (1987). Coordinate regulation and sensory transduction in the control of bacterial virulence. *Science*, **243**, 916–22.

Miller, S. I., Kukral, A. M. & Mekalanos, J. J. (1989a). A two-component regulatory system (*phoP phoQ*) controls *Salmonella typhimurium* virulence. *Proceedings of the National Academy of Sciences, USA*, **86**, 5054–8.

Miller, V. L., DiRita, V. J. & Mekalanos, J. J. (1989b). Identification of *toxS*, a regulatory gene whose product enhances ToxR-mediated activation of the cholera toxin promoter. *Journal of Bacteriology*, **171**, 1288–93.

Miller, V. L. & Mekalanos, J. J. (1988). A novel suicide vector and its use in construction of insertion mutations: osmoregulation of outer membrane proteins and virulence determinants in *Vibrio cholerae* requires *toxR*. *Journal of Bacteriology*, **170**, 2575–83.

Miller, V. L., Taylor, R. K. & Mekalanos, J. J. (1987). Cholera toxin transcriptional activator ToxR is a transmembrane DNA binding protein. *Cell*, **48**, 271–9.

Misra, R. & Reeves, P. R. (1987). Role of *micF* in the *tolC*-mediated regulation of OmpF, a major outer membrane protein of *Escherichia coli* K-12. *Journal of Bacteriology*, **169**, 4722–30.

Mizusawa, S. & Gottesman, S. (1983). Protein degradation in *Escherichia coli*: the *lon* gene controls the stability of *sulA* protein. *Proceedings of the National Academy of Sciences, USA*, **80**, 358–62.

Nassif, X., Honoré, N., Vasselon, T., Cole, S. T. & Sansonetti, P. J. (1989). Positive control of colanic acid synthesis in *Escherichia coli* by *rmpA* and *rmpB*, two virulence-plasmid genes of *Klebsiella pneumoniae*. *Molecular Microbiology*, **3**, 1349–59.

Neidhardt, F. C. & Van Bogelen, R. A. (1987). Heat shock response. In Escherichia coli *and* Salmonella typhimurium: *Cellular and Molecular Biology*, Neidhardt, F. C., Ingraham, J. L., Low, K. B., Magasanik, B., Schaechter, M. & Umbarger, H. E., eds, pp. 1334–1345. Washington: American Society for Microbiology.

Ní Bhriain, N., Dorman, C. J. & Higgins, C. F. (1989). An overlap between osmotic and anaerobic stress responses: a potential role for DNA supercoiling in the coordinate regulation of gene expression. *Molecular Microbiology*, **3**, 933–42.

Nikaido, H. & Vaara, M. (1987). Outer membrane. In Escherichia coli *and* Salmonella typhimurium: *Cellular and Molecular Biology*, Neidhardt, F. C., Ingraham, J. L., Low, K. B., Magasanik, B., Schaechter, M. and Umbarger, H. E., eds, 7–22. American Society for Microbiology: Washington, DC.

Nilsson, L., Vanet, A., Vijgenboom, E. & Bosch, L. (1990). The role of FIS in *trans* activation of stable RNA operons of *E. coli*. *EMBO Journal*, **9**, 727–34.

Ninfa, A. J. & Magasanik, B. (1986). Covalent modification of the *glnG* product, NR_I, by the *glnL* product, NR_{II}, regulates the transcription of the *glnALG* operon in *Escherichia coli*. *Proceedings of the National Academy of Sciences, USA*, **83**, 5909–13.

Ninfa, A. J., Ninfa, E. G., Lupas, A., Stock, A., Magasanik, B. & Stock, J. (1988). Crosstalk between bacterial chemotaxis signal transduction proteins and the regulators of transcription of the Ntr regulon: evidence that nitrogen assimilation and chemotaxis are controlled by a common phosphotransfer mechanism. *Proceedings of the National Academy of Sciences, USA*, **85**, 5492–6.

Nohno, T., Noji, S. & Saito, T. (1989). The *narX* and *narL* genes encoding the nitrate-sensing regulators of *Escherichia coli* are homologous to a family of prokaryotic two-component regulatory genes. *Nucleic Acids Research*, **17**, 2947–57.

Norioka, S., Ramakrishnan, G., Ikenaka, K. & Inouye, M. (1986). Interaction of a transcriptional activator, OmpR, with reciprocally osmoregulated genes, *ompF* and *ompC*, of *Escherichia coli*. *Journal of Biological Chemistry*, **261**, 17113–19.

Parsot, C. & Mekalanos, J. J. (1990). Expression of ToxR, the transcriptional activator of the virulence factors in *Vibrio cholerae*, is modulated by the heat shock response. *Proceedings of the National Academy of Sciences, USA*, **87**, 9898–902.

Pettijohn, D. E. (1982). Structure and properties of the bacterial nucleoid. *Cell*, **30**, 667–9.

Pettijohn, D. E. & Pfenninger, O. (1980). Supercoils in prokaryotic DNA restrained *in vivo*. *Proceedings of the National Academy of Sciences, USA*, **77**, 1331–5.

Pruss, G. J., Manes, S. H. & Drlica, K. (1982). *Escherichia coli* DNA topoisomerase mutants: increased supercoiling is corrected by mutations near gyrase genes. *Cell*, **31**, 35–42.

Raji, A., Zabel, D. J., Laufer, C. S. & Depew, R. E. (1985). Genetic analysis of mutations that compensate for loss of *Escherichia coli* DNA topoisomerase I. *Journal of Bacteriology*, **162**, 1173–9.

Rescei, P., Kreiswirth, B., O'Reilly, M., Schlievert, P., Gruss, A. & Novick, R. (1986). Regulation of exoprotein gene expression in *Staphylococcus aureus* by *agr*. *Molecular and General Genetics*, **202**, 58–61.

Richardson, S. M. H., Higgins, C. F. & Lilley, D. M. J. (1988). DNA supercoiling and the *leu500* promoter mutation of *Salmonella typhimurium*. *EMBO Journal*, **7**, 1863–9.

Roberts, I. S. & Coleman, M. J. (1991). The virulence of *Erwinia amylovora*: molecular genetic perspectives. *Journal of General Microbiology*, **137**, 1453–7.

Ronson, C. W., Nixon, B. T. & Ausubel, F. M. (1987). Conserved domains in bacterial regulatory proteins that respond to environmental stimuli. *Cell*, **49**, 579–81.

Rouvière-Yaniv, J., Bonnefoy, E., Huisman, O. & Almeida, A. (1990). Regulation of HU protein synthesis in *Escherichia coli*. In *The Bacterial Chromosome*, Drlica, K. & Riley, M., eds. pp. 247–257. American Society for Microbiology, Washington, DC.

Rouvière-Yaniv, J., Germond, J. & Yaniv, M. (1979). *E. coli* DNA binding protein HU forms nucleosome-like structures with circular double-stranded DNA. *Cell*, **17**, 265–74.

Roy, C. R., Miller, J. F. & Falkow, S. (1989). The *bvgA* gene of *Bordetella pertussis* encodes a transcriptional activator required for coordinate regulation of several virulence genes. *Journal of Bacteriology*, **171**, 6338–44.

Sanderson, K. E. & Hurley, J. A. (1987). Linkage map of *Salmonella typhimurium*. In Escherichia coli *and* Salmonella typhimurium: *Cellular and Molecular Biology*, Neidhardt, F. C., Ingraham, J. L., Low, K. B., Magasanik, B., Schaecter, M. & Umbarger, H. E., eds, pp. 877–918. American Society for Microbiology: Washington, DC.

Sanzey, B. (1979). Modulation of gene expression by drugs affecting deoxyribonucleic acid gyrase. *Journal of Bacteriology*, **138**, 40–7.

Scarlato, V., Prugnola, A., Aricó, B. & Rappuoli, R. (1990). Positive transcriptional feedback at the *bvg* locus controls expression of virulence factors in *Bordetella pertussis*. *Proceedings of the National Academy of Sciences, USA*, **87**, 6753–7.

Schwartz, M. (1987). The maltose regulon. In Escherichia coli *and* Salmonella typhimurium: *Cellular and Molecular Biology*, Neidhardt, F. C., Ingraham, J. L., Low, K. B., Magasanik, B., Schaecter, M. & Umbarger, H. E., eds, pp. 91–103. American Society for Microbiology: Washington, DC.

Shaw, D. J., Rice, D. W. & Guest, J. R. (1983). Homology between CAP and Fnr, a regulator of anaerobic respiration in *Escherichia coli*. *Journal of Molecular Biology*, **166**, 241–7.

Sheehan B. J., Foster, T. J., Dorman, C. J., Park, S. & Stewart, G. S. A. B. (1992).

Osmotic and growth-phase dependent regulation of the *eta* gene of *Staphylococcus aureus*: a role for DNA supercoiling. *Molecular and General Genetics*, **232**, 49–57.

Shellman, V. L. & Pettijohn, D. E. (1991). Introduction of proteins into living bacterial cells: distribution of labelled HU protein in *Escherichia coli*. *Journal of Bacteriology*, **173**, 3047–59.

Sinden, R. R. & Pettijohn, D. E. (1981). Chromosomes in living *Escherichia coli* cells are segregated into domains of supercoiling. *Proceedings of the National Academy of Sciences, USA*, **78**, 224–8.

Slauch, J. M. & Silhavy, T. J. (1991). *cis*-acting *ompF* mutations that result in OmpR-dependent constitutive expression. *Journal of Bacteriology*, **173**, 4039–48.

Smith, G. R. (1981). DNA supercoiling: another level for regulating gene expression. *Cell*, **24**, 599–600.

Smith, H. (1990). Pathogenicity and the microbe *in vivo*. *Journal of General Microbiology*, **136**, 377–93.

Spassky, A., Rimsky, S., Garreau, H. & Buc, H. (1984). H1a, an *E. coli* DNA-binding protein which accumulates in stationary phase, strongly compacts DNA *in vitro*. *Nucleic Acids Research*, **12**, 5321–40.

Srivenugopal, K. S., Lockshon, D. & Morris, D. R. (1984). *Escherichia coli* DNA topoisomerase III: purification and characterisation of a new type I enzyme. *Biochemistry*, **23**, 1899–906.

Stewart, V., Parales Jr., J. & Merkel, S. M. (1989). Structure of genes *narL* and *narX* of the *nar* (nitrate reductase) locus in *Escherichia coli* K-12. *Journal of Bacteriology*, **171**, 2229–34.

Stibitz, S. (1990). Genetic studies on the *vir* locus of *Bordetella pertussis*. In *Bacterial Protein Toxins*, Rappouli, R., Alouf, J. E., Falmagne, P., Fehrenbach, F. J., Freer, J., Gross, R., Jeljaszewicz, J., Montecucco, C., Tomasi, M., Wadström, T. & Witholt, B., eds, pp. 363–364. Gustav Fischer Verlag, Stuttgart.

Stibitz, S., Aaronson, W., Monack, D. & Falkow, S. (1989). Phase variation in *Bordetella pertussis* by a frameshift mutation in a gene for a novel two-component system. *Nature*, **338**, 266–9.

Stock, J. B., Ninfa, A. J. & Stock, A. M. (1989). Protein phosphorylation and regulation of adaptive responses in bacteria. *Microbiological Reviews*, **53**, 450–90.

Strauch, K. L. & Beckwith, J. (1988). An *Escherichia coli* mutation preventing degradation of abnormal periplasmic proteins. *Proceedings of the National Academy of Sciences, USA*, **85**, 1576–80.

Taylor, R. K. (1989). Genetic studies of enterotoxin and other potential virulence factors of *Vibrio cholerae*. In *Genetics of Bacterial Diversity*, Hopwood, D. A. & Chater, K. F., eds, pp. 309–329. Academic Press, London.

Thompson, J. F. & Landy, A. (1988). Empirical estimation of protein-induced bending angles: application to λ site-specific recombination complexes. *Nucleic Acids Research*, **16**, 9687–705.

Torres-Cabassa, A. S. & Gottesman, S. (1987). Capsule synthesis in *Escherichia coli* K-12 is regulated by proteolysis. *Journal of Bacteriology*, **169**, 981–9.

Tse-Dinh, Y.-C. & Beran, R. K. (1988). Multiple promoters for transcription of the *Escherichia coli* DNA topoisomerase I gene and their regulation by DNA supercoiling. *Journal of Molecular Biology*, **202**, 735–42.

Tsui, P., Helu, V. & Freundlich, M. (1988). Altered osmoregulation of *ompF* in integration host factor mutants of *Escherichia coli*. *Journal of Bacteriology*, **170**, 4950–3.

Tsui, P., Huang, L. & Freundlich, M. (1991). Integration host factor binds specifically to multiple sites in the *ompB* promoter of *Escherichia coli* and inhibits transcription. *Journal of Bacteriology*, **173**, 5800–7.

Van Bogelen, R. A., Kelley, P. M. & Neidhardt, F. C. (1987). Differential induction of heat shock, SOS, and oxidative stress regulons and accumulation of nucleotides in *Escherichia coli, Journal of Bacteriology*, **169**, 26–32.

Vinograd, J., Lebowitz, J., Radloff, R., Watson, R. & Laipis, P. (1965). The twisted circular form of polyoma viral DNA. *Proceedings of the National Academy of Sciences, USA*, **53**, 1104–11.

Vodkin, M. H. & Williams, J. C. (1988). A heat shock operon in *Coxiella burnetii* produces a major antigen homologous to a protein in both mycobacteria and *Escherichia coli. Journal of Bacteriology*, **170**, 1227–34.

Voellmy, R. & Goldberg, A. L. (1980). Guanosine 5'-diphosphate-3'-diphosphate (ppGpp) and the regulation of protein breakdown in *Escherichia coli. Journal of Biological Chemistry*, **255**, 1008–14.

Wang, J. C. (1971). Interaction between DNA and an *Escherichia coli* protein. *Journal of Molecular Biology*, **55**, 523–33.

Wang, J. C. (1985). DNA topoisomerases. *Annual Review of Biochemistry*, **54**, 665–97.

Wang, J. C. (1987). Recent studies of DNA topoisomerases. *Biochimica Biophysica Acta*, **909**, 1–9.

Wanner, B. L. (1983). Overlapping and separate controls on the phosphate regulon in *Escherichia coli. Journal of Molecular Biology*, **166**, 283–308.

Wanner, B. L. (1987). Phosphate regulation of gene expression in *Escherichia coli*. In Escherichia coli *and* Salmonella typhimurium: *Cellular and Molecular Biology*, Neidhardt, F. C., Ingraham, J. L., Low, K. B., Magasanik, B., Schaechter, M. & Umbarger, H. E., eds, pp. 1326–1333. American Society for Microbiology, Washington, DC.

Weiss, A. A. & Hewlett, E. L. (1986). Virulence factors of *Bordetella pertussis*. *Annual Review of Microbiology*, **40**, 661–86.

Winans, S. C., Ebert, P. R., Stachel, S. E., Gordon, M. P. & Nester, E. W. (1986). A gene essential for *Agrobacterium* virulence is homologous to a family of positive regulatory loci. *Proceedings of the National Academy of Sciences, USA*, **83**, 8278–82.

Winans, S. C., Kerstettler, R. A. & Nester, E. W. (1988). Transcriptional regulation of the *virA* and *virG* genes of *Agrobacterium tumefaciens. Journal of Bacteriology*, **170**, 4047–54.

Yamamoto, N. & Droffner, M. L. (1985). Mechanisms determining aerobic or anaerobic growth in the facultative anaerobe *Salmonella typhimurium. Proceedings of the National Academy of Sciences, USA*, **82**, 2077–81.

Young, D. B., Mehlert, A., Bal, V., Mendez-Samperio, P., Ivanyi, J. & Lamb, J. R. (1988). Stress proteins and the immune response to mycobacteria: antigens as virulence factors? *Antonie van Leeuwenhoek, International Journal of General and Molecular Microbiology*, **54**, 431–9.

YERSINIAE, FINELY TUNED PATHOGENS

GUY R. CORNELIS

Microbial Pathogenesis Unit
International Institute of Cellular and Molecular Pathology (ICP)
and Faculté de Médecine, Université Catholique de Louvain, UCL
54.90, B–1200 Brussels, Belgium

INTRODUCTION

Among the many bacterial species that recycle organic matter, a very small number are adapted to multiply at the expense of a living host. This lifestyle drastically simplifies the problem of nutrient acquisition but it poses the problem of escaping from the immune system. To succeed in this, bacterial pathogens use a number of specific functions which they have evolved or borrowed from other organisms, possibly from the host itself. A bacterial infection is a sequential process in which the infecting organism will encounter different environmental conditions and predators. To survive, the invading bacteria must thus express their virulence functions at the appropriate time. Hence, successful pathogens not only acquire specific virulence attributes but they also subject these to a regulatory network connected to specific environmental sensors to provide adequate expression. Temperature is one of the environmental signals commonly used by the pathogens. Yersiniae offer a particularly complex example of a virulence regulatory network partially controlled by temperature. After a brief description of the genus *Yersinia*, virulence functions and their genetic determinants will be reviewed and, finally, regulation of their expression will be discussed, in particular recent knowledge in the field of regulation by temperature and Ca^{2+}. Some virulence determinants such as the iron chelating system will not be dealt with. For a complete overview of yersiniae, the reader should refer to broader reviews (for instance, Butler, 1983; Cornelis *et al.*, 1987*a*).

THE GENUS *YERSINIA*

Three *Yersinia* species are pathogenic for humans and rodents. *Yersinia pestis* is the agent of plague which is an essentially zoonotic infection transmitted among urban and sylvatic rodents. After inoculation through flea bites or by inhalation, *Y. pestis* invades the lymph node corresponding to the inoculation point, converts it into the typical bubo and then colonizes deep organs via the blood stream. In the particularly severe pneumonic form, *Y. pestis* becomes airborne transmitted and it can then cause death of healthy individuals within three days (for review see Butler, 1983). Like *Y. pestis*, *Y. pseudotuberculosis* is also essentially a rodent pathogen. After oral inoculation, it causes diarrhoea, emaciation and death as a result of septicaemia. *Y. enterocolitica* is a common human pathogen which causes gastrointestinal syndromes of various severities, ranging from mild self-limited diarrhoea to mesenteric adenitis resembling appendicitis. Systemic involvement is unusual with *Y. enterocolitica* but reactive arthritis and erythema nodosum are common complications (Cover & Aber, 1989). Transmission of *Y. enterocolitica* generally occurs by consumption of contaminated food or water (Bottone, 1981; Cover & Aber, 1989) but several cases of septic shock due to transfusion of blood contaminated with *Y. enterocolitica* have recently been reported (Tipple *et al.*, 1990). In Europe, pork meat is a current contamination source of *Y. enterocolitica* (Tauxe *et al.*, 1987). Although not within the scope of this review, it is worth mentioning that only a few, well defined, serotypes of *Y. enterocolitica* are pathogenic. Also, by some cruel stroke of fate, the pathogenic strains isolated in the United States (serotypes O:4.32, O:8, O:13, O:18, O:20, O:21...) are markedly more virulent than the strains isolated anywhere else in the world (serotypes O:3, O:9, O:5.27) (for review see Cornelis *et al.*, 1987a).

Although the three yersiniae infect their host via different routes and cause diseases of very different severity, they share a common tropism for lymphoid tissue and a remarkable ability to resist the non-specific immune response of the host. According to recent research, their main strategy seems to consist of avoiding lysis by complement and avoiding phagocytosis by polymorphonuclear leucocytes or macrophages and of forming extra-cellular microcolonies in the infected tissues (Lian, Hwang & Pai, 1987; Hanski *et al.*, 1989, 1991; Simonet, Richard & Berche, 1990). As will appear from this review, the three yersiniae have common basic virulence functions. For the sake of clarity, attention will be essentially focused on *Y. enterocolitica* and the differences with the other species will be mentioned throughout.

THE CHROMOSOME-ENCODED THERMOREGULATED VIRULENCE FUNCTIONS OF *Y. ENTEROCOLITICA*

The enterotoxin Yst

The chromosome of *Y. enterocolitica* encodes a heat-stable enterotoxin, Yst, detectable in broth culture supernatant by the infant mouse test (Pai & Mors, 1978). It is a 30 amino acid peptide (Takao, Tominaga & Shimonishi, 1984) which has the same mode of action as the heat-stable toxin ST_a (also called STI) of *Escherichia coli* (Robins-Browne *et al.*, 1979). According to the nucleic acid sequence, the enterotoxin Yst is synthesized as a 70 amino acid polypeptide (Delor *et al.*, 1990). The C-terminal 30 amino acids correspond to the toxin extracted from culture supernatants. The N-terminal 18 amino acids have the properties of a signal sequence. The central 22 residues are removed during or after the secretion process. This organization in three domains (Pre, Pro and mature Yst) also resembles that of the enterotoxin ST_a of *E. coli* but Yst is much larger than ST_a. The degree of conservation between the *E. coli* and *Y. enterocolitica* toxins is much higher in the mature proteins than in the Pre and the Pro domains. In particular, the active site appears to be highly conserved in both enterotoxins (Takao *et al.*, 1985). Interestingly, only the *Y. enterocolitica* strains that belong to the pathogenic serotypes possess the *yst* gene (Delor *et al.*, 1990). While fresh isolates generally express Yst, many strains in collections no longer produce the enterotoxin (Delor *et al.*, 1990). It has been observed that one such Yst-negative strain resumed its production of Yst after mutation in a gene called *ymoA* and encoding a histone like protein (see below) (Cornelis *et al.*, 1991). Hence, the loss of production of the enterotoxin results not from mutations but rather from some silencing process (I. Delor & G. P. Cornelis, unpublished data). In view of the close resemblance between Yst and ST_a and of the association of this character with pathogenic serotypes, it is tempting to speculate that the production of Yst is responsible for the diarrhoeal manifestation associated with yersiniosis. However, one factor argues against this hypothesis: Yst is only detected in supernatants of cultures incubated at temperatures less than 30°C.

The invasins

Two chromosomal genes, *inv* (*inv*asion) and *ail* (*a*ccessory *i*nvasion *l*ocus) have been associated with the ability of *Y. enterocolitica* to enter various lines of cultured human epithelial cells (Miller & Falkow, 1988). When either gene is introduced into a laboratory *E.coli* strain, it confers the invasive phenotype on this normally non-invasive organism. The *inv* gene encodes a 92 kD outer membrane protein (Young *et al.*, 1990) that binds to mammalian cell receptors of the integrin family (Isberg & Leong, 1990). Gene *ail* encodes a 17 kD invasin that is also membrane associated (Miller, Bliska & Falkow, 1990). Only pathogenic strains of *Y. enterocolitica* have functional *ail* (Miller *et al.*, 1989) and *inv* loci (Pierson & Falkow, 1990). This observation suggests a role for the invasins in pathogenicity but this role has not yet been formally demonstrated. The invasins could be involved in the translocation of the bacterium through the intestinal epithelium, but histopathological analysis of infected mouse ileum suggests that crossing the epithelium occurs through the phagocytic M cells rather than through the normally non-phagocytic enterocytes (Grützkau *et al.*, 1990). *Y. pseudotuberculosis* also produces the invasin Inv (Isberg, Voorhis & Falkow, 1987) and the inactivation of the *inv* gene results in a slower rate of infection after oral challenge (Rosqvist, Bölin & Wolf-Watz, 1988a). Surprisingly, the

concomitant loss of both Inv and YadA (Yersinia *ad*hesin, see below) by
Y. pseudotuberculosis results in a severe increase (and not decrease) in
virulence. Consistent with this observation, *Y. pestis* makes neither Inv nor
YadA (Rosqvist, Skurnik & Wolf-Watz, 1988*b*). The temperature regulation of Inv production adds even more confusion as to its potential role:
more Inv is produced by *Y. enterocolitica* as well as *Y. pseudotuberculosis*
when the strains are grown at 30°C rather than at 37°C (Isberg, Swain &
Falkow, 1988; Young *et al.*, 1990). In contrast, *Y. enterocolitica* synthesizes
more Ail at 37°C than at 28°C (Miller *et al.*, 1990).

The lipopolysaccharide composition

The lipopolysaccharide composition of *Y. enterocolitica* also varies with
growth temperature. The modification consist of a change in the composition in fatty acids and in a reduction of the length of the O-side chains
(Wartenberg *et al.*, 1983). Recently, cloning of the *rfb* region determining
the synthesis of the O-side chain of the lipopolysaccharide from *Y. enterocolitica* O:3 was reported (Al-Hendy, Toivanen & Skurnik, 1991*a*). Interestingly, transcription of this region is dramatically repressed at 37°C (Al-
Hendy, Toivanen & Skurnik, 1991*b*).

THE PANOPLY OF THERMOREGULATED PROPERTIES AND VIRULENCE FUNCTIONS ENCODED BY THE pYV PLASMID

Ca^{2+} dependency

In vitro, virulent yersiniae restrict their growth within two generations when
they are shifted from 28°C to 37°C in the absence of millimolar concentrations of Ca^{2+} ions. Growth can be reinitiated if the cultures are returned
to 26°C or if Ca^{2+} is added to the medium. Surprisingly, a variety of
nucleotides, including ATP, can prevent growth restriction (Zahorchak &
Brubaker, 1982). Mg^{2+} ions, on the contrary, exacerbate the growth
restriction, with a maximal effect at a concentration of 20 mM. This growth
restriction phenomenon, called Ca^{2+} dependency, is associated with the
massive production of a set of about ten proteins called Yops (see below).
Although it is quite obvious that massive production of a given set of
proteins must divert the metabolic potential from other biosynthetic pathways, it is still not known whether Ca^{2+} dependency involves a specific
repression of growth.

  The physiopathological significance of regulation by Ca^{2+} is far from
clear. Brubaker (1983) suggested a correlation between Ca^{2+} dependency
and the known, distinct levels of Ca^{2+} in the mammalian intracellular (μM
range) and extracellular (about 2.5 mM) fluids. According to this suggestion, the Yops would essentially be produced in the intracellular environment. Although this hypothesis is very appealing, there is accumulating

evidence that yersiniae spread and multiply essentially outside cells (Lian *et al.*, 1987; Hanski *et al.*, 1989, 1991; Simonet *et al.*, 1990). There is thus a paradox: *in vivo*, yersiniae proliferate in conditions which are supposed to be non-permissive for Yops production, yet they do produce Yops (see below).

Secretion of the Yops

During their growth restriction, *Y. pestis*, *Y. pseudotuberculosis* and *Y. enterocolitica* produce a set of proteins now uniformly called Yop followed by a letter which is identical for the homologous proteins in the three species. The 11 Yops identified in *Y. enterocolitica* are YopB, D, E, H, M, N, O, P, Q, R and LcrV. YopB, D, E, H, M, N and LcrV are well characterized in *Y. pseudotuberculosis* (see, for instance, Rosqvist *et al.*, 1990) and in *Y. pestis* (Straley, 1988). In the latter species, Straley (1988) also described YopJ, K and L. It is not known whether these Yops are the homologues of YopP, Q, and R of *Y. enterocolitica* or whether they are unique to *Y. pestis*.

The Yops were initially described as outer membrane proteins (Portnoy, Moseley & Falkow, 1981; Straley & Brubaker, 1981; Bölin, Portnoy & Wolf-Watz, 1985). However, Yops are also released into the culture supernatant (Heesemann, Algermissen & Laufs, 1984; 1986). Some of the released Yops (LcrV, YopM, Q, R) are soluble in the culture supernatant but others (YopH, E, O, B, D, P, N) have a propensity to aggregate as visible filaments in the culture (Fig. 1) (Michiels *et al.*, 1990). This casts doubt on the outer membrane localization of the Yops. Michiels *et al.* (1990) studied the kinetics of transcription and appearance of the Yops in the different cellular compartments of yersiniae. They observed the following: (i) Yops are first detected in the supernatant and later in the membrane fraction; (ii) the appearance of Yops in the membrane fraction is concomitant with the decrease of the corresponding protein in the supernatant; (iii) disappearance of the less-soluble Yops from the supernatant is not a consequence of degradation; (iv) there is a correlation between the propensity of a given Yop to aggregate in the supernatant and the presence of that Yop in the membrane fraction; (v) Yops still accumulate in the membrane fraction 3 h after induction, whereas transcription of the *yop* genes at that time is dramatically reduced; (vi) Yops are separated from the cell fraction upon treatment with hydrophobic agents such as xylene or hexadecane, whereas chromosome-encoded membrane proteins and YadA are not; and (vii) according to the hydrophobicity/hydrophilicity analysis of their amino acid sequences, Yops do not have the characteristics of membrane proteins. On the basis of these observations, Michiels *et al.* (1990) concluded that Yops are not membrane-anchored proteins. Their detection in the membrane fraction during the early experiments presumably resulted either

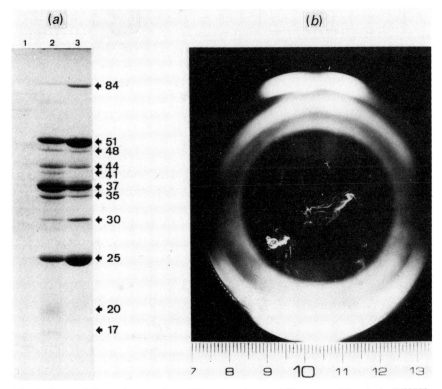

Fig. 1. Secreted Yops: Aggregation of Yops in cultures of *Y. enterocolitica* strain W22703. Filamentous aggregates become visible in cultures of *Y. enterocolitica* strain W22703 after 3 to 4 h of induction at 37°C. (*b*). Photograph of the culture in the conical flask, seen from above (scale in centimetres). These aggregates were collected with a glass rod and were run on SDS-PAGE (*a*, lane 3) together with supernatant proteins extracted from the same strain (lane 2) and from the plasmidless derivative (lane 1). Reprinted with permission from Michiels *et al.* (1990).

from co-purification of aggregated Yops with the membranes or from the adsorption of secreted Yops to the bacterial cell surface. The name Yop, introduced by H. Wolf-Watz (Bölin *et al.*, 1985) for *Yersinia outer mem-brane protein* could thus be questioned. However, it is so popular and widely used that it was decided, during a recent UCLA meeting on yersiniae (1990) to retain it but to write it as Yop(s) rather than YOPs, to indicate that it is not a set of initials.

Surprisingly, the secretion of Yops by *Yersinia* does not involve the cleavage of a classical signal sequence (Forsberg & Wolf-Watz, 1988; Michiels *et al.*, 1990). However, the 48 N-terminal residues of YopH contain all the information required for exportation (Michiels & Cornelis, 1991). Coupling the N-terminus of YopH to the α-peptide of β-galactosidase or to *E. coli* alkaline phosphatase (Michiels & Cornelis, 1991) or to the B subunit of cholera toxin (Sory *et al.*, 1990) results in efficient extracellular secretion of the hybrid protein. The recognition of YopE and YopQ by the exportation apparatus also involves the N-terminal region (Michiels & Cornelis, 1991). There is no similarity between the exportation domains of these proteins with respect to amino acid sequence, hydrophobicity profile, distribution of charged residues or prediction of secondary structure, suggesting a conformational recognition (Michiels & Cornelis, 1991). It is important to mention that the Yop proteins are not detected in the culture supernatant of *Y. pestis* cultures as they are in *Y. enterocolitica* and *Y. pseudotuberculosis* culture supernatants. This results from the degradation of the *Y. pestis* Yops by a membrane protease encoded by a small bacteriocinogenic plasmid that is present only in *Y. pestis* strains (Sodeinde *et al.*, 1988). This protease is either the plasminogen activator or the coagulase, both of which are encoded by the single *pla* gene with subsequent posttranslational processing (Sodeinde & Goguen, 1989).

Role of the Yops

The Yops are highly conserved in the genus *Yersinia* but no homology exists between different Yops in a single species (Forsberg & Wolf-Watz, 1988; Bölin & Wolf-Watz, 1988; Michiels & Cornelis, 1988; Michiels *et al.*, 1990). One of the Yops, LcrV, is the V antigen (38 kD) already described in the mid-1950s as a diffusible antigen differentiating virulent from avirulent strains of *Y. pestis* (Burrows & Bacon, 1956). The V antigen is protective by active immunisation and specific antibodies protect by passive immunization (Lawton, Erdman & Surgalla, 1963; Une & Brubaker, 1984).

Patients suffering an infection (Martinez, 1983) or mice artificially infected with *Y. enterocolitica* grown at low temperature (Sory & Cornelis, 1988) develop antibodies against the Yops, which clearly demonstrates that they are synthesized *in vivo* in the course of the infection. Mutants unable to express one or the other of the Yops have been constructed in the three *Yersinia* species and most of them are less virulent than the parental strain (Forsberg & Wolf-Watz, 1988; Sory & Cornelis, 1988; Bölin & Wolf-Watz, 1988; Mulder *et al.*, 1989; Straley & Cibull, 1989; Leung, Reisner & Straley, 1990). A few mutants, in particular those defective in YopB and YopD, can, however, not be validly tested because they are thermosensitive as an unexpected result of the mutation (see below).

The Yops constitute the major antihost components encoded by the pYV plasmid (see below). Whether they are produced inside or outside the phagocyte remains a matter of debate. Although yersiniae now tend to be considered as extracellular pathogens, it has been shown that the phagolysosomal environment of human macrophages allows the expression of the *yop* genes (Pollack, Straley & Klempner, 1986).

The functions of individual Yops are now emerging. YopE is cytotoxic for cultured HeLa cells (Rosqvist *et al.*, 1990). Interestingly, it is only active if it is produced by bacteria adhering at the cell surface or if it has been internalized by microinjection (Rosqvist, Forsberg & Wolf-Watz, 1991). YopH contributes to the ability of the bacteria to resist phagocytosis by peritoneal macrophages (Rosqvist, Bölin & Wolf-Watz, 1988). It is a tyrosine phosphatase (PTPase; EC 3.1.3.48) (Guan & Dixon, 1990) acting on multiple substrates in the cytoplasm of macrophages, which suggests that it interacts with the macrophage regulation (Bliska et al., 1991). Interestingly, the catalytic site of YopH is very similar to that of eukaryotic protein tyrosine phosphatases, which raises the appealing hypothesis that the *yopH* gene could be of eukaryotic origin. YopM is a 41 kD protein containing a 14-amino-acid sequence repeated six times. It shares some significant similarity with the von Willebrand factor and thrombin-binding domain of the alpha chain of human platelet membrane glycoprotein Ib, GPIbα (Leung & Straley 1989). This similarity obviously suggests a possible role in coagulation and YopM was indeed shown to inhibit platelet aggregation (Leung et al., 1990). Accordingly, YopM may prevent platelet-mediated host defense events and serve yersiniae at some stage of the infection. No cytotoxic or enzymatic activity has been reported for YopB and YopD, two other major Yops. Interestingly enough, Rosqvist et al. (1991) observed that a mutant affected in YopD looses its cytotoxicity in spite of the fact that it still produces YopE. These authors logically concluded that YopD could act as an internalization factor for YopE, a phenomenon reminiscent of the exotoxins PA and EF of *Bacillus anthracis* (Leppla, 1982). In accordance with the suggestion of Rosqvist et al., YopD and YopB are predicted by the Eisenberg plot (Eisenberg et al., 1984) to have a transmembrane domain (S. Häkansson, J.-C. Vanooteghem, G. R. Cornelis and H. Wolf-Watz, unpublished data). This is not true for YopE (Michiels et al., 1990), YopH (Michiels & Cornelis, 1988), YopM (Leung et al., 1990) and LcrV (Bergman et al., 1991). It is thus intriguing to speculate that the set of Yops is composed of two groups: one of toxin-like elements (YopE, YopM, YopH,...) and one of internalization proteins (YopD, YopB). Table 1 gives the main properties of the Yops as well as their previous designations.

The adhesin YadA

Protein YadA, formerly called P1 or YopA (Skurnik & Wolf-Watz, 1989) is a major outer membrane protein that forms a fibrillar matrix on the surface of *Y. enterocolitica* and *Y. pseudotuberculosis* (Kapperud et al., 1987) when they are cultivated at 37°C irrespective of the Ca^{2+} concentration. In SDS–PAGE analysis, YadA can be seen as a band migrating with an apparent molecular weight of between 200 and 240 kD. This band dissociates into 45 to 52.5 kDa subunits upon prolonged boiling in sample buffer or by addition of 8 M urea in the sample (Skurnik et al., 1984). Analysis of the amino acid sequence of YadA reveals the presence of a typical 25 amino acid signal sequence, which suggests that YadA is translocated via the classical Sec export pathway (Skurnik & Wolf-Watz, 1989). The name YadA was given for Yersinia *adh*esin, because it makes the bacteria adherent to eucaryotic cells (Heesemann & Grüter, 1987) and it promotes colonization of the mouse intestine (Kapperud *et al.*, 1987). YadA is also responsible for a marked autoagglutination (Skurnik *et al.*, 1984). In addition, it protects *Y. enterocolitica* (Balligand, Laroche & Cornelis, 1985) but apparently not *Y. pseudotuberculosis* (Perry & Brubaker, 1983) against the bactericidal action of human serum. YadA from *Y. enterocolitica*, expressed in *E. coli*, also confers resistance to the bactericidal action of serum (Martinez, 1989). In agreement with these observations, inactivation of *yadA* decreases the virulence of *Y. enterocolitica* but not that of *Y. pseudotuberculosis* (Kapperud *et al.*, 1987; Bölin & Wolf-Watz, 1984). The *yadA* gene is not functional in *Y. pestis* and, surprisingly, the introduction of a functional allele reduces the virulence of the strain (Rosqvist *et al.*, 1988*b*). The role of YadA in determining virulence is thus variable among the three *Yersinia* species. One could speculate that YadA has a positive role for its *Yersinia* host, not by increasing its virulence but by allowing its adaptation to another form of parasitism that offers, in the long run, better survival chances.

The lipoprotein YlpA

YlpA is a 29 kDa lipoprotein expressed by *Y. enterocolitica* at 37°C, in absence of Ca^{2+} (China, Michiels & Cornelis, 1990). YlpA is related to the TraT proteins encoded by the F plasmid, by various resistance plasmids and by the virulence plasmid of *Salmonella typhimurium*. although several TraT proteins have been shown to be involved in resistance to the bactericidal activity of human serum (for review see Sukupolvi & O'Connor, 1990), no evidence has been obtained for such a role in *Y. enterocolitica*. So far, the only element that pleads for a role in pathogenesis is of a genetic order: the expression of *ylpA* is regulated like that of the *yop* genes (see below). Gene *ylpA* is present in *Y. pseudotuberculosis* (China *et al.*, 1990).

THE pYV PLASMID

General organization

Growth restriction and the production of YadA, YlpA and the Yops are governed by a 70 kb plasmid called pYV (Ben Gurion & Shafferman, 1981; Gemski, Lazere & Casey 1980*a*; Gemski *et al.*, 1980*b*; Portnoy *et al.*, 1981). The pYV plasmids are non-conjugative and incompatible with the F plasmid as a result of the presence of an *incD* determinant which is part of their partition system (Biot & Cornelis, 1988). The replicon is of the

Table 1. *pYV encoded proteins.*

Nomenclature used in this review	Old or alternative nomenclature	M_r in kD	Main characteristics
LcrD	ORF5	77	Inner membrane protein. Regulation or export of Yops
LcrG	—	11	Regulatory role, cytoplasmic
LcrH	—	19	Negative regulator, cytoplasmic
LcrR	—	16	Regulatory role
LcrV	Yop41[a]	37–41	V antigen, secreted
RepA	—	33	Replicase of the RepFIIA replicon
RepB	—	9	Replicase repressor
VirF	LcrF[b,d]	31	Transcriptional activator of the *yop* regulon
YadA	Yop 1[b], Protein 1[c], P1[a]	>200	Adhesin, serum resistance determinant (45–50 kD)$_n$
YlpA	—	29	TraT-like lipoprotein
YopB	Yop44[a], Yop3[b]	41–44	Secreted
YopD	Yop37[a], Yop4a[b]	34–37	Secreted, translocation of YopE
YopE	Yop25[a], Yop5[b]	23–26	Secreted, cytotoxin
YopH	Yop51[a], Yop2b[b]	45–51	Secreted. Protein tyrosine phosphatase
YopJ	—	31	Only described in *Y. pesis* so far
YopK	—	21	Only described in *Y. pestis* so far
YopL	—	14	Only described in *Y. pestis* so far
YopM	Yop48[a], Yop2a[b]	44–48	Secreted, inhibitor of platelet aggregation
YopN	Yop35[a], Yop4b[b], LcrE[b]	33	Secreted, hypothetical Ca^{2+} sensor
YopO	Yop84[a]	84	Secreted, only described in *Y. enterocolitica* so far
YopP	Yop30[a]	30	Secreted, only described in *Y. enterocolitica* so far
YopQ	Yop20[a]	20	Only described in *Y. enterocolitica* so far
YopR	Yop17[a]	17	Only described in *Y. enterocolitica* so far
YscC		64	Putative secretion factor, similarity to PulD and gene IV product
YscD		46	Putative secretion factor
YscJ	YlpB[a]	27	Lipoprotein, putative secretion factor
YscL	—	25	Putative secretion factor
YscM	—	12	Similarity to central domain of YopH

Notes to Table 1 on opposite page.

RepFIIA type, like that of many large antibiotic resistance and virulence plasmids (Vanooteghem & Cornelis, 1990). The genetic maps of the pYV plasmids from the three species are very similar except for a quadrant containing at least *yopE* and *yadA*, which has been rearranged during evolution (Biot & Cornelis, 1988; Forsberg *et al.*, 1987; Forsberg & Wolf-Watz, 1990). The map of the pYV plasmid from *Y. enterocolitica* is given in Fig. 2.

The yop, yadA *and* ylpA *genes*

Genes *yadA, ylpA* and several *yop* genes, namely *yopE, yopH, yopQ* and *yopM* are scattered around pYV. Genes *yopO* and *yopP* form an operon located near the origin of replication. Transcription of the *Y. enterocolitica yop* genes was initially monitored by assaying the expression of various *yop–lacZ* or *yop–cat* operon fusion generated by transposon mutagenesis. Consistent with the fact that the Yops are only produced at 37 °C, transcription of the *yop* genes is strongly thermodependent. When exponential cultures in rich medium are shifted from 28 °C to 37 °C, transcription of the *yop* genes is increased by a factor of more than 100 (Cornelis *et al.*, 1986, Cornelis, Vanooteghem & Sluiters 1987*b*, Mulder *et al.*, 1989). By contrast, the presence of Ca^{2+} at 37 °C has a negative influence but transcription is only reduced by a factor of 3 to 10, as measured by the expression of *yop–lacZ* operon fusions (Cornelis *et al.*, 1986). Gene *ylpA* is regulated like the *yop* genes (China *et al.*, 1990). The same observation was made for the *yop* genes of *Y. pestis*, also analysing mini-Mud*lac* insertion mutants (Straley & Bowmer, 1986). *Y. pseudotuberculosis* also regulates the *yop* genes in the same way (Forsberg & Wolf-Watz, 1988).

In contrast to the *yop* genes, the expression of *yadA* in *Y. enterocolitica* and *Y. pseudotuberculosis* is regulated by temperature only, with no effect of Ca^{2+} (Bölin, Norlander & Wolf-Watz, 1982; Skurnik & Wolf-Watz, 1989).

Mutations in all these *yopE,H,Q,M,O,P,ylpA* and *yadA* genes simply result in the loss of the corresponding protein without alteration of the

---

Notes to Table 1

The actual nomenclature was agreed during the 1990 UCLA Symposium on *Yersinia*. The letters identifying the Yops have been introduced by Straley and co-workers.
[a] Nomenclature previously used by Cornelis and co-workers.
[b] Nomenclature previously used by Wolf-Watz and co-workers.
[c] Nomenclature used by Martinez.
[d] Nomenclature used by Straley and co-workers.

Note: the references are too numerous to be included in this consolidated table. The reader should refer to the text.

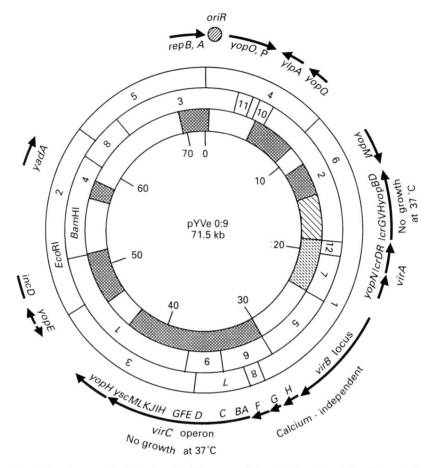

Fig. 2. Genetic map of a pYV plasmid of *Y. enterocolitica*. The *Eco*RI and *Bam*HI restriction map as well as most of the genetic map is from *Y. enterocolitica* strain W22703 (serotype O:9). Genes *yopN, lcrG, lcrH* and the *lcrDR* operon have been characterized in other systems namely *Y. enterocolitica* serotype O:3 (Viitanen *et al.*, 1990), *Y. pestis* strain KIM (Perry *et al.*, 1986; Barve & Straley, 1990; Plano *et al.*, 1991) and *Y. pseudotuberculosis* strain YPIII (Bergman *et al.*, 1991; Häkansson *et al.*, unpublished data). The arrows give the direction of transcription. *incD* is the stabilization system and *repBA* encodes the plasmid replication machinery. The shaded areas identify available nucleic acid sequences.

general phenotype: the mutant remains Ca^{2+} dependent and produces all the proteins save the one(s) encoded by the mutated gene(s) (operon).

The car *operon, or the* lcrGVHyopBD *operon*

The genes encoding the V antigen, YopB and YopD are clustered on pYV. The analysis of *Y. enterocolitica* mutants generated by mini-Mud*lac* or Tn*2507*, both generating operon fusions, showed that the three genes are transcribed in the same direction and that the insertions in *lcrV* (encoding the V antigen, see below) and *yopB* exert polar effects on *yopD* (Cornelis *et al.*, 1986; Mulder *et al.*, 1989). Since the three genes are transcribed as a messenger of about 4 kb, it was concluded that these three genes, *lcrV*, *yopB* and *yopD* form an operon. The transcription of this operon is temperature controlled and subjected to Ca^{2+} regulation (Mulder *et al.*, 1989). The special status of this operon is apparent from the fact that the insertion mutants are unable to grow at 37°C (TS phenotype), which is not the case for the other *yop* mutants. Even more, the distal polar mutants in the operon die at 37°C in the presence of Ca^{2+}. This indicates that some element encoded by the operon exerts a regulatory role, presumably in the Ca^{2+} regulation pathway. The name *car* for, *Ca^{2+} r*egulation, was thus proposed for this operon (Mulder *et al.*, 1989). In spite of the fact that the *car* mutants are unable to grow at 37°C in the presence of Ca^{2+}, they nevertheless do not secrete the Yops under these conditions. At 37°C, in the absence of Ca^{2+}, their phenotype is quite normal: the bacteria exhibit restricted growth and secrete the Yops, save YopD (YopB LcrV). The fact that the *car* mutants show restricted growth even under conditions in which they do not secrete the Yops suggests that growth restriction is not a consequence of over-production of the Yops *per se* but that some specific growth restriction mechanism is involved.

In *Y. pestis*, it was also noticed very early on that a mini-Mud*lac* insertion mutant defective in the production of the V antigen had a reduced growth rate at 37°C, irrespective of the presence of Ca^{2+} (Perry *et al.*, 1986). These authors also noted that the regulation by Ca^{2+} was disturbed and they concluded that the locus encoding the V antigen had a regulatory function. Hence, they called these genes *lcr* for '*low Ca^{2+} r*esponse'. Since V antigen is an exported protein which acts as a virulence factor, Straley and co-workers reasoned that the locus could encode other proteins. They analysed the locus in minicells and discovered that the gene encoding the V antigen was flanked by two small genes encoding a 13 kD protein (LcrG) and a 18 kD protein (LcrH) (Perry *et al.*, 1986). Going back to the original *Y. pestis lcrV* insertion mutant, they noticed that it did not produce any of the three LcrG, V and H factors. Hence, they concluded that the genes *lcrGVH* form an operon. In the wild-type, this operon is regulated by temperature and Ca^{2+} at the transcriptional level (Price *et al.*, 1989). By analogy with the situation in *Y. enterocolitica* and *Y. pseudotuberculosis* (see below), this operon presumably also contains *yopB* and *yopD* (Plano, Barve & Straley, 1991).

The corresponding locus in *Y. pseudotuberculosis* has been thoroughly investigated by the group of H. Wolf-Watz. The DNA sequence analysis reveals a great similarity with *Y. pestis*. The phenotype of polar mutants and the existence of a 4.6 kb transcript hybridizing with an *lcrH* probe led these authors to conclude that the *lcrGVHyopBD* genes form a single large operon (Bergman *et al.*, 1991). As in *Y. enterocolitica* and in *Y. pestis*, transcription of this operon is regulated by temperature and by Ca^{2+}. Mutants in the operon are thermosensitive and impaired by Ca^{2+} in regulation of transcription of the *car* operon itself and of *yopE* (Bergman *et al.*, 1991).

The vir *or* lcr *region*

A contiguous 20 kb region of the pYV plasmid, called the Ca^{2+} dependence region is required for the production of all the Yops. Insertion mutagenesis in this region defined a series of loci called *vir* (because they condition the *vir*ulence) in the *Y. enterocolitica* strain W22703 system and *lcr* (because they condition the '*low calcium response*') in *Y. pseudotuberculosis* and *Y. pestis*. Some of these loci have been characterized but the information is not yet complete. These loci are *virA* (subdivided in *lcrE* and *lcrD* in *Y. pestis*), *virB*, (*lcrB*), *virC* (*lcrC*), *virF* (*lcrF*), and *virG* (Goguen, Yother & Straley, 1984; Yother & Goguen, 1985; Yother, Chamness & Goguen, 1986; Cornelis *et al.*, 1986, 1987*b*, 1989, unpublished data; Hoe, Minion & Goguen, 1990). Locus *virA* is transcribed anticlockwise on the map presented in Fig. 2, like the neighbouring *car* operon, while all the other *vir* (*lcr*) genes are transcribed clockwise. Taking into account that *yopH*, located downstream from *virC* is also transcribed clockwise, it appears that about one contiguous third of the pYV plasmid is transcribed in the same direction (Goguen *et al.*, 1984; Cornelis *et al.*, 1986; Michiels *et al.*, 1991). Like transcription of the *yop* genes, transcription of the *vir* (*lcr*) genes is thermodependent (Goguen *et al.*, 1984; Cornelis *et al.*, 1986; Yother *et al.*, 1986, Cornelis *et al.*, 1989).

Mutations in *virA, virB, virF, virG* or the corresponding *lcr* loci completely abolish the extracellular Yops production and make the host strain independent of Ca^{2+} for growth at 37°C (Ca^{2+} independent, CI, phenotype). They also reduce the level of transcription of the *yop* genes (Cornelis *et al.*, 1986). The reduction of transcription of the *yop* genes is particularly drastic in the *virF* (*lcrF*) mutants, suggesting that *virF* encodes a transcriptional activator (Yother *et al.*, 1986; Cornelis *et al.*, 1987*b*). Mutations in the *virB* locus, reduce the transcription of *yop* genes though not to the same extent (Cornelis *et al.*, 1986; Michiels *et al.*, 1991). It is hypothesized that *virB* is also involved in the regulation of transcription of the *yop* genes (Michiels *et al.*, 1991) but this locus still needs to be analysed in detail. The *virF, virA* and *virC* loci are described in the following sections.

VirF and the yop *regulon*

Gene *virF* of *Y. enterocolitica* strain W22703 encodes a 30 kD transcriptional activator that belongs to the AraC family of regulators (Fig. 3). This family includes regulators of degradative pathways in *E. coli* and *Pseudomonas*

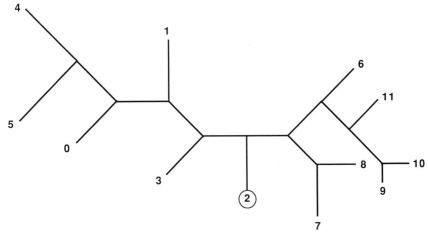

Fig. 3. The regulators of the AraC family and their phylogeny. The 12 members are (0) MelR from *E. coli* (Webster *et al.*, 1987); (1) AraC from *E. coli* (Wallace, Lee & Fowler., 1980); (2) VirF from *Y. enterocolitica* (Cornelis *et al.*, 1989); (3) RhaS and (4) RhaR from *E. coli* (Tobin & Schleif, 1987); (5) XylS from *Pseudomonas putida* (Inouye, Nakazawa & Nakazawa, 1988); (6) EnvY from *E. coli* (Lundrigan, Friedrich & Kadner, 1989); (7) Rns from *E. coli* (Caron, Coffield & Scott, 1989); (8) VirF from *Shigella flexneri* (Sakai *et al.*, 1986); (9) TetD from Tn*10* (Schollmeier & Hillen, 1984); (10) SoxS from *E. coli* (Wu & Weiss, 1991); (11) AppY from *E. coli* (Atlung, Nielsen & Hansen, 1989). The tree has been computed according to the method of Hein (1990) on the total proteins. The lengths of the branches are proportional to the evolutionary distances.

putida as well as regulators involved in the control of virulence of *Shigella*, *Yersinia* and enterotoxigenic *E. coli*. The similarity between VirF and other transcriptional activators, as well as the presence of a potential DNA-binding motif in its amino acid sequence (Cornelis *et al.*, 1989), are indications that VirF acts as a DNA-binding protein. Indeed, *in vitro*, a specific binding of VirF to *yopH* and *yopE* promoter regions was observed. DNase I-footprinting experiments on the *yopH* gene identified a protected region spanning 31 bp immediately upstream of the RNA polymerase binding site. This VirF-binding sequence is located in an AT-rich region and only comprises an imperfect 8 bp inverted repeat (Lambert *et al.*, 1992). Protection assays of several promoters of the *yop* regulon are currently being carried out in order to define a consensus VirF-binding sequence. VirF is, so far, the only *Yersinia* regulator which has been shown to bind to DNA.

Transcription of *virF* itself is regulated by temperature but not by Ca^{2+} (Cornelis *et al.*, 1989). VirF controls the expression of the *yop* genes (Cornelis *et al.*, 1987b; 1989), of *ylpA* (China *et al.*, 1990), of *yadA* (Martinez, 1989; Michiels *et al.*, 1991) and also of the *virC* operon (Lambert *et al.*, 1992). The *yop*, *ylpA*, *yadA* and *virC* genes thus constitute what is referred to as the *yop* regulon. By contrast, VirF is not essential for the induction of *virA* and *virB* (Lambert *et al.*, 1992).

The virA *locus:* lcrD, lcrR *and* lcrE

The *lcrA (virA)* locus was described during the early mini-Mud*lac* mutagenesis experiments in *Y. pestis* and *Y. enterocolitica* (Goguen *et al.*, 1984; Cornelis *et al.*, 1986). The mini-Mud*lac virA* and *lcrA* mutants are independent of Ca^{2+} for their growth (CI phenotype), while the mini-Mud*lac* mutants in the *car* operon (see above) or in *virC* (see below) are growth thermosensitive (TS phenotype).

Simultaneously Yother & Goguen (1985) selected TS mutants of *Y. pestis* strain KIM5 after mutagenesis by ethyl methanesulphonate. Presumably because of the non-polar character of these mutations, the authors did not come across *virC (lcrC)* mutants or *car* mutants but they isolated mutants mapping in the left part of *lcrA (virA)*. Interestingly, these new TS mutants produce the V antigen at 37°C, irrespective of the Ca^{2+} concentration. Yother & Goguen called them Ca^{2+}-blind or Lcr^c (constitutive). Hence, the authors divided the *virA* locus in two new loci: *lcrD* (right of *lcrA*) and *lcrE* (left of *lcrA*). Since the mutants in *lcrE* are Ca^{2+}-blind, the authors concluded that this gene encodes a diffusible element of the Ca^{2+} regulation pathway (Yother & Goguen, 1985). Later, Forsberg *et al.* (1987) showed that, at variance with the other *yop* genes, *yopN* was located in the Ca^{2+} region, close to *lcrE*. This suggested that YopN is involved in Ca^{2+} regulation and raised the hypothesis that YopN could be LcrE. The complete *lcrE* locus of *Y. enterocolitica* was then sequenced (Viitanen, Toivanen & Skurnik, 1990). The sequenced region consists of six tightly packed open-reading frames and it contains the site where transcription diverges in the Ca^{2+} region. The first ORF, which is transcribed anticlockwise, encodes a 32.9 kD protein which turned out indeed to be both LcrE and YopN (Viitanen *et al.*, 1990; Forsberg *et al.*, 1991).

The locus *lcrD* of *Y. pestis* strain KIM encodes a 77 kD inner membrane protein (Plano *et al.*, 1991). According to the structure prediction models, LcrD contains eight amino-terminal transmembrane helices and a large carboxy-terminal cytoplasmic domain. LcrD mutants possess a CI phenotype. Expression of LcrV, YopB and YopD proteins is severely reduced in an *lcrD* mutant but, surprisingly, the expression of YopH and YopM is enhanced (Plano *et al.*, 1991) which suggests that there could be two different post-transcriptional pathways for the Yops. The same observation was made with a *virA* mutant of *Y. enterocolitica* strain W22703 (Michiels *et al.*, 1991). LcrD could be involved in the secretion of the Yops or in the sensing or transmembrane signalling of the environmental Ca^{2+} or nucleotide concentrations.

The gene located downstream of *lcrD* is called *lcrR* (Barve & Straley, 1990). The two genes form a tight operon with an overlap between the translational stop of *lcrD* and the translational start of *lcrR* (Plano *et al.*, 1991). A *lcrR* mini-Mud*lac* insertion mutant of *Y. pestis* has a TS phenotype and it produces the V antigen in the defined liquid medium called TMH, even in the presence of 2.5 mM Ca^{2+}. However, growth restriction and V antigen production are still relieved by the presence of 18 mM ATP. In rich medium, growth restriction of the mutant is relieved by 2.5 mM Ca^{2+}. Hence, the *lcrR* mutant of *Y. pestis* is only blind to Ca^{2+} and only in a synthetic medium. According to Barve & Straley (1990), the *lcrR* mutant is also impaired in its ability to synthesize LcrG (but not LcrV and LcrH encoded by the same *car* operon). Its phenotype is thus different from that of *yopN* (= *lcrE*) mutants. Gene *lcrR* encodes a 16 kD acidic protein (Barve & Straley, 1990).

The virC *operon :* yscA-M

As mentioned previously, the Yops are secreted by a specialized transport system. It is very likely that this system is encoded by the *vir* region of the pYV plasmid. However, the identification of the genes involved in this process turned out to be difficult because none of the many pYV mutants that have been constructed, accumulates large amounts of Yops intracellularly. This probably results from a feedback inhibition of the Yops synthesis when export is compromised. Thus, distinguishing between regulatory genes and genes involved in secretion must make use of sensitive detection methods for the Yops. We reasoned that mutations in the Yop secretion pathway would prevent the production of the Yops but not of YadA, since YadA possesses the structure of a protein exported via the classical Sec pathway. This turned out to be the case for the *virC* mutants. These mutants transcribe the *yop* genes, yet produce small amounts of intracellular Yops but do not secrete them. Hence, it was concluded that the locus *virC* encodes at least some of the components of the Yops secretion machinery (Michiels *et al.*, 1991). This hypothesis is consistent with the fact that the *virC* locus is controlled by VirF, like the *yop* genes (Lambert *et al.*, 1992). The 8.5-kb *virC* locus constitutes a single large operon composed of 13 genes called *yscA* to *yscM* (for Yop secretion) (Michiels *et al.*, 1991). ORFs *yscB* to *yscK* are all contiguous with four cases of overlap between a stop codon and the start of the next gene. Complementation experiments were only carried out on the distal half of the operon (Michiels *et al.*, 1991). According to the results of these experiments, at least genes *yscD, yscJ* and *yscL* are required to secrete the Yops. The role of *yscA, B, C* could not be evaluated genetically but the putative *YscC* gene product has a signal sequence and it shares significant similarity with outer membrane proteins known to be involved in the secretion of pullulanase by *Klebsiella pneumoniae* (PulD) (d'Enfert *et al.*, 1989) or in the release of filamentous bacteriophages (gene IV product) (Brissette & Russel, 1990). YscJ is a 27 kD lipoprotein that was initially described as YlpB (China *et al.*, 1990). The presence of this lipoprotein among the *virC* encoded proteins is also compatible with the hypothesis that the *virC* operon is part of the secretion machinery (Michiels *et al.*, 1991).

Mutations in the *virC* operon confer on their host a TS phenotype (Cornelis *et al.*, 1986). However, the phenotype differs from that of the *car* mutants because the latter do secrete the Yops at 37 °C in absence of Ca^{2+} (see above) while the *virC* mutants do not. The region in *Y. pseudotuberculosis* corresponding to the distal part of the *virC* operon has been called *lcrK* by the group of H. Wolf-Watz (Rosqvist *et al.*, 1990).

GENETIC REGULATION BY TEMPERATURE

The previous sections have demonstrated that growth restriction and Yops secretion only occur at 37 °C and in absence of Ca^{2+} ions. None of these two regulatory networks is perfectly understood so far but it is quite clear that they are independent from each other. This section will deal with temperature regulation. Regulation by Ca^{2+} will be discussed afterwards.

Effect of temperature

As seen previously, transcription of the *yop*, *yadA* and *vir* genes, including *virF*, is thermo-induced. The *yop* and *ysc* genes (*virC* operon) require the presence of VirF for their transcription. The temperature regulatory system is therefore of the positive type. The *virA* and *virB* genes do not require VirF for their transcription in spite of the fact that they are thermoregulated. Accordingly the thermoregulation of Yops secretion involves a regulatory network of which VirF is only one piece. The transcription of *virF* is not regulated by Ca^{2+}. VirF is thus an element of the temperature regulatory system but not of the Ca^{2+} regulatory pathway. For the genes depending on VirF, two levels of regulation must be considered: the level of *virF* expression and the level of expression of the subordinated genes. To try to reconstitute the regulatory network, the following questions were posed: (i) how is *virF* thermoregulated; (ii) is *virF* the key to the thermoregulation of the *yop* and *yadA* genes?

Control of virF *by chromatin structure*

Transcription of a cloned *virF* is thermodependent in a *Y. enterocolitica* strain cured of pYV (Cornelis *et al.*, 1989). This observation indicates that *virF* is itself regulated by a chromosomal gene. In order to identify the chromosomal regulator, *Y. enterocolitica* strain W22703(pGC1152), which carries *lacZ* fused to *yopH*, was mutagenized. This strain produces very little *β*-galactosidase at 28°C and the production is strongly induced upon transfer to 37°C (Cornelis *et al.*, 1987*b*). On a modified McConkey medium, the colonies are white at 28°C and they subsequently turn red upon incubation at 37°C. Using this system, 33000 transposants were screened (Cornelis et al., 1991). Only a few colonies remained white after transfer to 37°C and all of them carried the transposon in *virF* or in the neighbouring *virG* gene or in *lacZ*. Hence, no chromosomal mutant was obtained that failed to induce *yopH* at high temperature. This suggests that the thermoregulation of *virF* does not involve a classical transcription activator.

In contrast, two chromosomal mutants expressing *yopH–lacZ* strongly at 28°C were isolated (Cornelis et al., 1991). These mutants transcribe not only *yopH* but also *yopE* and *yadA* at low temperature. However, they do not secrete the Yops. Transcription of the regulatory gene *virF* is itself increased at 28°C, which may account for the increased transcription of the genes of the regulon. Although the elements of the *yop* regulon are overexpressed at low temperature in the mutants, there is still an increase of transcription upon transfer to 37°C. Hence, the thermal response is not abolished but rather 'modulated'. The phenotype is thus not that of a classical repressor minus mutant. The copy number of the pYV plasmid is not altered and the increase in the transcription rate at low temperature is not a generalized phenomenon. Thus, the mutations specifically affect a component of the temperature response of yersiniae. In both mutants, the transposon is inserted in the same gene that has been designated *ymoA* for '*Yersinia* mo*dulator*'. The cloned *ymoA* gene complements the mutants and encodes a 8064 D protein extremely rich in positively and negatively charged residues. Although there is no similarity between YmoA and HU, IHF or H1 (H-NS) (see Dorman & Ní Bhriain, this symposium), it is very likely that YmoA is an histone like protein. The lack of YmoA does not modify the superhelicity of reporter plasmids but it results in a greater *in vitro* fragility of chromosomal DNA. Interestingly, the expression of the *yopH–lacZ* fusion becomes osmoregulated in the *ymoA* mutants (Cornelis et al., 1991). All these properties of the *ymoA* mutants are strikingly reminiscent of the mutants of *E. coli* and *Shigella flexneri* lacking H1 (Higgins et al., 1988; Dorman, Ní Bhriain & Higgins, 1990; Göransson et al., 1990).

In conclusion, the only chromosomal regulatory mutants isolated so far are affected in a protein presumably involved in the chromatin structure. This does, of course, not rule out the existence of a classical repressor, but it is striking that the search for a thermoregulator in *Yersinia*, as in *Shigella* and in uropathogenic *E. coli*, converged on histone-like proteins (see below). It thus seems reasonable to postulate that DNA compaction, and maybe the YmoA histone itself, does indeed play a major role in thermoregulation. This conclusion is supported by the fact that DNA supercoiling is known to vary in response to temperature changes. It is strengthened by the observation that, in *ymoA* mutants, the *yop* regulon becomes regulated by osmolarity, an environmental stimulus known to modify DNA supercoiling (Higgins et al., 1988; see Dorman & Ní Bhriain, this symposium).

VirF and temperature are both required to induce the yop *regulon*

The fact that *virF* is thermoregulated can explain why the Yops are only produced at 37°C. However, it does not demonstrate that temperature is only involved in the regulation of *virF*. Would the Yops secretion and growth restriction be induced if gene *virF* was expressed at low temperature? To answer this question, *virF* was placed downstream from a *tac* promoter and the response of the *Y. enterocolitica* recombinant strain to IPTG was measured. In spite of the fact that *virF* was strongly transcribed at 25°C, no Yops were produced at that temperature. In contrast, at 37°C, the response to IPTG mimicked the normal response to thermal induction. The lack of Yops secretion at 25°C can be explained by the observation that *virA* and *virB* are activated by temperature but not by VirF. At 25°C, upon IPTG induction, YadA was undetectable as well (Lambert *et al.*, 1992). This is more surprising since the export of YadA occurs via the chromosome-encoded Sec export pathway. However, in a pYV⁺ strain, the full expression of YadA requires *virB* (Michiels *et al.*, 1991). Hence, the lack of a *virB*-encoded factor could explain the absence of YadA at low temperature.

Although the lack of production of Yops and YadA at 25°C could be explained by the absence of *virA*- or *virB*-encoded factors, the *yop* and *yadA* genes on pYV are nevertheless poorly transcribed at 25°C upon induction of *virF* by IPTG. Thus, full transcription of the *yop* and *yadA* genes requires both VirF and temperature. This is strikingly reminiscent of the regulation of the invasion phenomenon by *Shigella flexneri*: transcription of gene *virB* from *S. flexneri* depends on both temperature and on the presence of the transcriptional activator VirF (Tobe *et al.*, 1991).

The poor transcription of the *yop* regulon at 25°C in the presence of VirF could be due (i) to some feedback inhibition caused by a defect in the export system, (ii) to the action of a pYV-encoded repressor, or (iii) simply to a lack of activity of VirF resulting from an inadequate conformation of either the activator or the promoter at 25°C. Since VirF has been observed to bind *in vitro* to *yop* promoter regions even at 25°C (Lambert *et al.*, 1992), the hypothesis of a temperature-mediated conformational change in the protein is improbable. To check whether the conformation of the promoter influences transcription of *yop* genes, a *yopH–cat* reporter construct was introduced into the *ymoA* mutant *Y. enterocolitica* strain W22711. It was found that the *yopH* promoter is active at 37°C in the absence of VirF and that it is extremely active at 25°C in the presence of VirF (Lambert *et al.*, 1992). This result indicates that chromatin structure is involved in transcription activation of the *yop* genes. Hence, this suggests that, *in vivo*, the promoters of the *yop* regulon are more susceptible to VirF activation at 37°C. Thus, chromatin structure would have an influence on *yop* promoters, in addition to its effect on the transcription of *virF* itself.

The regulation of the production of YadA was also studied in a simplified system. The expression of the *yadA* gene is strictly dependent on VirF in pYV⁺ strains of *Y. enterocolitica* (Michiels *et al.*, 1991) as well as in *E. coli* (Martinez, 1989). On the contrary, when cloned on vector plasmid pACYC184, *yadA* is expressed in a pYV⁻ strain of *Y. enterocolitica* even in absence of VirF (Balligand *et al.*, 1985). This discrepancy cannot be attributed to the copy number of the vector for two reasons: first, plasmid pACYC184 has a medium copy number; second, transcription of the *yopH* promoter on another reporter plasmid also derived from pACYC184, remains completely silent in absence of VirF (Cornelis *et al.*, 1989). Interestingly, although *yadA* seems to be less tightly regulated by *virF* in the absence of the whole pYV plasmid, its expression remains thermodependent. This observation reinforces the conclusion that temperature and VirF act distinctly to activate the expression of the genes constituting the *yop* regulon (Lambert *et al.*, 1992). An observation on *Y. pestis* from Goguen's laboratory also points to the same conclusion: in *E. coli*, induction of *lcrF* transcription is not required for temperature-dependent activation of *yopE* transcription (Hoe, Minion & Goguen, 1990). In conclusion, temperature appears somehow to modify the structure of chromatin, making the promoters more accessible to VirF. However, no information is currently available on the nature of the DNA structural change occurring during the temperature shift. One can only speculate that some histone-like protein, possibly YmoA, is involved in this change. This observation is just another one pointing to the role of chromatin structure in transcriptional regulation. This phenomenon, which seems now to be classical in the regulation of bacterial virulence functions, is reviewed and discussed in detail by Higgins *et al.* (1990) and by Dorman (1991) (see also Dorman & Ní Bhriain, this symposium).

GENETIC REGULATION BY Ca²⁺

While temperature controls nearly all the virulence functions of yersiniae, the presence of Ca²⁺ only regulates the production of the Yops. This second regulation appears to be very complex and, in spite of many efforts, it is far from being understood, yet. Although the *in vivo* occurrence of the Ca²⁺ regulation and its significance is not perfectly understood, the phenomenon is extremely clearcut *in vitro*.

The facts

A number of facts regarding the effect of Ca²⁺ are well established. These are summarized here to permit presentation of the current working model.

- Radioactive Ca²⁺ is not actively accumulated by yersiniae but yersiniae may possess a surface ligand capable of recognizing Ca²⁺ (Perry & Brubaker, 1987).

- Transcription of the *yop* and *vir* genes is reduced in the presence of Ca^{2+} ions (Cornelis *et al.*, 1987*b*; Leung *et al.*, 1990; Forsberg & Wolf-Watz, 1988; Bölin *et al.*, 1988; Price *et al.*, 1989; Mulder *et al.*, 1989). The negative influence of Ca^{2+} ions is probably not as drastic as the positive effect of temperature. This is especially clear when one analyses transcription by measuring enzymic activities encoded by operon fusions (Cornelis *et al.*, 1987*b*; Mulder *et al.*, 1989).
- The polar mutants in the *lcrGVHyopBD* operon (*car* operon for simplicity) are thermosensitive and they transcribe the *yop* genes even in the presence of Ca^{2+} (Price & Straley, 1989; Bergman *et al.*, 1991).
- Overexpression of gene *lcrH* in a polar *car* mutant shuts off the expression of the *car* operon itself as well as of the *yopE* gene (taken as a reporter) (Bergman *et al.*, 1991).
- Overexpression of *lcrH* in a wild-type strain has no influence on the phenotype (Bergman *et al.*, 1991).
- Distal mutants in the *car* operon die at 37°C in the presence of Ca^{2+} (Mulder *et al.*, 1989).
- Non-polar mutants in *yopN* are Ca^{2+} blind: they secrete the Yops at 37°C, even in the presence of Ca^{2+} (Yother & Goguen, 1985; Forsberg *et al.*, 1991). YopN is a secreted protein.

The negative loop model

The simplest interpretation of the facts listed above can be summarized in the six following points:

1. At least part of the regulation by Ca^{2+} occurs at the transcriptional level.
2. The *car* operon is involved in this regulation.
3. LcrH is a repressor of transcription of the *yop* genes.
4. An unidentified product, mic RNA or protein, counteracts *lcrH* or its product, only in the presence of Ca^{2+}. This product could be encoded by a distal gene of the *car* operon.
5. Ca^{2+} acts at the cytoplasmic membrane as an environmental regulatory signal.
6. YopN is a secreted Ca^{2+} sensor (Forsberg *et al.*, 1991).

The first model of Ca^{2+} regulation was presented by Price & Straley (1989), when only part of the information was available. Basically, this model confers to LcrH a negative regulatory role in response to the presence of Ca^{2+}. Later, the group of H. Wolf-Watz proposed a more sophisticated model (Forsberg *et al.*, 1991). In agreement with the previous model, it also attributes a key negative regulatory role to LcrH. It includes an antagonist to LcrH, tentatively called LcrX, and YopN, the Ca^{2+} sensor. This model is appealing and it has the great merit of presenting the first coherent explanation for Ca^{2+} regulation. It fits with the facts listed above and it also

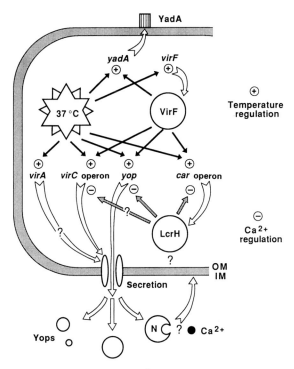

Fig. 4. A model for the temperature and Ca^{2+} regulation of *Yersinia* virulence functions. Curved lines symbolize synthesis and movement. Straight lines symbolize regulation.

provides an explanation for the TS phenotype of the *virC* mutants. According to the model, the mutants in the secretion machinery no longer secrete the Ca^{2+} sensor and, hence, are deregulated in the presence of Ca^{2+}. In spite of the fact that they are deregulated, they do not secrete the Yops because they lack an intact export system. The model is presented, in a simplified version, in Fig. 4.

The grey area

Although appealing, the negative loop model is based on some assumptions which still need to be demonstrated. Moreover, several points remain unaccounted for. These questions are discussed here. They certainly do not invalidate the model but they rather suggest experiments and in doing so, confirm the utility of the model.

The role of YopN remains unclear. First, YopN has never been shown to bind Ca^{2+}. Secondly, YopN has also not been shown conclusively to bind to the bacterial surface. According to its weak hydrophobicity, YopN would probably not insert into the outer membrane but rather bind to a receptor.

Not only has this receptor not yet been identified but there is not the slightest clue for its existence.

The negative loop model proposes that LcrH is a repressor of transcription. Although this is based on sound genetic arguments, LcrH has not been shown to bind to DNA. The link between the sensor (YopN) and the repressor (LcrH) is still completely missing. One should, however, not lose hope: many genes of the pYV plasmid remain orphans! One also requires to explain some other odd phenotypes such as the CI phenotype of the non-polar mutants in the gene encoding the V antigen (Bergman *et al.*, 1991): although the V antigen has the properties of a virulence determinant, there is genetic evidence that V antigen also has a regulatory role. What is then the major function of the V antigen? Should it be called YopV, emphasizing its possible role in virulence or rather LcrV to suggest a major regulatory role? In spite of a decade of intensive genetic research on yersiniae, this factor, which was identified in the mid-fifties, still awaits a status.

Finally, there could be more than one Ca^{2+} regulatory pathway. Although transcription of the *yop* genes is reduced in the presence of Ca^{2+}, the degree of reduction does not explain the complete lack of secretion of the Yops (Cornelis *et al.*, 1987*b*). The Ca^{2+} shutdown system could involve some kind of post-transcriptional regulation.

A PUZZLING SIMILARITY WITH *SHIGELLA*

As seen above, the *Yersinia* virulence machinery consists of several functions, including a secretion function, that appears to be quite unique among procaryotes. The number of functions shared with other bacterial pathogens is very limited but if one focuses on the similarities with other bacteria, the number of convergences with *Shigella* is surprising. They occur in the ancillary functions, in the regulatory functions and, finally, in the pathogenicity proteins themselves.

In both species, major virulence functions are encoded by a large plasmid. The *Shigella* virulence plasmids are larger (over 200 kb) than the *Yersinia* plasmids but both groups contain a RepFIIA replicon and thus are derived from a common ancestral plasmid. To be complete, this also applies to *Salmonella typhimurium* and enterotoxigenic *E. coli* (for review see Couturier *et al.*, 1988).

The regulatory cascade of the virulence functions is also strikingly similar in *Yersinia* and *Shigella*. In both genera, a 37°C temperature is the signal which triggers the expression of specialized virulence proteins: the Yops in *Yersinia* and the Ipa proteins in *Shigella* (for a review on *Shigella* virulence, see Hale, 1991; see also Sansonetti, this symposium). In both systems, the search for a high-rank temperature-sensitive regulatory locus ended up with the discovery of a chromosomal gene encoding a histone-like protein: YmoA in *Yersinia* (Cornelis *et al.*, 1991) and H1 in *Shigella* (Maurelli &

Sansonetti, 1988; Dorman *et al.*, 1990). H1 turned out also to be the product of gene *drdX*, involved in thermoregulation of the *E. coli* Pap pili (Görans-son *et al.*, 1990; Hulton *et al.*, 1990). All these observations lead to the conclusion that DNA topology and chromatin structure play a major role in the temperature-dependent induction process (for review see Dorman, 1991; see Dorman & Ní Bhriain, this symposium).

In *Yersinia*, one of the targets of temperature regulation is *virF*, the gene encoding the transcriptional activator (Cornelis *et al.*, 1989). The same applies to *Shigella* where, by mere chance, the activator gene is also called *virF* (Sakai *et al.*, 1988). Surprisingly, both VirF proteins turned out to be members of the same AraC family. The comparison between all these transcription activators is given in Fig. 3. In *Yersinia*, VirF activates directly the transcription of the *yop* genes, at least in a simplified system consisting only of the regulatory gene and a reporter gene cloned downstream from a *yop* promoter (Cornelis *et al.*, 1989). In the physiological genetic environment, i.e. in the presence of a complete pYV plasmid, the situation could be more complex. There are clues for the existence of another regulatory gene neighbouring *virF*, which has been designated *virG* (G. R. Cornelis & C. Lambert, unpublished data). In *Shigella*, VirF controls directly only *virG* (= *icsA*), a gene associated with intercellular bacterial spread (Bernardini *et al.*, 1989) and *virB* (= *ipaR*), a vassal regulator (Adler *et al.*, 1989). The latter, in turn, regulates positively the *ipa* operon. Consequently, expression of the invasive phenotype of shigellae is under the control of the dual activation system directed by the *virF* and *virB* genes (Adler *et al.*, 1989).

In both species, the activation by *virF* (of *yop* genes or of *virB*) requires, not only VirF but also temperature. Thus, temperature-induced chromatin modifications are required at two levels: first the induction of the activator and, second, the induction of its targets (Tobe *et al.*, 1991; Lambert *et al.*, 1992).

The regulatory networks offer another similarity between *Yersinia* and *Shigella*: LcrH (18 kD) has 26% amino acid sequence identity with the putative protein encoded by *ippI*, the gene immediately adjacent to *ipaB* (Fig. 5). The role of this protein has not yet been determined. By analogy with *Yersinia*, one would predict a regulatory role, possibly in a pathway responding to a stimulus different from temperature.

Finally, there is some similarity at the level of the virulence proteins themselves. First IpaH, like YopM, contains several evenly spaced leucine rich glycoprotein (LRG) motifs and the two proteins share 50% amino acid sequence identity in a 93-amino acid overlap. Both proteins could serve the same role: inhibition of thrombus formation and recruitment of inflammatory cells in the lamina propria during the early stages of the infection (Hartman *et al.*, 1990). A second similarity has recently been identified at the level of the virulence proteins: two domains of YopB appear in IpaB

```
      1        10        20        30        40        50        60
IPPI  MSLNITENESISTAVIDAINSGATLKDINAIPDDMMDDIYSYAYDFYNKGRIEEAEVFFR
      *    *       *    *  *   *  *   * * *   ** *   *   *  * *   *
LCRH  MQQETTDTQEYQLAMESFLKGGGTIAMLNEISSDTLEQLYSLAFNQYQSGKYEDAHKVFQ
      1        10        20        30        40        50        60

               70        80        90       100       110       120
IPPI  FLCIYDFYNVDYIMGLAAIYQIKEQFQQAADLYAVAFALGKNDYTPVFHTGQCQLRLKAP
      **   *  *     ** *   *   *   *     *             **    *  *
LCRH  ALCVLDHYDSRFFLGLGACRQAMGQYDLAIHSYSYGAIMDIKEPRFPFHAAECLLQKGEL
               70        80        90       100       110       120

               130       140       150  155
IPPI  LKAKECFELVIQHSNDE....KLKIKAQSYLDAIQDIKE
       *      *        *    *     * * **    **
LCRH  AEAESGLFLAQELIADKTEFKELSTRVSSMLEAIKLKKEMEHECVDNP
               130       140       150       160       168
```

Fig. 5. Alignment between LcrH from *Y. pestis* (Price *et al.*, 1989) and IppI (Ipa 18) from *Shigella flexneri* (Venkatesan, Buysse & Kopecko, 1988). Note that LcrH from *Y. pseudotuberculosis* strain YPIII differs from its *Y. pestis* counterpart, only at amino acid 138 (Bergman *et al.*, 1991). The alignment was achieved using the 'MULTIALIGN' program (Corpet, 1988).

(S. Häkansson, J.-C. Vanooteghem, H. Wolf-Watz, G. R. Cornelis, unpublished data). Interestingly, the same domains also occur in HlyA, the *E. coli* hemolysin and LtP, the *Pasteurella* leucotoxin (Fig. 6). The link is particularly appealing because all these proteins are involved in processes occurring at the eucaryotic cell membrane: HlyA and LtP are membrane damaging

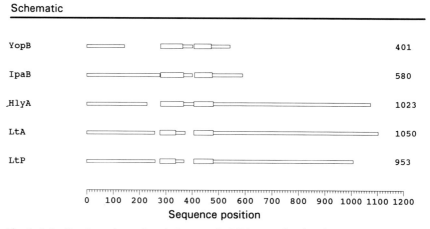

Fig. 6. A family of membrane-interfering proteins? Diagram showing the conserved domains (larger blocks) between YopB (Häkansson *et al.*, unpublished data), the *Shigella flexneri* invasion protein IpaB (Venkatesan, Buysse & Kopecko, 1988), the *E. coli* haemolysin HlyA (Felmlee, Pellett & Welch, 1985) and the leukotoxins LtA from *Actinobacillus actinomycetemcomitans* (Lally *et al.*, 1989) and LtP from *Pasteurella hemolytica* (Highlander *et al.*, 1989). This figure was constructed using the 'MACAW' program of Schuler, Altschul & Lipman (1991). There is similarity not only within the two groups of aligned blocks but also between some blocks of each group. Sometimes, the best pairwise fit even occurs between non-aligned blocks.

Table 2. *Similarities in pathogenicity functions and regulatory functions between Yersinia and Shigella.*

	Pathogenicity functions		Regulators		Ancillary
Shigella	IpaH	IaB	VirF	IppI	RepFIIA
Yersinia	YopM	YopB	VirF	LcrH	RepFIIA
Other homologues		HlyA			
		LtA			
		LtP			
Putative role	Coagulation interference	Membrane translocation	Activator	Repressor	Replicons

and IpaB is essential for the invasion of the cells by *Shigella*. One might hypothesize that YopB is an internalization factor for an unidentified Yop having an intracellular target.

In spite of the fact that yersiniae and shigellae elicit quite different syndromes, these repeated convergent observations (summarized in Table 2) suggest that both genera build up their pathogenicity with some identical basic pieces. Could not the virulence plasmids in both genera derive from a common ancestor that was distributed to the *E. coli* ancestor of *Shigella* and to the avirulent *Yersinia* ancestor of *Y. pestis, Y. pseudotuberculosis* and *Y. enterocolitica*? This ancestor plasmid probably contained the basic ancillary and regulatory functions as well as a few pathogenicity functions. It could then have progressively gained, possibly from a host, the new pathogenicity functions that led to the differentiation of the two pathogens.

ACKNOWLEDGEMENTS

I thank S. Straley and H. Wolf-Watz for communicating data prior to publication. I also thank M.-P. Sory, J.-C. Vanooteghem, P. Wattiau, B. China and T. Biot for a careful reading and critical discussion of the manuscript and G. Gobert for handling the references. The compilation of VirF homologues is due to C. Lambert. The computer searches for sequence similarity and the alignments were done by J.-C. Vanooteghem. Our research was supported by grant Nr 3.4514.83 from the Belgian Fund for Medical Scientific Research (FRSM) and by an 'Action Concertée' (Nr 86-91/86) from the Belgian Ministry for Science (SPPS).

REFERENCES

Adler, B., Sasakawa, C., Tobe, T., Makino, S., Komatsu, K. & Yoshikawa, M. (1989). A dual transcriptional activation system for the 230 kb plasmid genes coding for virulence-associated antigens of *Shigella flexneri*. *Molecular Microbiology*, **3**, 627–35.

Al-Hendy, A., Toivanen, P. & Skurnik, M. (1991a). Expression cloning of *Yersinia enterocolitica* O:3 *rfb* gene cluster in *Escherichia coli* K12. *Microbial Pathogenesis*, **10**, 47–59.

Al-Hendy, A., Toivanen, P. & Skurnik, M. (1991b). The effect of growth temperature on the biosynthesis of *Yersinia enterocolitica* O:3 lipopolysaccharide: temperature regulates the transcription of the *rfb* but not of the *rfa* region. *Microbial Pathogenesis*, **10**, 81–6.

Atlung, T., Nielsen, A. & Hansen, F. G. (1989). Isolation, characterization, and nucleotide sequence of *appY*, a regulatory gene for growth-phase-dependent gene expression in *Escherichia coli*. *Journal of Bacteriology*, **171**, 1683–91.

Balligand, G., Laroche, Y. & Cornelis, G. (1985). Genetic analysis of virulence plasmid from a serotype 9 *Yersinia enterocolitica* strain: role of outer membrane protein P1 in resistance to human serum and autoagglutination. *Infection and Immunity*, **48**, 782–6.

Barve, S. S. & Straley, S. C. (1990). *lcrR*, a low-Ca^{2+}-response locus with dual Ca^{2+}-dependent functions in *Yersinia pestis*. *Journal of Bacteriology*, **172**, 4661–71.

Ben Gurion, R. & Shafferman, A. (1981). Essential virulence determinants of different *Yersinia* species are carried on a common plasmid. *Plasmid*, **5**, 183–7.

Bergman, T., Håkansson, S., Forsberg, A. *et al.* (1991). Analysis of the V antigen *lcr*GVH-*yopBD* operon of *Yersinia pseudotuberculosis*: evidence for a regulatory role of LcrH and LcrV. *Journal of Bacteriology*, **173**, 1607–16.

Bernardini, M. L., Mounier, J., D'Hauteville, H., Coquis-Rondon, M. & Sansonetti, P. J. (1989). Identification of *icsA*, a plasmid locus of *Shigella flexneri* that governs bacterial intra- and intercellular spread through interaction with F-actin. *Proceedings of National Academy of Sciences USA*, **86**, 3867–71.

Biot, T. & Cornelis, G. (1988). The replication, partition and *yop* regulation of the pYV plasmids are highly conserved in *Yersinia enterocolitica* and *Y. pseudotuberculosis*. *Journal of General Microbiology*, **134**, 1525–34.

Bliska, J. B., Guan, K., Dixon, J. E. & Falkow, S. (1991). A mechanism of bacterial pathogenesis: tyrosine phosphate hydrolysis of host proteins by an essential *Yersinia* virulence determinant. *Proceedings of National Academy of Sciences USA*, **88**, 1187–91.

Bölin, I., Forsberg, A., Norlander, L., Skurnik, M. & Wolf-Watz, H. (1988). Identification and mapping of the temperature-inducible, plasmid-encoded proteins of *Yersinia* spp. *Infection and Immunity*, **56**, 343–8.

Bölin, I., Norlander, L. & Wolf-Watz, H. (1982). Temperature-inducible outer membrane protein of *Yersinia pseudotuberculosis* and *Yersinia enterocolitica* is associated with the virulence plasmid. *Infection and Immunity*, **37**, 506–12.

Bölin, I., Portnoy, D. A. & Wolf-Watz, H. (1985). Expression of the temperature-inducible outer-membrane proteins of yersiniae. *Infection and Immunity*, **48**, 234–40.

Bölin, I. & Wolf-Watz, H. (1984). Molecular cloning of the temperature-inducible outer membrane protein 1 of *Yersinia pseudotuberculosis*. *Infection and Immunity*, **43**, 72–8.

Bölin, I. & Wolf-Watz, H. (1988). The plasmid-encoded Yop2b protein of *Yersinia pseudotuberculosis* is a virulence determinant regulated by calcium and temperature at the level of transcription. *Molecular Microbiology*, **2**, 237–45.

Bottone, E. J. (ed.). (1981). *Yersinia enterocolitica*. Florida, Boca Raton: CRC Press Inc.

Brissette, J. L. & Russel, M. (1990). Secretion and membrane integration of a filamentous phage-encoded morphogenetic protein. *Journal of Molecular Biology*, **211**, 565–80.

Brubaker, R. R. (1983). The Vwa$^+$ virulence factor of *Yersinia*: the molecular basis of the attendant nutritional requirement for Ca^{2+}. *Reviews of Infectious Diseases*, 5, S748–58.

Burrows, T. W. & Bacon, G. A. (1956). The basis of virulence in *Pasteurella pestis*: an antigen determining virulence. *British Journal of Experimental Pathology*, 37, 481–93.

Butler, T. (1983). Plague and other *Yersinia* infections. In *Current Topics in Infectious Disease*, W. B. Greenough III and T. C. Merigan, eds, New York and London: Plenum Medical Book Company.

Caron, J., Coffield, L. M. & Scott, J. R. (1989). A plasmid-encoded regulatory gene, *rns*, required for expression of the CS1 and CS2 adhesins of enterotoxigenic *Escherichia coli*. *Proceedings of National Academy of Sciences, USA*, 86, 963–7.

China, B., Michiels, T. & Cornelis, G. R. (1990). The pYV plasmid of *Yersinia* encodes a lipoprotein YlpA, related to TraT. *Molecular Microbiology*, 4, 1585–93.

Cornelis, G., Laroche, Y., Balligand, G., Sory, M.-P. & Wauters, G. (1987a). *Y. enterocolitica*, a primary model for bacterial invasiveness. *Reviews of Infectious Diseases*, 9, 64–87.

Cornelis, G., Sluiters, C., Delor, I. *et al.* (1991). *ymoA*, a *Yersinia enterocolitica* chromosomal gene modulating the expression of virulence functions. *Molecular Microbiology*, 5, 1023–34.

Cornelis, G., Sluiters, C., Lambert De Rouvroit, C. & Michiels, T. (1989). Homology between VirF, the transcriptional activator of the *Yersinia* virulence regulon, and AraC, the *Escherichia coli* arabinose operon regulator. *Journal of Bacteriology*, 171, 254–62.

Cornelis, G., Sory, M.-P., Laroche, Y. & Derclaye, I. (1986). Genetic analysis of the plasmid region controlling virulence in *Y. enterocolitica* O:9 by mini-Mu insertions and *lac* gene fusions. *Microbial Pathogenesis*, 1, 349–59.

Cornelis, G., Vanooteghem, J.-C. & Sluiters, C. (1987b). Transcription of the *yop* regulon from *Y. enterocolitica* requires trans-acting pYV and chromosomal genes. *Microbial Pathogenesis*, 2, 367–79.

Corpet, F. (1988). Multiple sequence alignment with hierarchical clustering. *Nucleic Acids Research*, 16, 10881–90.

Couturier, M., Bex, F., Bergquist, P. L. & Maas, W. K. (1988). Identification and classification of bacterial plasmids. *Microbiological Reviews*, 52, 375–95.

Cover, T. L. & Aber, R. C. (1989). *Yersinia enterocolitica. The New England Journal of Medicine*, 6, 16–24.

d'Enfert, C., Reyss, I., Wandersman, C. & Pugsley, A. P. (1989). Protein secretion by Gram-negative bacteria. *Journal of Biological Chemistry*, 264, 17462–8.

Delor, I., Kaeckenbeeck, A., Wauters, G. & Cornelis, G. R. (1990). Nucleotide sequence of *yst*, the *Yersinia enterocolitica* gene encoding the heat-stable enterotoxin, and prevalence of the gene among the pathogenic and non-pathogenic yersiniae. *Infection and Immunity*, 58, 2983–8.

Dorman, C. J. (1991). DNA supercoiling and environmental regulation of gene expression in pathogenic bacteria. *Infection and Immunity*, 59, 745–9.

Dorman, C. J., Ní Bhriain, N. & Higgins, C. F. (1990). DNA supercoiling and environmental regulation of virulence gene expression in *Shigella flexneri*. *Nature*, 344, 789–92.

Eisenberg, D., Schwarz, E., Komaromy, M. & Wall, R. (1984). Analysis of membrane and surface protein sequences with the hydrophobic moment plot. *Journal of Molecular Biology*, 179, 125–142.

Felmlee, T., Pellett, S. & Welch, R. A. (1985). Nucleotide sequence of an *Escherichia coli* chromosomal hemolysin. *Journal of Bacteriology*, 163, 94–105.

Forsberg, Å., Bölin, I., Norlander, L. & Wolf-Watz, H. (1987). Molecular cloning and expression of calcium-regulated, plasmid-coded proteins of *Y. pseudotuberculosis. Microbial Pathogenesis*, **2**, 123–37.

Forsberg, Å, Viitanen, A.-M., Skurnik, M. & Wolf-Watz, H. (1991). The surface-located YopN protein is involved in calcium signal transduction in *Yersinia pseudotuberculosis. Molecular Microbiology*, **5**, 977–86.

Forsberg, Å. & Wolf-Watz, H. (1988). The virulence protein Yop5 of *Yersinia pseudotuberculosis* is regulated at the transcriptional level by plasmid-pIB1-encoded trans-acting elements controlled by temperature and calcium. *Molecular Microbiology*, **2**, 121–33.

Forsberg, Å. & Wolf-Watz, H. (1990). Genetic analysis of the *yopE* region of *Yersinia* spp.: identification of a novel conserved locus, *yerA*, regulating *yopE* expression. *Journal of Bacteriology*, **172**, 1547–55.

Gemski, P., Lazere, J. R. & Casey, T. (1980*a*). Plasmid associated with pathogenicity and calcium dependency of *Yersinia enterocolitica. Infection and Immunity*, **27**, 682–5.

Gemski, P., Lazere, J. R., Casey, T. & Wohlieter, J. A. (1980*b*). Presence of a virulence-associated plasmid in *Y. pseudotuberculosis. Infection and Immunity*, **28**, 1044–7.

Goguen, J. D., Yother, J. & Straley, S. C. (1984). Genetic analysis of the low calcium response in *Yersinia pestis* Mu d1(Ap *lac*) insertion mutants. *Journal of Bacteriology*, **160**, 842–8.

Göransson, M., Sonden, B., Nilsson, P. *et al.* (1990). Transcriptional silencing and thermoregulation of gene expression in *Escherichia coli. Nature*, **344**, 682–5.

Grützkau, A., Hanski, C., Hahn, H. & Riecken, E. O. (1990). Involvement of M cells in the bacterial invasion of Peyer's patches: a common mechanism shared by *Yersinia enterocolitica* and other enteroinvasive bacteria. *Gut*, **31**, 1011–15.

Guan, K. & Dixon, J. E. (1990). Protein tyrosine phosphatase activity of an essential virulence determinant in *Yersinia. Science*, **249**, 553–6.

Hale, T. L. (1991). Genetic basis of virulence in *Shigella* species. *Microbiological Reviews*, **55**, 206–24.

Hanski, C., Kutschka, U., Schmoranzer, H. P. *et al.* (1989). Immunohistochemical and electron microscopic study of interaction of *Yersinia enterocolitica* serotype O:8 with intestinal mucosa during experimental enteritis. *Infection and Immunity*, **57**, 673–8.

Hanski, C., Naumann, M., Grutzkau, A., Pluschke, G., Friedrich, B., Hahn, H. & Riecken, E. O. (1991). Humoral and cellular defense against intestinal murine infection with *Yersinia enterocolitica. Infection and Immunity*, **59**, 1106–11.

Hartman, A. B., Venkatesan, M., Oaks, E. V. & Buysse, J. M. (1990). Sequence and molecular characterization of a multicopy invasion plasmid antigen gene, *ipaH*, of *Shigella flexneri. Journal of Bacteriology*, **172**, 1905–15.

Heesemann, J., Algermissen, B. & Laufs, R. (1984). Genetically manipulated virulence of *Y. enterocolitica. Infection and Immunity*, **46**, 105–10.

Heesemann, J., Gross, U., Schmidt, N. & Laufs, R. (1986). Immunochemical analysis of plasmid-encoded proteins released by enteropathogenic *Yersinia* sp. grown in calcium-deficient media. *Infection and Immunity*, **54**, 561–7.

Heesemann, J. & Grüter, L. (1987). Genetic evidence that the outer membrane protein Yop1 of *Yersinia enterocolitica* mediates adherence and phagocytosis resistance to human epithelial cells. *FEMS Microbiology Letters*, **40**, 37–41.

Hein, J. (1990). Unified approach to alignment and phylogenies. In "Methods in Enzymology" vol 183, ed. R. Doolittle, pp. 626–645. Academic Press: London.

Higgins, C. F., Dorman, C. J., Stirling, D. A. *et al.* (1988). A physiological role for

DNA supercoiling in the osmotic regulation of gene expression in *S. typhimurium* and *E. coli. Cell*, **52**, 569–84.

Higgins, C. F., Hinton, J. C. D., Hulton, C. S. J. *et al.* (1990). Protein H1: a role for chromatin structure in the regulation of bacterial gene expression and virulence? *Molecular Microbiology* **4**, 2007–12.

Highlander, S. K., Chidambaram, M., Engler, M. J. & Weinstock, G. M. (1989). DNA sequence of the *Pasteurella haemolytica* leukotoxin gene cluster. *DNA*, **8**, 15–28.

Hoe, N. P., Minion, F. C. & Goguen, J. D. (1990). Analysis of the *lcrF* gene of *Y. pestis. Journal of Cellular Biochemistry* (Suppl. 14C), 183.

Hulton, C. S. J., Seirafi, A., Hinton, J. C. D. *et al.* (1990). Histone-like protein H1 (H-NS), DNA supercoiling, and gene expression in bacteria. *Cell*, **63**, 631–42.

Inouye, S., Nakazawa, A. & Nakazawa, T. (1988). Nucleotide sequence of the regulatory gene *xylR* of the TOL plasmid from *Pseudomonas putida. Gene*, **66**, 301–6.

Isberg, R. R. & Leong, J. M. (1990). Multiple β_1 chain integrins are receptors for invasin, a protein that promotes bacterial penetration into mammalian cells. *Cell*, **60**, 861–71.

Isberg, R. R., Swain, A. & Falkow, S. (1988). Analysis of expression and thermo-regulation of the *Yersinia pseudotuberculosis inv* gene with hybrid proteins. *Infection and Immunity*, **56**, 2133–8.

Isberg, R. R., Voorhis, D. L. & Falkow, S. (1987). Identification of invasin: a protein that allows enteric bacteria to penetrate cultured mammalian cells. *Cell*, **50**, 769–78.

Kapperud, G., Namork, E., Skurnik, M. & Nesbakken, T. (1987). Plasmid-mediated surface fibrillae of *Yersinia pseudotuberculosis* and *Yersinia enterocolitica*: relationship to the outer membrane protein YOP1 and possible importance for pathogenesis. *Infection and Immunity*, **55**, 2247–54.

Lally, E. T., Folub, E. E., Kieba, I. R., Taichman, N. S., Rosenbloom, J., Rosenbloom, J. C., Gibson, C. W. & Demuth, D. R. (1989). Analysis of the *Actinobacillus actinomycetemcomitans* leukotoxin gene: delineation of unique features and comparison to homologous toxins. *Journal of Biological Chemistry*, **264**, 15451–6.

Lambert de Rouvroit, C., Sluiters, C. & Cornelis, G. R. (1982). Role of the transcriptional activator, VirF, and temperature in the expression of the pYV plasmid genes of *Yersinia enterocolitica. Molecular Microbiology*, **6**, 395–409.

Lawton, W. D., Erdman, R. L. & Surgalla, M. J. (1983). Biosynthesis and purification of V and W antigens in *Pasteurella pestis. Journal of Immunology*, **91**, 179–84.

Leppla, S. H. (1982). Anthrax toxin edema factor: a bacterial adenylate cyclase that increases cAMP concentration in eucaryotic cells. *Proceedings of National Academy of Sciences, USA*, **72**, 2284–8.

Leung, K. Y., Reisner, B. S. & Straley, S. C. (1990). YopM inhibits platelet aggregation and is necessary for virulence of *Yersinia pestis* in mice. *Infection and Immunity*, **58**, 3262–71.

Leung, K. Y. & Straley, S. C. (1989). The *yopM* gene of *Yersinia pestis* encodes a released protein having homology with the human platelet surface protein GPIbα. *Journal of Bacteriology*, **171**, 4623–32.

Lian, C-J., Hwang, W. S. & Pai, C. H. (1987). Plasmid-mediated resistance to phagocytosis in *Yersinia enterocolitica. Infection and Immunity*, **55**, 1176–83.

Lundrigan, M. D., Friedrich, M. J. & Kadner, R. J. (1989). Nucleotide sequence of the *Escherichia coli* porin thermoregulatory gene *envY*. *Nucleic Acids Research*, **17**, 800.

Martinez, R. J. (1983). Plasmid-mediated and temperature-regulated surface properties of *Yersinia enterocolitica*. *Infection and Immunity*, **41**, 921–30.

Martinez, R. J. (1989). Thermoregulation-dependent expression of *Yersinia enterocolitica* protein 1 imparts serum resistance to *Escherichia coli* K-12. *Journal of Bacteriology*, **171**, 3732–9.

Maurelli, A. T. & Sansonetti, P. J. (1988). Identification of a chromosomal gene controlling temperature-regulated expression of *Shigella* virulence. *Proceedings of National Academy of Sciences, USA*, **85**, 2820–4.

Michiels, T. & Cornelis, G. (1988). Nucleotide sequence and transcription analysis of *yop51* from *Yersinia enterocolitica* W22703. *Microbial Pathogenenis*, **5**, 449–59.

Michiels, T. & Cornelis, G. R. (1991). Secretion of hybrid proteins by the *Yersinia* Yop export system. *Journal of Bacteriology*, **173**, 1677–85.

Michiels, T., Vanooteghem, J.-C., Lambert de Rouvroit, C. *et al.* (1991). Analysis of *virC*, an operon involved in the secretion of Yop proteins by *Yersinia enterocolitica*. *Journal of Bacteriology*, **173**, 4994–5009.

Michiels, T., Wattiau, P., Brasseur, R., Ruysschaert, J.-M. & Cornelis, G. (1990). Secretion of Yop proteins by Yersiniae. *Infection and Immunity*, **58**, 2840–9.

Miller, V. L. & Falkow, S. (1988). Evidence for two genetic loci in *Yersinia enterocolitica* that can promote invasion of epithelial cells. *Infection and Immunity*, **56**, 1242–8.

Miller, V. L., Bliska, J. B. & Falkow, S. (1990). Nucleotide sequence of the *Yersinia enterocolitica ail* gene and characterization of the Ail protein product. *Journal of Bacteriology*, **172**, 1062–9.

Miller, V. L., Farmer III, J. J., Hill, W. E. & Falkow, S. (1989). The *ail* locus is found uniquely in *Yersinia enterocolitica* serotypes commonly associated with disease. *Infection and Immunity*, **57**, 121–31.

Mulder, B., Michiels, T., Sory, M.-P., Simonet, M. & Cornelis, G. (1989). Identification of additional virulence determinants on the pYV plasmid of *Y. enterocolitica* W227. *Infection and Immunity*, **57**, 2534–41.

Pai, C. H. & Mors, V. (1978). Production of enterotoxin by *Yersinia enterocolitica*. *Infection and Immunity*, **19**, 908–11.

Perry, R. D. & Brubaker, R. R. (1983). Vwa$^+$ phenotype of *Yersinia enterocolitica*. *Infection and Immunity*, **40**, 166–71.

Perry, R. D. & Brubaker, R. R. (1987). Transport of Ca^{2+} by *Yersinia pestis*. *Journal of Bacteriology*, **169**, 4861–4.

Perry, R. D., Harmon, P. A., Bowmer, W. S. & Straley, S. C. (1986). A low-Ca^{2+} response operon encodes the V antigen of *Yersinia pestis*. *Infection and Immunity*, **54**, 428–34.

Pierson, D. & Falkow, S. (1990). Nonpathogenic isolates of *Yersinia enterocolitica* do not contain functional *inv*-homologous sequences. *Infection and Immunity*, **58**, 1059–64.

Plano, G. V., Barve, S. S. & Straley, S. C. (1991). LcrD, a membrane-bound regulator of the *Yersinia pestis* low-calcium response. *Journal of Bacteriology*, **173**, 7293–303.

Pollack, C., Straley, S. C. & Klempner, M. S. (1986). Probing the phagolysosomal environment of human macrophages with a Ca^{2+} responsive operon fusion in *Yersinia pestis*. *Nature*, **322**, 834–6.

Portnoy, D. A., Moseley, S. L. & Falkow, S. (1981). Characterization of plasmids and plasmid-associated determinants of *Yersinia enterocolitica* pathogenesis. *Infection and Immunity*, **31**, 775–82.

Price, S. B., Leung, K. Y., Barve, S. S. & Straley, S. C. (1989). Molecular analysis of *lcrGVH*, the V antigen operon of *Yersinia pestis*. *Journal of Bacteriology*, **171**, 5646–53.

Price, S. B. & Straley, S. C. (1989). *lcrH*, a gene necessary for virulence of *Yersinia pestis* and for the normal response of *Y. pestis* to ATP and calcium. *Infection and Immunity*, **57**, 1491–8.

Robins-Browne, R. M., Still, C. S., Miliotis, M. D. & Koornhof, H. J. (1979). Mechanism of action of *Yersinia enterocolitica* enterotoxin. *Infection and Immunity*, **25**, 680–4.

Rosqvist, R., Bölin, I. & Wolf-Watz, H. (1988*a*). Inhibition of phagocytosis in *Yersinia pseudotuberculosis*: a virulence plasmid-encoded ability involving the Yop2b protein. *Infection and Immunity*, **56**, 2139–43.

Rosqvist, R., Forsberg, A., Rimpilainen, M., Bergman, T. & Wolf-Watz, H. (1990). The cytotoxic protein YopE of *Yersinia* obstructs the primary host defence. *Molecular Microbiology*, **4**, 657–7.

Rosqvist, R., Forsberg, A. & Wolf-Watz, H. (1991). Intracellular targeting of the *Yersinia* YopE cytotoxin in mammalian cells induces actin microfilament disruption. *Infection and Immunity*, **59**, 4562–9.

Rosqvist, R., Skurnik, M. & Wolf-Watz, H. (1988a). Increased virulence of *Yersinia pseudotuberculosis* by two independent mutations. *Nature*, **334**, 522–5.

Sakai, T., Sasakawa, C., Makino, S. & Yoshikawa, M. (1986). DNA sequence and product analysis of the *virF* locus responsible for Congo red binding and cell invasion in *Shigella flexneri* 2a. *Infection and Immunity*, **54**, 395–402.

Sakai, T., Sasakawa, C. & Yoshikawa, M. (1988). Expression of four virulence antigens of *Shigella flexneri* is positively regulated at the transcriptional level by the 30 kiloDalton *virF* protein. *Molecular Microbiology*, **2**, 589–97.

Schollmeier, K. & Hillen, W. (1984). Transposon *Tn10* contains two structural genes with opposite polarity between *tetA* and IS*10*$_R$. *Journal of Bacteriology*, **160**, 499–503.

Schuler, G. D., Altschul, S. F. & Lipman, D. J. (1991). A workbench for multiple alignment construction and analysis. *Prot. Stuct. Funct. Genet.* **9**, 180–90.

Simonet, M., Richard, S. & Berche, P. (1990). Electron microscopic evidence for *in vivo* extracellular localization of *Yersinia pseudotuberculosis* harboring the pYV plasmid. *Infection and Immunity*, **58**, 841–5.

Skurnik, M., Bölin, I., Heikkinen, H., Piha, S. & Wolf-Watz, H. (1984). Virulence plasmid-associated autoagglutination in *Yersinia* spp. *Journal of Bacteriology*, **158**, 1033–6.

Skurnik, M. & Wolf-Watz, H. (1989). Analysis of the *yopA* gene encoding the Yop1 virulence determinants of *Yersinia* spp. *Molecular Microbiology*, **3**, 517–29.

Sodeinde, O. A. & Goguen, J. D. (1989). Nucleotide sequence of the plasminogen activator gene of *Yersinia pestis*: relationship to *ompT* of *Escherichia coli* and gene E of *Salmonella typhimurium*. *Infection and Immunity*, **57**, 1517–23.

Sodeinde, O. A., Sample, A. K., Brubaker, R. R. & Goguen, J. D. (1988). Plasminogen activator/coagulase gene of *Yersinia pestis* is responsible for degradation of plasmid-encoded outer membrane protein 3. *Infection and Immunity*, **56**, 2749–52.

Sory, M.-P. & Cornelis, G. (1988). *Yersinia enterocolitica* O:9 as a potential live oral carrier for protective antigens. *Microbial Pathogenesis*, **4**, 431–42.

Sory, M.-P., Hermand, P., Vaerman, J. P. & Cornelis, G. R. (1990). Oral immunization of mice with a live recombinant *Yersinia enterocolitica* O:9 strain that produces the cholera toxin B subunit. *Infection and Immunity*, **58**, 2420–8.

Straley, S. C. (1988). The plasmid-encoded outer-membrane proteins of *Yersinia pestis*. *Reviews of Infectious Diseases*, **10**, S323–6.

Straley, S. & Bowmer, W. (1986). Virulence genes regulated at the transcriptional level by Ca^{2+} in *Yersinia pestis* include structural genes for outer membrane proteins. *Infection and Immunity*, **51**, 445–54.

Straley, S. C. & Brubaker, R. R. (1981). Cytoplasmic and membrane proteins of yersiniae cultivated under conditions simulating mammalian intracellular environment. *Proceedings of National Academy of Sciences, USA*, **78**, 1224–8.

Straley, S. C. & Cibull, M. L. (1989). Differential clearance and host-pathogen interaction of YopE⁻ and YopK⁻ YopL⁻ *Yersinia pestis* in Balb/c micc. *Infection and Immunity*, **57**, 1200–10.

Sukupolvi, S. & O'Connor, D. (1990). TraT lipoprotein, a plasmid-specified mediator of interactions between Gram-negative bacteria and their environment. *Microbiological Reviews*, **54**, 331–41.

Takao, T., Tominaga, N. & Shimonishi, Y. (1984). Primary structure of heat-stable enterotoxin produced by *Yersinia enterocolitica*. *Biochemical and Biophysical Research Communications*, **125**, 845–51.

Takao, T., Tominaga, N., Yoshimura, S. *et al.* (1985). Isolation, primary structure and synthesis of heat-stable enterotoxin produced by *Yersinia enterocolitica*. *European Journal of Biochemistry*, **152**, 199–206.

Tauxe, R. V., Vandepitte, J., Wauters, G. *et al.* (1987). *Yersinia enterocolitica* infections and pork: the missing link. *Lancet*, **ii**, 1129–32.

Tipple, M. A., Bland, L. A., Murphy, J. J. *et al.* (1990). Sepsis associated with transfusion of red cells contaminated with *Yersinia enterocolitica*. *Transfusion*, **30**, 207–13.

Tobe, T., Nagai, S., Okada, N., Adler, B., Yoshikawa, M. & Sasakawa, C. (1991). Temperature-regulated expression of invasion genes in *Shigella flexneri* is controlled through the transcriptional activation of the *virB* gene on the large plasmid. *Molecular Microbiology*, **5**, 887–93.

Tobin, J. F. & Schleif, R. F. (1987). Positive regulation of the *Escherichia coli* L-rhamnose operon is mediated by the products of tandemly repeated regulatory genes. *Journal of Molecular Biology*, **196**, 789–99.

Une, T. & Brubaker, R. R. (1984). Roles of V antigen in promoting virulence and immunity in yersiniae. *Journal of Immunology*, **133**, 2226–30.

Vanooteghem, J.-C. & Cornelis, G. R. (1990). Structural and functional similarities between the replication region of the *Yersinia* virulence plasmid and the RepFIIA replicons. *Journal of Bacteriology*, **172**, 3600–8.

Viitanen, A.-M., Toivanen, P. & Skurnik, M. (1990). The *lcrE* gene is part of an operon in the *lcr* region of *Yersinia enterocolitica*. *Journal of Bacteriology*, **172**, 3152–62.

Venkatesan, M. M., Buysse, J. M. & Kopecko, D. J. (1988). Characterization of invasion plasmid antigen genes (*ipaBCD*) from *Shigella*. *Proceedings of National Academy of Sciences, USA*, **85**, 9317–21.

Wallace, R. G., Lee, N. & Fowler, A. V. (1980). The *araC* gene of *Escherichia coli*: transcriptional and translational start-points and complete nucleotide sequence. *Gene*, **12**, 179–90.

Wartenberg, K., Knapp, W., Ahamed, N. M., Wideman, C. & Mayer, H. (1983). Temperature-dependent changes in the sugar and fatty acid composition of lipopolysaccharides from *Yersinia enterocolitica* strains. *Zentralblatt für Bakteriologie Mikrobiologie und Hygiene*, **253**, 523–30.

Webster, C., Kempsell, K., Booth, I. & Busby, S. (1987). Organisation of the regulatory region of the *Escherichia coli* melibiose operon. *Gene*, **59**, 253–63.

Wu, J. & Weiss, B. (1991). Two divergently transcribed genes, *soxR* and *soxS*, control a superoxide response regulon of *Escherichia coli*. *Journal of Bacteriology*, **173**, 2864–71.

Yother, J., Chamness, T. W. & Goguen, J. D. (1986). Temperature-controlled plasmid regulon associated with low calcium response in *Yersinia pestis*. *Journal of Bacteriology*, **165**, 443–7.

Yother, J. & Goguen, J. D. (1985). Isolation and characterization of Ca^{2+} blind mutants in *Yersinia pestis*. *Journal of Bacteriology*, **164**, 704–11.

Young, V. B., Miller, V. L., Falkow, S. & Schoolnik, G. K. (1990). Sequence, localization and function of the invasin protein of *Yersinia enterocolitica*. *Molecular Microbiology*, **4**, 1119–28.

Zahorchak, R. J. & Brubaker, R. R. (1982). Effect of exogenous nucleotides on Ca^{2+} dependence and V antigen synthesis in *Yersinia pestis*. *Infection and Immunity*, **38**, 953–9.

BACTERIAL FIMBRIAE: VARIATION AND REGULATORY MECHANISMS

C. J. SMYTH AND S. G. J. SMITH

Department of Microbiology,
Moyne Institute,
Trinity College,
Dublin,
Republic of Ireland

INTRODUCTION

The capacities of bacteria to cause disease relate to the sums of expression of a variety of virulence attributes at different stages in the pathogenesis of infections. There is often a complex series of interactions between the susceptible host and the invading pathogen. For those pathogens which colonize mucosal surfaces or invade through them, the ability to adhere to epithelial cells in sufficient numbers to establish an infection is a crucial event.

Knowledge of the nature and structure of bacterial adhesins and of their receptors on target cells has steadily increased through advances in scientific technology and inventiveness. Not least over the past decade, the advent of molecular genetic techniques and their application to bacterial pathogenesis has led to a considerable body of information on the genetic determinants encoding the adhesins of a broad spectrum of bacteria, especially on those specifying the fimbrial adhesins of *Escherichia coli*. The amino acid sequences of the subunits of many types of bacterial fimbriae have been revealed by cloning and nucleotide sequencing of the pilin genes. However, the status of knowledge of each fimbrial determinant differs widely in terms of operon structure, the numbers of individual genes involved in biogenesis, the natures and roles of individual gene products and the regulatory mechanisms underlying biogenesis, antigenic modulation and phase variation. Indeed, the recognition that sets of virulence factors are regulated in a coordinated manner in response to changes in the environment has expanded the intellectual horizons of required understanding of mechanisms of bacterial pathogenicity, not just in terms of primary host tissue–pathogen interaction but also within the host tissues at different body sites.

This chapter will focus principally on variation and regulation of expression of bacterial fimbriae. For in-depth treatment of other aspects of bacterial adhesion there is no shortage of excellent reviews (Birkbeck & Penn, 1986; Donachie, Griffiths & Stephen, 1988; Doyle & Rosenberg,

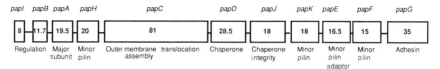

Fig. 1. Structure and function of the *papI* and *papB* operons. The established and postulated functions of the gene products are indicated. The numbers refer to molecular mass in kD. (Adapted from Hultgren *et al.*, 1991 and Uhlin *et al.*, 1985.)

1990; Iglewski & Clark, 1990; Jann & Jann, 1990; Krogfelt, 1991; Lark, 1986; Mirelman, 1986; Proctor, 1987; Ron & Rottem, 1991; Roth, 1988; Switalski, Höök & Beachey, 1988; see Smyth, 1988 for compilation of other reviews).

ESCHERICHIA COLI

P Fimbriae

The majority of *E. coli* isolates associated with acute infantile pyelonephritis express mannose-resistant adhesins that bind to the P blood group antigens of uroepithelial cells. These adhesins recognize the $Gal(\alpha1-4)Gal$ moiety of these glycolipid antigens and are termed P fimbriae. There are eight serological variants of P fimbriae and these have been organized into different F serogroups (Hacker, 1990). The best characterized serovars are F13, $F7_1$, $F7_2$, and F11 which are encoded by the *pap, fso, fst,* and *fel* determinants, respectively.

The chromosomally located determinant for Pap (*p*yelonephritis *a*ssociated *p*ili) fimbriae (serogroup F13) has been cloned and its nucleotide sequence determined (Tennent *et al.*, 1990; Uhlin *et al.*, 1985; Normark *et al.*, 1985, 1986). Using a combination of transposon mutagenesis in conjunction with minicell analysis and DNA sequence analysis, 11 genes have been implicated in the expression of Pap fimbriae. These genes are organized into two divergently transcribed operons, namely the *papI* and *papB* operons. The *papB* operon comprises ten cistrons, whereas the *papI* operon is monocistronic. The products of the *papB* and *papI* genes regulate Pap fimbriae expression, whereas the remaining nine genes (*papA–papK*) are involved in fimbrial assembly, subunit translocation and adhesion (Hultgren, Normark & Abraham, 1991). A diagrammatic representation of the organization of the *papB* and *papI* operons is shown in Fig. 1.

The regulation of Pap fimbriae expression is complex, requiring several other proteins in addition to PapB and PapI (Fig. 2). Expression is thermoregulated and is subject to catabolite repression and phase variation. The stoichiometry of polypeptides within the fimbrial structure dictates that the major pilin should be present in a 1000-fold excess with respect to the

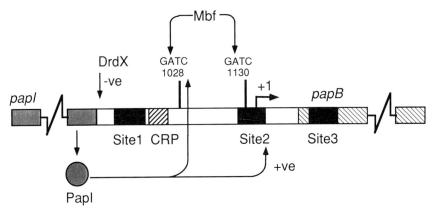

Fig. 2. The repertoire of proteins acting on the UAS region. The three PapB-binding sites and the binding site for the cAMP–CRP complex are indicated. DrdX reduces transcription from the *papI* promoter whilst PapI has a stimulatory effect on transcription from the *papB* promoter. The GATC sites which Mbf alone or Mbf and PapI protect from the action of Dam methylase are indicated. It is not known whether the methylation blocking activity of PapI and the stimulatory effect of PapI are one and the same.

adhesin and anchor proteins. This presents the bacterium with the problem of differential expression of genes within the same operon.

The regulatory region of the *pap* gene cluster is located upstream of the *papA* gene. The PapB and PapI proteins encoded in this region appear to act as transcriptional activators and have molecular weights of 11.7 kD and 8–8.5 kD, respectively (Båga *et al.*, 1985). The *papI* gene is weakly expressed and its gene product stimulates *papB* transcription. Transcription from the *papB* promoter and hence expression of the *papB*, *papA* and subsequent genes also requires the PapB protein. Transcription from the *papB* promoter is also dependent on cAMP and the cAMP receptor protein (CRP) (see Dorman & Ní Bhriain, this symposium). Production of β-galactosidase from a *papA–lacZ* transcriptional fusion plasmid is 10- to 20-fold lower in a *crp* or a *cya* mutant compared to wild-type strains. The wild-type level of β-galactosidase is restored by providing exogenous cAMP. A possible binding site for the cAMP–CRP complex is located upstream of the *papB* promoter, and this complex has been shown to bind to this region in gel mobility shift assays (Forsman, Göransson & Uhlin, 1989).

The *papI/papB* intercistronic region, also known as the upstream activating sequence (UAS), contains three binding sites for PapB (Forsman *et al.*, 1989). These sites are consecutively occupied at increasing concentrations of the PapB activator. Site 1, which is adjacent to the cAMP–CRP binding site and upstream of the *papB* promoter, appears to be occupied when the concentration of PapB is low. It is likely that PapB and cAMP–CRP act in concert at this site to stimulate expression from both the *papB* and *papI* promoters. Site 2 is localized in the *papB* promoter region and is occupied with increasing levels of PapB. Binding of PapB to this site is likely to prevent RNA polymerase from initiating transcription. A further binding site, site 3, lies within the *papB* gene.

The existence of an operator within the coding region of the *papB* gene is reminiscent of the organization of the *gal* operon in *E. coli* (Irani, Orsoz & Adhya, 1983). It has been proposed that the cooperative binding of GalR to operators within and outside the coding region of the *galE* gene efficiently reduces transcription from the *galE* promoter. The formation of DNA loops brought about by the interaction of two proteins bound at two different DNA sites is seen in the regulation of the *araCBAD* operon (Huo, Martin & Schleif, 1988). The *araCBAD* regulatory region, like the *pap* UAS, contains three binding sites for its regulator, AraC. In the absence of arabinose, a repressive DNA loop is formed. With increasing concentrations of arabinose, an alternative DNA loop is formed by AraC which precludes the formation of the repressive DNA loop and allows expression of the operon. It may be possible that the formation of alternative DNA loops is involved in the PapB-modulated expression of the *papB* operon. The overproduction of the PapB protein in *trans* causes reduced expression from the *papB* operon. The differential binding efficiencies of these regulator-binding sites offers an explanation for the autoregulatory mode of action of PapB.

The major pilin gene, *papA*, is transcribed within a polycistronic mRNA. This message is subject to post-transcriptional processing to maintain the stoichiometry of components required for fimbrial integrity (Båga *et al.*, 1988). Transcription commences at the *papB* promoter and proceeds through the *papA* gene. In the *papA/papH* intercistronic region there exists a region of dyad symmetry which forms a stem-loop structure and efficiently reduces transcription into the remainder of the operon, thereby ensuring that the *papBA* transcript is relatively more abundant. The *papBA* message is then subject to endonucleolytic cleavage in the *papB/papA* intercistronic region; the 5' half of the mRNA, encoding the *papB* gene, is rapidly degraded by 3'–5' exoribonucleases. The *papA*-encoding 3' end of the message is not degraded. This stability may be due to the previously mentioned stem-loop structure located downstream of the *papA* termination codon. The overall result of these processes is the selective accumulation of the *papA* transcript with respect to other transcripts encoded in this operon. This accumulation of the *papA* message is a major contributing factor to the high level of expression of the pilin polypeptide.

Pap fimbriae undergo phase variation (Low *et al.*, 1987). This is mediated through the differential methylation states of two Dam (*deoxyadenosine methylase*) methylation sites ($GATC_{1130}$ and $GATC_{1028}$) within the UAS region (Blyn, Braaten & Low, 1990; Blyn *et al.*, 1989). When fimbriae are expressed, $GATC_{1028}$ is unmethylated and $GATC_{1130}$ is methylated; conversely, when fimbriae are not expressed, $GATC_{1028}$ is methylated and

$GATC_{1130}$ is unmethylated. The protection of these sites from the action of Dam requires Mbf (*M*ethylation *b*locking *f*actor) encoded by the *mbf* locus which maps at 19.6 minutes on the *E. coli* chromosome (Braaten *et al.*, 1991). The protection of $GATC_{1028}$ requires PapI and Mbf, whereas the protection of $GATC_{1130}$ requires only Mbf. A model explaining Pap fimbriae phase variation is dealt with by Robertson & Meyer (this symposium). It is important to note that phase variation is not seen when cells harbour multicopy plasmids encoding the *pap* determinant (Blyn *et al.*, 1989).

The expression of Pap fimbriae, like that of many adhesins, is subject to thermoregulation. Transcription from both the *papI* and *papB* operons is essentially turned off at 26 °C. The thermoregulation of fimbrial expression is due to the limiting amount of *papI* mRNA. The overexpression of *papI* from an expression vector leads to fimbrial expression at low temperatures (Göransson, Forsman & Uhlin, 1989). The expression of β-galactosidase at 20 °C in a strain bearing a *pap–lacZ* fusion was derepressed in a spontaneously occurring mutant. The mutant locus mapped at 27.5 minutes on the *E. coli* chromosome and was designated *drdX*, for *d*erepressed *ex*pression. This locus was cloned and sequenced and was shown to have one open-reading frame encoding a 15.5 kD protein (Göransson *et al.*, 1990). The amino acid sequence of this protein was shown to be identical to that of H-NS, a DNA-binding protein (see Dorman & Ní Bhriain, this symposium). The *drdX* mutation maps to a similar location on the *E. coli* chromosome as *bglY* (Lejeune & Danchin, 1990), *osmZ* (Higgins *et al.*, 1988), *pilG* (Kawula & Orndorff, 1991) and *virR* (Dorman, Ní Bhriain & Higgins, 1990). Further to this, *drdX, osmZ*, and *bglY* demonstrate similar Bgl phenotypes.

White-Ziegler *et al.* (1990) have identified a locus, *tcp* (*t*hermoregulatory *c*ontrol of *p*ap), which is important in the thermoregulatory control of Pap pilin expression. This locus is located at 23.4 minutes on the *E. coli* chromosome. Transposon Tn*10* insertion mutations in this region lead to derepressed expression of Pap fimbriae and a higher Pap pilin phase OFF → phase ON transition frequency at 37 °C. It is probable, although not proven, that *drdX* and *tcp* are identical. Two separate phenotypes were recognized for these mutants, i.e. those that showed elevated Pap pilin phase OFF → phase ON transition frequency independent of temperature and those that demonstrated elevated Pap pilin phase OFF → phase ON transition frequency at 37 °C only. The chromosomal site of transposon insertion corresponding to these two mutant phenotypes is different but they are closely linked (within 5 kb).

An analogous situation occurs with the expression of type 1 fimbriae, since mutations in the gene for PilG cause higher DNA inversion rates of the *pilA* promoter (Kawula & Orndorff, 1991). A mutant allele termed *pilG2–tetR*, with an insertion mutation within the *pilG* coding region, shows an elevated DNA inversion rate irrespective of temperature. In contrast to the second type of *tcp* mutant, a T-to-G base change within the *pilG* promoter, gives rise to the mutant phenotype at low temperature only. Recently, the *osmZ* locus was shown to consist of three genetically active regions with respect to their effects on *proU* expression, i.e. regions I, II and III, with region I being equivalent to *hns* (Barr, Ní Bhriain & Dorman, 1992). If the *tcp* locus is identical to the *drdX/osmZ* locus, it may be possible that the different mutant phenotypes outlined above are due to insertions in different genetically active regions of the *tcp* operon.

Cloned determinants for P fimbrial expression from strains producing serotype F7$_2$ and F11 fimbriae have sequences homologous to the *papI* and *papB* genes of serotype F13 pili. These cloned determinants can complement mutations in *papI* and *papB*. The *papI* and *papB* genes are also homologous to the *sfaB* and *sfaC* genes which are involved in the regulation of the S fimbrial adhesin determinant (Göransson, Forsman & Uhlin, 1988). The *sfa* determinant shares many characteristics with the *pap* determinant. For example, its expression is subject to catabolite repression and thermoregulation (Schmoll *et al.*, 1990*b*), there is a divergent arrangement of promoters (Schmoll *et al.*, 1990*a*), and there are overall similarities in the organization of cistrons within the *pap* and *sfa* determinants (Hacker, 1990). The regulators of these two fimbrial systems appear to be interchangeable since *papI* and *papB* mutations can be complemented by a cloned S fimbrial determinant (Göransson *et al.*, 1988).

In contrast to the *pap* determinant, the *sfa* determinant has a third promoter (P$_A$) located in front of the *sfaA* gene for the major subunit. However, this promoter is weak and cannot be solely responsible for high level expression of *sfaA* (Schmoll *et al.*, 1990*a*). The *sfaC* and *sfaB* genes are organized in a manner analogous to the *papI* and *papB* genes, respectively. SfaC positively influences the *sfaB* promoter, P$_B$, and has no effect on its own promoter, P$_C$. The SfaB protein stimulates expression from both of these promoters. Neither SfaC nor SfaB have any effect on P$_A$. The *sfa* determinant shows multiple gene homologies with another gene cluster (*foc*) encoding F1C fimbriae (Ott *et al.*, 1987). However, F1C fimbriae do not agglutinate erythrocytes. Six genes involved in the biogenesis of F1C fimbriae have been identified (Riegman *et al.*, 1990). To date, no regulatory genes have been identified for expression of the *foc* determinant but a *sfaC/ sfaB* probe did hybridize to a cloned *foc* fimbrial operon.

Human ETEC Fimbriae

Adhesion of enterotoxigenic *Escherichia coli* (ETEC) to the mucosa of the small intestine is mediated by adhesins termed colonization factor antigens (CFAs) of which there are several types (for review see Smyth, 1986; Smyth *et al.*, 1991). CFA/II and CFA/IV fimbriae represent multiple fimbrial types. The CFA/II complex is composed of three coli-surface-associated antigens, i.e. CS1, CS2 and CS3 fimbriae. CS3 fimbriae may be expressed alone or with either CS1 or CS2 fimbriae. Occasionally CS2 fimbriae are expressed alone (Smyth, 1986). The CFA/IV complex consists of three antigens, viz., CS4, CS5 and CS6. The CS6 antigen can be expressed on its own or with either CS4 or CS5 fimbriae or CS5 fimbriae can be expressed alone (Smyth *et al.*, 1991).

Expression of CFA/II fimbriae is plasmid-mediated. Mobilization of CS-fimbriae-associated plasmids from their wild-type serotype O6: K15: H16 or H-background into other O6 serotypes or other O serovars or *E. coli* K-12 gives rise to expression of CS3 fimbriae only (Boylan & Smyth, 1985; Twohig, Boylan & Smyth, 1988). This suggested that expression of CS1 or CS2 fimbriae or both required genetic information from the host chromosome and from the CS-associated plasmids (Smyth, 1986; Smyth *et al.*, 1991).

CS3 fimbriae

The genetic determinant for the expression of CS3 fimbriae has been cloned (Boylan, Coleman & Smyth, 1987; Manning, Timmis & Stevenson, 1985). Regions associated with the expression of specific gene products have been identified by deletion mutagenesis and transposon mutagenesis coupled with minicell analysis. It became apparent from these experiments that more DNA was required to encode the proteins identified than was available in the cloned CS3 fimbrial determinants. Nucleotide sequencing of the entire cloned determinant has revealed that the genes for four proteins (63 kD, 48 kD, 33 kD and 20 kD) are encoded within the reading frame of a fifth protein (104 kD). For synthesis of the 104 kD protein, suppression of an internal amber codon is required. The genes encoding two further proteins, a periplasmic protein and the major CS3 fimbrial subunit, are located upstream and downstream, respectively, of the open-reading frame for the 104 kD protein (Jalajakumari *et al.*, 1989).

Rns

CS1/CS2 fimbrial determinants cloned and subcloned from plasmids in ETEC strains gave rise to the same host-related restriction of expression of these fimbriae in different host backgrounds as their parental CS-associated-plasmids (Boylan *et al.*, 1988; Boylan & Smyth, 1985; Willshaw *et al.*, 1988). DNA sequence analysis of the recombinant plasmid pCS200, which gives rise to CS1 or CS2 fimbrial expression in appropriate backgrounds (Boylan *et al.*, 1988), showed that there were no sequences present for fimbrial proteins (Caron, Coffield & Scott, 1989). Instead, the DNA insert in plasmid pCS200 contained the gene for a regulatory protein that acts positively on the genes for the production of CS1 and CS2 fimbriae. Its gene product was termed Rns (*Regulation of CS1 and CS2*). Recombinant plasmids containing DNA inserts with only the *rns* gene give rise to expression of CS1 and CS2 fimbriae in appropriate backgrounds.

The Rns protein is a 265 amino acid polypeptide with a molecular mass of 30 kD and a pI of 10.1. Significant amino acid homology exists between the C-terminal halves of the amino-acid sequences of Rns and the *E. coli* AraC protein, the RhaR and RhaS regulators of the *E. coli* rhamnose operon (Caron *et al.*, 1989), VirF, a plasmid-encoded transcriptional activator of the *Yersinia* virulence regulon (Cornelis *et al.*, 1989), and also the plasmid-encoded regulator of virulence factors of *Shigella flexneri*, VirF (Sakai *et al.*, 1986*a*,*b*; Sakai, Sasakawa & Yoshikawa, 1988). More importantly, the strongest amino acid homology between the Rns and AraC proteins is in the AraC DNA-binding domain (Caron *et al.*, 1989). The *Yersinia* VirF protein has been shown to be a DNA-binding protein (Lambert de Rouvroit, Sluiters & Cornelis, 1992). These data when taken together infer that the Rns protein is a transcriptional activator with DNA-binding properties, although the binding of the Rns protein to its target DNA *in vitro* has not yet been demonstrated. The G + C content of the *rns* gene (28%) is low for an *E. coli* gene (50%). This suggests that the *rns* gene originated from another genus. However, no hybridization was obtained with genomic DNA from other bacterial species using a *rns* DNA probe (Caron *et al.*, 1990).

Interestingly, a positive regulator of expression of the 987P operon in ETEC of porcine origin, designated FapR (*fimbria* 987*P*), shows an overall similarity to Rns (Klaasen & de Graaf, 1990; de Graaf, van der Woude & Klaasen, 1991). It has a molecular mass of 30.4 kD, comprises 260 amino acids, and has a pI of 9.6.

CS1/CS2 fimbriae

The CS1 fimbrial operon appears to be plasmid-located. Such plasmids are distinct from those containing the *rns* gene. Two groups have cloned the *cso* determinant (for *CS* one fimbriae, also known as *coo* [for *coli*-surface antigen *one*]) from different serogroups of *E. coli* (Perez-Casal, Swartley & Scott, 1990; Willshaw *et al.*, 1990*b*). The nucleotide sequences of the first two genes of the *cso* determinant, *csoA* and *csoB*, have been determined (Jordi *et al.*, 1991; Perez-Casal *et al.*, 1990; Scott *et al.*, 1992). The CsoA protein is the CS1 pilin, whereas the CsoB protein appears to be needed for the assembly but not the transport of the CS1 pilin. The *csoB* promoter region has been fused to the promoterless *phoA* gene of a promoter-probe vector. When the *rns* gene was supplied in *trans* on a compatible plasmid, phosphatase activity was enhanced approximately two-fold suggesting that Rns acts on the *csoB* promoter (Jordi *et al.*, 1991). Scott *et al.* (1992), using *in vitro* transcription-translation analysis on a cloned *cso* determinant, have shown that the determinant encodes an additional protein with a molecular mass of 43 kD. The function of this protein is currently unknown. Cloning of the CS2 fimbrial operon has not yet been described, although its location is likely to be chromosomal. Totally plasmidless CS2-fimbriae-negative strains, such as strain C91f-b (Smyth, 1982), which have lost their CS-associated-plasmid, express CS2 fimbriae upon introduction of a *rns*-containing recombinant plasmid (Boylan *et al.*, 1988; Caron *et al.*, 1989).

CFA/I fimbriae

Restriction endonuclease mapping and transposon mutagenesis of a CFA/I-associated plasmid has revealed two widely separated regions required for the expression of CFA/I fimbriae (Smith *et al.*, 1979). These two regions, designated regions 1 and 2, have been cloned into compatible plasmid vectors and both are required for the expression of CFA/I fimbriae (Willshaw, Smith & Rowe, 1983). DNA sequence analysis of region 1 revealed the presence of five open-reading frames, i.e. *cfaA*, *cfaB*, *cfaC*, *cfaE* and *cfaD'* (Gaastra *et al.*, 1990; Hamers *et al.*, 1989; Karjalainen *et al.*, 1989). The first four open-reading frames form an operon. The product of the *cfaB* gene is the major CFA/I fimbrial subunit, the gene product of the *cfaC* gene is likely to act as a membrane anchor and the CfaA protein is likely to have an analogous function to the CsoB protein. The major pilin subunits of CS1 and CFA/I fimbriae show 55% amino-acid sequence identity and the CsoB and CfaA proteins show 63% similarity. Both *csoA* and *cfaA* have in common a region of dyad symmetry downstream of their respective termination codons. This is analogous to the region of dyad symmetry found downstream of the *papA* gene. Accordingly, these motifs are likely to function in a similar manner.

Nucleotide sequencing of CFA/I region 2 has revealed the existence of an open-reading frame encoding a protein with a molecular mass of 30 kD (Savelkoul *et al.*, 1990). The amino acid sequence of this protein, CfaD, is highly homologous to Rns with only 11 amino acid differences due to 28 nucleotide differences between their respective DNA sequences. The Rns and CfaD proteins are functionally interchangeable with respect to their effects on the expression of CS1, CS2 and CFA/I fimbriae (Caron & Scott, 1990; Savelkoul *et al.*, 1990). Nucleotide sequencing of region 1 revealed the existence of a sequence homologous to the *cfaD* gene of region 2 (Gaastra *et al.*, 1990). This sequence, *cfaD'*, produces a truncated protein with no regulatory function in the production of CFA/I fimbriae. An analogous situation occurs in the expression of CS1 fimbriae, where strains may encode one or two *rns* gene homologues located on different plasmids (Smith, Coleman & Smyth, 1991).

To analyse the function of the CfaD protein, the promoter region of CFA/I region 1 (upstream of *cfaA*) was cloned in front of a promoterless β-galactosidase gene on a promoter vector. Introduction of a compatible plasmid containing the *cfaD* gene into a strain bearing this construct resulted in an approximately 10-fold increase in β-galactosidase activity (Savelkoul *et al.*, 1990). CFA/I region 1 lacks a good promoter, but a DNA-bending site (CAAAAAAAAAT) exists in front of the *cfaA* gene (Hamers *et al.*, 1989). Bending of promoter regions enhances the efficiency of RNA polymerase interaction. It may be possible that CfaD bends the promoter region to enhance expression of the *cfa* determinant (Savelkoul *et al.*, 1990).

Recently, a role for iron has been described in the regulation of expression of CFA/I fimbriae. The expression of CFA/I fimbriae by an ETEC strain was reduced in the presence of iron and this repressive effect was relieved by the addition of iron chelators. A role for the *E. coli* metalloregulatory protein Fur (*f*erric *u*ptake *r*egulation) was proposed, since in *fur* mutants there is no difference in the β-galactosidase activity of bacteria harbouring a *cfaB* gene promoter–*lacZ* fusion in medium containing iron, or iron with iron chelators. There are several Fur-binding sites in the *cfaB* promoter region (Karjalainen, Evans & Evans, 1991). It has also been proposed that CFA/I fimbriae are subject to catabolite repression, as a possible binding site for the cAMP–CRP complex exists upstream of the proposed Fur-binding site.

CFA/IV

The genetic determinant for the biogenesis of CS5 fimbriae is plasmid located since loss of a 78 MD plasmid in a serotype O167:H5 ETEC strain leads to loss of expression of CS5 and CS6 fimbriae (McConnell *et al.*, 1988, Hibberd *et al.*, 1991). This 78 MD plasmid also hybridizes with the *cfaD* probe, albeit weakly. The CS5 determinant has been cloned into a cosmid vector. The introduction of this recombinant plasmid into *E. coli* K-12 led to expression of CS5 fimbriae (Heuzenroeder *et al.*, 1989). The minimal amount of DNA required for expression of CS5 fimbriae is approximately 7 kb. If CS5 fimbriae require a positive regulator for expression, it too must be encoded on the cloned DNA insert (Heuzenroeder *et al.*, 1989). Minicell analysis has demonstrated that the cloned determinant encodes six polypeptides. One of these polypeptides, with a molecular mass of 31 kD, is a possible candidate CfaD-type protein on the basis of its size. The nucleotide sequence of the CS5 subunit gene has recently been determined and its deduced amino acid sequence shows strong homology to the amino acid sequence of the *E. coli* F41 pilin (Clark, Heuzenroeder & Manning, 1992).

The determinant for the plasmid-encoded CS6 antigen has been cloned and the minimal amount of DNA required for expression is 3 kb (Willshaw *et al.*, 1988). Expression of the CS6 antigen in *E. coli* K-12 bearing the recombinant plasmid pDEP5 is similar to expression directed from a wild-type plasmid. Minicell analysis has revealed that the DNA insert of plasmid pDEP5 encodes three polypeptides. This simplicity of organization is paralleled in the organization of the *cso* operon (Scott *et al.*, 1992).

In ETEC of serotype O25:H42 expressing the CS4 and CS6 antigens the genes required for expression of these fimbriae are located on separate plasmids. The plasmid encoding the CS6 antigen hybridized with a *cfaD* probe and loss of this plasmid resulted in loss of CS4 fimbrial expression. Expression was restored if a cloned *cfaD* gene was introduced into those strains lacking this plasmid (Willshaw *et al.*, 1990*a*). This scenario resembles expression of CS1 fimbriae where the regulator and fimbrial operon genes are on separate plasmids. Cloning of the CS4 regulator gene has been accomplished and its gene product can functionally substitute for CfaD or Rns in the expression of CFA/I, CS1 and CS2 fimbriae (Willshaw *et al.*, 1991).

A DNA sequence which hybridizes weakly with a *cfaD* probe has been cloned from an ETEC strain of serogroup O167 which expressed CS5 fimbriae (Willshaw *et al.*, 1991). The cloned regulator gene could promote expression of CFA/I, CS1, CS2 and CS4 antigens. However, the level of expression was lower than that promoted by *cfaD* or the CS4 regulator and functional fimbriae were not assembled. This gene, designated *csvR* for *c*oli *s*urface *v*irulence *r*egulator, has been sequenced (de Haan *et al.*, 1991). The amino acid sequence of CsvR is 87% similar to that of CfaD but it is 34 amino acids longer.

BORDETELLA PERTUSSIS

Molecular genetics has added much to our knowledge of the virulence factors of *B. pertussis* and their roles in disease and potential as candidates for an acellular vaccine (Coote, 1991; Parton, 1989, 1991: Rappuoli *et al.*, 1991; Wardlaw & Parton, 1988; Weiss & Hewlett, 1986). However, how *B. pertussis* adheres to cells of the respiratory tract is not known, although several candidate adhesins have been recognized. In addition to filamentous haemagglutinin, which is a high-M_r rod-like surface structure, and a 69 kD outer membrane protein termed protactin, which is a non-fimbrial agglutinogen, the sero-specific agglutinogens 2 and 3, which are fimbrial in nature, have been proposed as candidate adhesins.

B. pertussis exhibits antigenic variation by two mechanisms, viz. antigenic modulation, which is a freely reversible phenotypic change, and phase variation, which is a genotypic change (Robinson *et al.*, 1986; Coote & Brownlie, 1988). The expression of the fimbrial 2 and 3 antigens is regulated by both of these processes. Moreover, these surface associated factors can be lost and regained independently of each other at frequencies of 10^{-3} to 10^{-4} in a process termed serotype variation (Stanbridge & Preston, 1974).

The ground-breaking research of Weiss *et al.* (1983, 1984) and Weiss & Falkow (1983, 1984) led to the identification of a central regulatory locus, originally termed *vir* and since redesignated *bvg* (*Bordetella v*irulence *g*ene), encoding *trans*-acting factors, the expression of which is influenced by modulating conditions. The *bvg* locus has two genes, *bvgA* and *bvgS*, which constitute an operon and encode proteins with predicted M_r values of 23 kD and 135 kD, respectively (Stibitz & Yang, 1991). These proteins belong to a family of regulatory proteins that transduce sensory signals in a two-component sensor–regulator system (Gross, Aricò & Rappuoli, 1989; Coote, 1991; see also Dorman & Ní Bhriain, this symposium for review). These systems comprise a sensory transmembrane protein which has kinase activity and a cytoplasmic DNA-binding protein, which when phosphorylated can activate gene expression. The current data support this concept in *B. pertussis* where BvgS acts as the transmembrane sensor and BvgA, the phosphorylatable regulator (Stibitz & Yang, 1991; Coote, 1991). However, in addition, a hierarchy of regulation exists with different requirements for the control of expression of *bvg*-dependent loci. For example, in the case of the *fha* locus for expression of the filamentous haemagglutinin, the *bvg* locus is sufficient for sensory transduction and transcriptional activation of *fha* genes and expression of *bvg* and *fha* loci is sensitive to modulation. In contrast, the influence of the *bvg* locus on *ptxA* and *cyaA*, the promoter-proximal genes of the operons for pertussis toxin and adenylate cyclase production, respectively, is indirect and auxiliary regulatory factors appear to be required (Coote, 1991).

The *bvg* locus itself has been shown to be subject to autoregulation from studies on gene fusions involving a single copy *lacZYA* operon fused to the *bvg* promoter region integrated into the *E. coli* chromosome and a *bvg* locus supplied in *trans* on a multicopy plasmid, and from studies on *bvg* gene fusions created by allelic exchange in the chromosome of *B. pertussis* (Coote, 1991).

Although *bvg*-regulation of the *fha* locus has been most intensively investigated, the promoter regions of several other virulence-associated genes have also been sequenced (Coote, 1991). The promoter regions of the *fha, bvg, ptx* and *cya* operons contain inverted or direct repeat sequences which appear to be sites for transcriptional activation through the binding of the regulatory protein. In contrast, the promoter region of the *fim3* gene for the synthesis of fimbrial agglutinogen 3 lacks the repeat sequences characteristic of these other virulence-associated loci, yet transcription of the *fim3* gene requires functional Bvg activity (Willems *et al.*, 1990; MacGregor *et al.*, 1991). However, a 29 bp region containing an 11 bp inverted repeat has been found in the *fim2* promoter region between 167 and 140 bp upstream of the translational start codon (Walker *et al.*, 1990, 1991). A similarly positioned inverted repeat was present in the *fim3* gene promoter region. It is, however, unknown at present if the transcriptional regulators interact with these repeat sequences in the *fim* promoter regions.

Neither the nature of the *in vivo* signals in the human respiratory tract that influence modulation via the *bvg* regulon nor how such signals might affect the properties of BvgS, the proposed sensory element of this regulatory system, is known. Antigenic modulation probably involves specific intermolecular reactions between the modulating molecules and BvgS. The effects of osmolarity and temperature are unclear, although these factors have been shown to be important in the *toxRS* regulon system in *Vibrio cholerae* (Miller & Mekalanos, 1985; Miller, Taylor & Mekalanos, 1987; Miller, DiRita & Mekalanos, 1989). A model for sensory activation of the *bvg* regulon has been proposed which relies on dimerization of BvgS to an active kinase and promotion of active BvgA in the form of a dimer in the cytoplasm (Coote, 1991; Fig. 3).

Phase variation, the genotypic change from a virulent to an avirulent form, is a reversible process which has been associated with insertion or deletion of a cytosine residue (a frameshift mutation) in a run of six such residues within the *bvgS* gene (Stibitz *et al.*, 1989) and with small deletions or additions to the *bvg* locus (McGillivray, Coote & Parton, 1989; Monack *et al.*, 1989).

Although expression of the fimbrial 2 and 3 agglutinogens is controlled by the *bvg* locus, these antigens can be lost or regained independently during serotype variation. This has been shown to be dependent on *cis*-acting elements associated with the *fim* genes rather than *trans*-acting molecules (Willems *et al.*, 1990; MacGregor *et al.*, 1991). As with phase variation due to a frameshift event in the *bvg* locus, serotype variation may also be due to insertion or deletion events within cytosine-rich regions, but in this case these regions are located 70 bp upstream of the coding regions of the *fim* genes. These cytosine-rich regions contain 15 and 13 cytosine residues, respectively, in the *fim2* and *fim3* genes from strains producing high levels of these fimbrial antigens, and between 12 and 9 cytosine residues, respectively, from strains producing low levels of these fimbriae. Thus, the numbers of cytosine residues may influence the ability of the transcriptional apparatus to express the *fim* genes.

Reiterated base sequences are known to cause duplications and deletions during DNA replication and underlie antigenic variation documented for several bacterial pathogens, e.g. opacity proteins (protein II) of *Neisseria gonorrhoeae* (Stern & Meyer, 1987; Meyer & van Putten, 1989) and oligosaccharide epitopes of lipopolysaccharide of *Haemophilus influenzae* (Weiser, Love & Moxon, 1989; Moxon & Weiser, 1991) (see also Robertson & Meyer, this symposium and Moxon & Maskell, this symposium). In these instances, however, as with *bvgS* in *B. pertussis*, translation is affected by the frameshift mutations. The significance of the frameshifts outside of the coding regions of the *fim* genes is unclear. Such changes may affect the distance between the DNA region involved in binding of RNA polymerase and a positive regulator.

N-METHYLPHENYLALANINE FIMBRIAE

Members of the type 4 class of bacterial fimbriae are found on a wide variety of Gram-negative bacteria including *Moraxella bovis*, *Moraxella lacunata*,

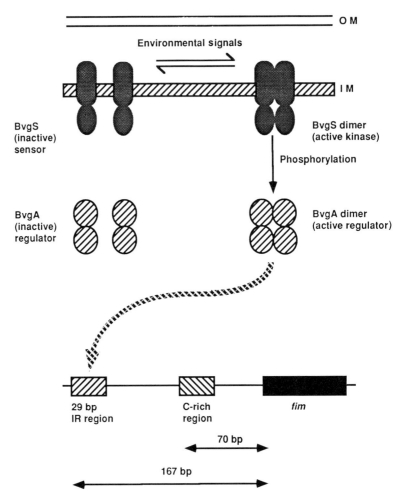

Fig. 3. A model for control of expression of fimbrial agglutinogens 2 and 3 of *Bordetella pertussis*. The *bvg* locus encodes two proteins, BvgS, the sensor (signal transducing) element located in the cytoplasmic membrane that responds to modulating environmental signals and BvgA, the regulatory element in the cytoplasm which is activated by BvgS. Appropriate environmental signals promote the BvgS protein to adopt either a monomeric (OFF) or dimeric (ON) form. The dimeric form of BvgS is an active kinase phosphorylating BvgA monomers, thereby promoting the formation of an active dimeric form of BvgA in the cytoplasm. Active BvgA is able to bind to the promoter regions of appropriate loci within the *bvg* regulon.

The *fim2* promoter region has a 29 bp region, about 167 to 140 bp upstream of its translational start site, comprising of an 11 bp inverted repeat (the promoter region and *fim* gene are not drawn to scale). A similar region is found in the *fim3* promoter region. BvgA or other transcriptional activators in a control hierarchy may interact with these repeat sequences. This mechanism underlies antigenic modulation. A cytosine-rich region 70 bp upstream of the coding regions of both the *fim2* and *fim3* genes is the site for insertion or deletion events leading to serotype variation by frameshift mutations. These may alter the distance between the DNA region involved in binding of RNA polymerase and a positive regulator. *cis*-Acting elements are also involved in serotype variation. (Adapted after Coote, 1991.)

Pseudomonas aeruginosa, *N. gonorrhoeae*, *Neisseria meningitidis*, *Bacteroides nodosus* and *Vibrio cholerae* (Paranchych & Frost, 1988; Paranchych, 1990). These fimbriae are composed of structural subunits which share extensive N-terminal amino-acid sequence homology and all, except the fimbriae of *V. cholerae*, contain the modified amino acid *N*-methylphenylalanine as the first residue of the mature protein (Dalrymple & Mattick, 1987; Paranchych, 1990). The subunit proteins are synthesized as prepilins containing similar N-terminal six or seven amino-acid cationic sequences (except in the case of *V. cholerae*) which are absent from the mature pilin subunits. Thus, the leader sequences are notably shorter than the typical signal sequences associated with *E. coli* pilins (Oudega & De Graaf, 1988). The export and assembly of type 4 pilins is accompanied by cleavage of the short leader peptide and methylation of the newly formed N-terminal phenylalanine residue.

The functional significance of this degree of conservation is attested to by the ability of *Ps. aeruginosa* containing a plasmid bearing a cloned *B. nodosus* pilin gene to process and assemble authentic *B. nodosus* fimbriae on the *Ps. aeruginosa* cell surface (Elleman *et al.*, 1986; Mattick *et al.*, 1987) and the ability of *Ps. aeruginosa* containing a *M. bovis* pilin gene to express chimaeric fimbriae containing both *M. bovis* and *Ps. aeruginosa* pilins (Beard *et al.*, 1990).

Type 4 fimbriate bacteria display several phenotypic differences from non-fimbriate bacteria including colony morphology, pitting of agar, autoagglutination, pellicle formation, haemagglutination, twitching motility, and increased competence for DNA transformation.

Similarities and differences in the genetic organization of type 4 pilin genes occur in different species. *Ps. aeruginosa*, *N. gonorrhoeae* and *M. bovis* pilin genes all appear to use *rpoN(glnF, ntrA)*-dependent promoters. RpoN has been shown to be the alternative sigma factor, σ^{54}, required for transcriptional activation of some genes (Kustu *et al.*, 1989). Differences exist in the copy numbers of pilin genes among type 4 fimbriate bacteria. For example, *Ps. aeruginosa* strains and most serotypes of *B. nodosus* have only a single copy of their respective pilin genes in each chromosome. In contrast, *N. gonorrhoeae* contains multiple pilin gene loci in each strain, with a single pilin expression locus and multiple silent variant pilin sequences lacking the common N-terminal coding sequence of pilin (Seifert & So, 1988).

Moraxella species

M. bovis is the primary cause of infectious bovine keratoconjunctivitis. *M. lacunata* is an occasional causative agent of human conjunctivitis and keratitis. These organisms are very closely related on the basis of DNA–DNA hybridization analysis and transformation studies. A single strain of *M. bovis* is capable of producing one or other of two pilin types, designated

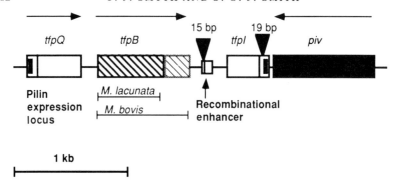

Fig. 4. Diagrammatic representation of the prepilin-encoding inversion regions of *Moraxella bovis* and *Moraxella lacunata*. The directions of transcription of these genes are shown by arrows. The *tfpQ* gene encoding Q prepilin is shown in the expression locus. The region 5′ of the expressed pilin has a *rpoN*-binding sequence (GG*CACAGACAA*TGC). The 6 amino-acid leader sequence and first 36 amino acids of mature pilins Q and I are identical. The inversion sites occur within the codon for amino acid residue 12 (isoleucine) of each pilin. The boxed areas encoding the N-terminal regions of the pilins indicate the conserved regions of *tfpQ* and *tfpI*. The silent *tfpI* gene is shown. The inverted repeats, which are shown as shaded areas within the conserved regions, are 26 bp long. The 19 bp perfect tandem repeat in the *tfpI* gene of *M. lacunata* is indicated. The 15 bp insert found in *M. lacunata* in the region equivalent to the 60 bp putative *M. bovis* recombinational enhancer is also indicated. The relative sizes of the *tfpB* genes of *M. bovis* (formerly ORF1) and *M. lacunata* are shown. The *piv* gene (putative inversion gene) which is thought to encode invertase in both *M. lacunata* and *M. bovis* (formerly ORF2) is located outside the pilin gene inversion region. (Adapted from Fulks *et al.*, 1990, Marrs *et al.*, 1990, 1991.)

Q pilin (previously β) and I pilin (previously α) (Marrs *et al.*, 1985; Ruehl *et al.*, 1988). Transitions from the production of Q pilin to production of I pilin and vice versa occur directly without the strain necessarily going through an afimbriate state.

This phase variation or switching between Q and I pilin gene expression is due to the inversion of an 2.1 kbp region of DNA, the end-points of which occur within the coding region of the expressed pilin gene (Fulks *et al.*, 1990; Marrs *et al.*, 1988). The inversion recombinational sites of the *M. bovis* pilin gene show sequence similarity to those of the Hin family of invertible segments, among the best characterized of which are the Hin system of *Salmonella typhimurium*, Gin and Cin of bacteriophages Mu and P1, and Pin of *E. coli* (Glasgow, Hughes & Simon, 1989). A region within the *M. bovis* pilin gene inversion site has DNA sequence similarity (21 out of 36 base-pairs) with the left inverted repeat of the *S. typhimurium* flagellar *hin* control region (Fulks *et al.*, 1990). In addition to the inversion region, a potential recombinational enhancer was identified which was similar to those found in the bacteriophage Mu *gin*, *E. coli pin*, bacteriophage P1 *cin* and *S. typhimurium hin* systems (Fulks *et al.*, 1990).

A representation of the chromosomal DNA region encoding Q and I pilin (*type four pilin genes tfpQ and tfpI*, respectively) is shown in Fig. 4. The inversion region contains one complete gene (*tfpQ*) in the expression locus and one partial gene (*tfpI*) capable of switching into the expression locus. In addition, two open-reading frames of unknown function were identified in the cloned *M. bovis* pilin determinant (Falks *et al.*, 1990). A similar type 4 pilin gene inversion region was subsequently identified in and cloned from chromosomal DNA of *M. lacunata* using the *M. bovis* pilin determinant as a hybridization probe (Marrs *et al.*, 1990). DNA sequencing revealed that the partial *tfpI* gene of *M. lacunata* corresponding to the non-expressed I pilin gene of *M. bovis* possessed a perfect 19-bp tandem repeat near the middle of the 26-bp inversion junction site. This 19-bp duplication causes a frameshift in the *M. lacunata tfpI* gene that disrupts translation of the transcript of this gene when in the expression locus (Rozsa & Marrs, 1991). In addition, it was shown that there was a 15-bp insertion relative to the putative *M. bovis* recombination enhancer, such that the critical spacing requirement of 48 bp between the centres of two potential Fis (*factor for inversion stimulation*) binding sites was disrupted (Hübner & Arber, 1989). This may account for the normal low inversion frequency of this pilin gene region.

DNA sequencing of the *M. lacunata* pilin determinant revealed an open-reading frame with over 98% DNA sequence homology to one of the ORFs previously found in the cloned *M. bovis* pilin determinant (termed ORF2) which lies downstream of the inversion region (Rozsa & Marrs, 1991; Marrs *et al.*, 1988; Fulks *et al.*, 1990). The *M. bovis* ORF2 supplied in *trans* could supply the inversion function missing in *M. lacunata* pilin-phase-blocked mutants, making these *Moraxella* loci good candidates for invertase-encoding genes (termed *putative invertase* [*piv*] genes).

The other identified ORF, ORF1 in the *M. bovis* determinant (Fulks *et al.*, 1990), now named *tfpB* (Rozsa & Marrs, 1991), may be involved in pathogenicity. Pathogenicity studies have demonstrated that *M. bovis* producing Q pilin establish eye infections more readily and produce more clinically severe disease than those producing I pilin, while organisms expressing I pilin are more frequently isolated from experimentally infected eyes, even although Q-pilin-producing bacteria may have been inoculated (Ruehl *et al.*, 1988; Lepper & Barton, 1987). Since the *tfpB* gene may be cotranscribed with the *tpfQ* gene, but would either not be expressed or would be transcribed by the promoter of the *piv* gene when the *tfpI* gene is in the expression locus, it is conceivable that differences in the amounts of TfpB rather than the change in pilin type may in some way be responsible for the observed differences in pathogenicity. It would be interesting to determine if the environmental conditions on the conjunctiva affect either expression of fimbriae or rates of inversion of the pilus locus or both.

Pseudomonas aeruginosa

The genetic determinants for the biogenesis of several classes of fimbriae of *E. coli*, including those for P fimbriae, K99 fimbriae and common type I fimbriae, contain several accessory genes and minor subunit genes involved in the stabilization and translocation of pilin subunits and their assembly into functional supramolecular structures on the bacterial cell surface (see above). In the case of type 4 fimbriae of *Ps. aeruginosa* three accessory genes, termed *pilB*, *pilC* and *pilD*, have so far been identified and shown to be required for the biogenesis of *Ps. aeruginosa* fimbriae from PilA subunits (Nunn, Bergman & Lory, 1990). These accessory genes are located on a 4.0 kbp region adjacent to the *pilA* gene. PilC and PilB may be integral membrane proteins (on the basis of hydrophobicity analysis of amino-acid sequences) involved in the processing of pilin precursors during membrane translocation. Mutations in *pilB* and *pilC* allow synthesis of mature pilin which is equally partitioned between the inner and outer membranes. Mutations in the *pilD* gene prevent processing of prepilin to pilin, suggesting that PilD is the prepilin leader peptidase (Strom, Nunn & Lory, 1991).

The roles of PilB and PilC appear to be restricted to fimbrial biogenesis. In contrast, mutations in the *pilD* gene prevented export of several extracellular proteins of *Ps. aeruginosa*, viz. alkaline phosphatase, phospholipase C, elastase and exotoxin A. In PilD mutants these extracellular proteins accumulated in the periplasmic space (Strom *et al.*, 1991). Thus, in addition to processing prepilin, PilD appears to have a role in secretion of proteins into the surrounding medium. It was proposed that PilD functions indirectly as a protease that may be involved in processing and assembly of one or more components of the membrane machinery or chaperones necessary for protein translocation. Export of true periplasmic proteins and outer membrane proteins is *pilD*-independent. The pleiotropic effect on secretion of *Ps. aeruginosa* extracellular proteins is similar to that described for mutants in several *xcp* (*extracellular proteins*) genes (Lazdunski *et al.*, 1990).

Whether or not *Ps. aeruginosa* possesses a global regulatory gene system equivalent to *bvg* of *B. pertussis* or *tox* of *V. cholerae* that could affect fimbrial expression remains to be determined. However, the production and export of alginate by *Ps. aeruginosa* appears to be controlled by a two-component *algRS* regulatory system (Deretic & Konyecsni, 1989; Deretic *et al.*, 1989; see also Dorman & Ní Bhriain, this symposium). Whether this *algRS* system regulates other virulence determinants in *Ps. aeruginosa* merits investigation.

Vibrio cholerae

A gene cluster encoding functions involved in the biogenesis of *V. cholerae* Tcp (*T*oxin *c*o-regulated *p*ilus) fimbriae has been described (Shaw *et al.*, 1988; Taylor *et al.*, 1987). Mutants which are unable to synthesize Tcp fimbriae are not able to colonize the small intestine successfully. The Tcp pilin and *N*-methylphenylalanine group of pilins share limited N-terminal homology, particularly in the regions immediately surrounding the cleavage

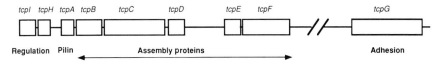

1 kbp

Fig. 5. Organization of the gene cluster determining expression of the Tcp fimbriae of *V. cholerae* O1 of both classical and El Tor biotypes. The *tcpH* and *tcpI* genes encode positive and negative regulators, respectively. The *pcpA* gene encodes the Tcp pilin. The *tcpB–tcpF* genes are involved in transport and assembly of Tcp fimbriae. The *tcpG* gene product may be a minor fimbrial subunit involved in adhesion or receptor recognition. The *tcpG* gene is unlinked to the *tcp* gene cluster. The *tcpA* gene is coregulated by ToxR, a positive regulator that is required for expression of the *ctxAB* operon (encoding cholera toxin). When Tcp fimbriae are highly expressed, the bacteria autoagglutinate. A further locus designated *tcpJ* (not shown on map) has been tentatively mapped to a 3 kb region at the right-hand end of the gene cluster. This locus may be involved in processing of the major pilin subunit. (Adapted from Shaw *et al.*, 1988.)

sites of *Ps. aeruginosa* and *V. cholerae* prepilins. Mutations in regions flanking the structural pilin gene (*tcpA*) resulted in phenotypes similar to those described for mutations in the *E. coli pap* determinant, pointing to the involvement of multiple genes in biogenesis (Shaw *et al.*, 1988).

The *tcpA* gene is positively co-regulated with the *ctx* (cholera *to*xin) operon by the two-component signal transduction regulatory operon *toxRS*, which plays a central role in the control of expression of multiple virulence properties of *V. cholerae* (Miller & Mekalanos, 1988; Miller *et al.*, 1987, 1989; see also Dorman & Ní Bhriain, this symposium). Tn*phoA* mutagenesis was used to characterize the genes and gene products of the *tcp* determinant. Ten genes have so far been identified (Fig. 5), indicating a determinant with a complexity similar to that described for the genetic determinant for P fimbriae. Interestingly mutants with lesions in the *tcpG* gene, the only gene unlinked to the *tcpA* gene thus far identified, produced wild-type amounts of morphologically normal fimbriae but did not possess phenotypes associated with expression of Tcp fimbriae. These mutants also showed decreased colonization ability. These findings suggest that the *tcpG* gene product may be a minor fimbrial subunit involved in adhesion, analogous to the PapG and FimH adhesin subunits of Pap fimbriae and common type 1 fimbriae, respectively (Krogfelt & Klemm, 1988). If this is the case, there must be a regulatory mechanism to ensure the correct stoichiometry of TcpA and TcpG pilin subunits in mature Tcp fimbriae.

Mutants with lesions in the *tcpH* and *tcpI* genes over-ride ToxRS-mediated regulation of *tcpA* expression, suggesting that these two genes encode positive and negative regulators, respectively, that interact either with the *tcpA* promoter or possibly with ToxR/ToxS.

Neisseria gonorrhoeae

The genetics of expression of fimbriae in *N. gonorrhoeae* have been extensively reviewed (Meyer & Haas, 1988; Meyer & van Putten, 1989; Meyer, 1990; Meyer, Gibbs & Haas, 1990; see also Robertson & Meyer, this symposium). Variability of the antigenicity of gonococcal fimbriae is based on variation in pilin. This pilin variation is not only implicated in immune evasion by gonococci but also contributes to modulation of adherence properties of these bacteria (Haas *et al.*, 1990). The gonococcal genome usually contains a single structural gene which is responsible for pilin expression (*pilE*). The *pilE* locus encodes production of the pilin precursor, prepilin, carrying a seven amino acid N-terminal export signal. Post-translational processing of the prepilin involves replacement of the signal peptide with an *N*-methyl group on the terminal phenylalanine residue. Variations between gonococcal pilins occur within six short regions toward the central and C-terminal domains, termed minicassettes. In addition to the expression locus, gonococci have many silent gene loci (*pilS*). A silent locus carries only one or more partial pilin gene copies which are tandemly arranged and connected by repetitive sequences.

Changes in the expression of pilins result from intragenic recombination events between *pilS* loci and the *pilE* locus which occur in the conserved regions flanking the variable minicassettes and are *recA*-dependent. Two pathways have now been identified, non-reciprocal recombination (intragenomic exchange) between two *pil* loci on the same chromosome and transformation-mediated recombination involving uptake of DNA released by bacteria in the same culture (see Robertson & Meyer, this symposium). As a consequence of these recombinations, *pilE* loci with altered combinations of minicassette encoding regions are generated. The repertoire of silent variant pilin genes in a single strain (MS11) has been investigated (Haas, Veit & Meyer, 1992). Seventeen silent copies were identified, which are truncated to varying degrees at their 5′ coding regions. They could be grouped into seven distinct *pil* loci. The *pilS1*, *pilS2* and *pilS6* loci contain six, two and three tandemly arranged silent copies, respectively. The locations of these loci on physical maps of the chromosome have been reported recently (Bihlmaier *et al.*, 1991; Dempsey *et al.*, 1991). The *pilS5* and *pilS7* loci each contain only one copy. Additionally, two silent copies are associated with each of the two *pilE* loci in this strain.

Fimbriation phase is affected by post-translational processing of prepilin to normal pilin, or S pilin (a truncated soluble secreted pilin), or L pilin (an extra long pilin that is not secreted or assembled) (Haas, Schwartz & Meyer, 1987; Manning *et al.*, 1991). In addition, changes in the expression of the assembly locus *pilC* by frameshift mutation can affect fimbriation phase (Jonsson, Nyberg & Normark, 1991).

Deletion and point mutants lacking fimbriae have been described (Meyer

& van Putten, 1989). Deletion mutants lacking the promoter and 5' coding sequences of the *pilE* structural gene show a complete loss of pilin expression. Point or frameshift mutations at distinct sites in the *pilE* gene affect the translational reading frame giving rise to cells producing assembly-deficient, truncated or unstable pilin. In contrast to deletion mutants, frameshift mutants can revert to a pilin-positive phenotype at very low frequency.

The *pilE* gene is not only subject to DNA rearrangements which give rise to antigenic and phase variation, but is also subject to regulatory control (Meyer & van Putten, 1989). The *pilE* promoter region contains nucleotide sequences which show homology with known DNA-binding sites for *Klebsiella* NtrA and NifA. *Klebsiella* NifA will activate gonococcal pilin expression in *trans* suggesting that the *pilE* gene is controlled by an activator similar to the *nifA* gene product.

Two genes, *pilA* and *pilB*, have been identified which affect transcription of the *pilE* gene in *trans* (Taha *et al.*, 1988). These are located downstream of the *pilE1* locus. The *pilA* gene product has an activating function and possesses a putative DNA-binding motif at its N-terminus. PilB is a repressor. These two genes are arranged in opposite orientations and appear to have overlapping regions (Meyer & van Putten, 1989). Whether gonococci possess a global virulence regulatory system which might be responsive to changes in the environment within the host (male or female) and inside phagocytic cells remains an intriguing question.

CONCLUSIONS

The application of molecular genetics to the study of the biogenesis of bacterial fimbriae has intensified research on regulation of expression, regulatory hierarchies and phase and antigenic variation not only in *E. coli* but in an ever-increasing number of Gram-negative bacteria, which, while not having the detailed genetic pedigree of *E. coli* or *S. typhimurium*, are none the less amenable to molecular genetic analysis (see Finlay, this symposium and Foster, this symposium).

While this chapter concentrates on fimbrial adhesins for host receptors, one must not lose sight of the fact that these morphological entities are but one class of adhesins on the surface of bacterial pathogens. Transposon mutagenesis, allele replacement, cloning strategies and DNA sequencing have led to the identification, in a small numbers of pathogens, of global regulatory systems which are involved in regulation of fimbrial biogenesis directly or indirectly. Almost certainly more such regulatory systems will be discovered in the future as interest focuses on the genetic basis of the influence of the host environment on expression of bacterial virulence factors. At the present rate of progress, one might predict that by the end of this century an understanding of the effects of host environment on expression of bacterial virulence factors, including fimbriae, will emerge. There is a need to define the host molecules that modulate fimbrial expression and the nature of their intermolecular interactions with sensor components of regulatory systems in the bacterial envelope.

The elegant resolution of morphological details of the tip adhesin structures of Pap fimbriae by electron microscopy techniques has added a new dimension to regulatory control of fimbrial assembly (Hultgren *et al.*, 1991; Kuehn *et al.*, 1992). How a bacterium controls ordered assembly of complex supramolecular structures comprising several different subunits in ordered arrays remains to be fully explained in terms of gene expression. Understanding regulation of fimbrial biogenesis in bacterial pathogens under conditions pertaining *in vivo* at the site of infection remains the challenge for future research. Knowledge of the host molecules which modulate expression may allow the development of novel strategies to combat bacterial infections.

ACKNOWLEDGEMENTS

Work in the authors' laboratory is supported by the Health Research Board of Ireland and the Wellcome Trust. S.G.J.S. is the holder of a research assistantship from the Wellcome Trust.

REFERENCES

Båga, M., Göransson, M., Normark, S. & Uhlin, B. E. (1985). Transcriptional activation of a Pap pilus virulence operon from uropathogenic *Escherichia coli*. *EMBO Journal*, **4**, 3887–93.

Båga, M., Göransson, M., Normark, S. & Uhlin, B. E. (1988). Processed mRNA with differential stability in the regulation of *E. coli* pilin gene expression. *Cell*, **52**, 197–206.

Barr, G. C., Ní Bhriain, N. & Dorman, C. J. (1992). Identification of two new genetically active regions associated with *osmZ* locus of *Escherichia coli*: role in regulation of *proU* expression and mutagenic effect at *cya*, the structural gene for adenylate cyclase. *Journal of Bacteriology*, **174**, 998–1006.

Beard, M. K. M., Mattick, J. S., Moore, L. J., Mott, M. R., Marrs, C. F. & Egerton, J. R. (1990). Morphogenetic expression of *Moraxella bovis* fimbriae (pili) in *Pseudomonas aeruginosa*. *Journal of Bacteriology*, **172**, 2601–7.

Bihlmaier, A., Römling, U., Meyer, T. F., Tümmler, B. & Gibbs, C. P. (1991). Physical and genetic map of the *Neisseria gonorrhoeae* strain M511–N198 chromosome. *Molecular Microbiology*, **5**, 2529–39.

Birkbeck, T. H. & Penn, C. W. (1986). *Antigenic Variation in Infectious Diseases*, Special Publications of the Society for General Microbiology, vol. 19, Oxford: IRL Press.

Blyn, L. B., Braaten, B. A. & Low, D. A. (1990). Regulation of *pap* pilin phase variation by a mechanism involving differential Dam methylation states. *EMBO Journal*, **9**, 4045–54.

Blyn, L. B., Braaten, B. A., White-Ziegler, C. A., Rolfson, D. H. & Low, D. A. (1989). Phase-variation of pyelonephritis-associated pili in *Escherichia coli*: evidence for transcriptional regulation. *EMBO Journal*, **8**, 613–20.

Boylan, M., Coleman, D. C., Scott, J. R. & Smyth, C. J. (1988). Molecular cloning of the plasmid-located determinants for CS1 and CS2 fimbriae of enterotoxigenic *Escherichia coli* of serotype O6:K15:H16 of human origin. *Journal of General Microbiology*, **134**, 2189–99.

Boylan, M., Coleman, D. C. & Smyth, C. J. (1987). Molecular cloning and characterization of the genetic determinant encoding CS3 fimbriae of enterotoxigenic *Escherichia coli*. *Microbial Pathogenesis*, **2**, 195–209.

Boylan, M. & Smyth, C. J. (1985). Mobilization of CS fimbriae-associated plasmids of enterotoxigenic *Escherichia coli* serotype O6:K15:H16 or H- into various wild-type hosts. *FEMS Microbiology Letters*, **29**, 83–9.

Braaten, B. A., Blyn, L. B., Skinner, B. S. & Low, D. A. (1991). Evidence for a methylation-blocking factor (*mbf*) locus involved in *pap* pilus expression and phase variation in *Escherichia coli*. *Journal of Bacteriology*, **173**, 1789–800.

Caron, J., Coffield, L. M. & Scott, J. R. (1989). A plasmid-encoded regulatory gene, *rns*, required for the expression of the CS1 and CS2 adhesins of enterotoxigenic *Escherichia coli*. *Proceedings of the National Academy of Sciences, USA*, **86**, 963–7.

Caron, J., Maneval, D. R., Kaper, J. B. & Scott, J. R. (1990). Association of *rns* homologs with colonization factor antigens in clinical *Escherichia coli* isolates. *Infection and Immunity*, **58**, 3442–4.

Caron, J. & Scott, J. R. (1990). A *rns*-like regulatory gene for colonization factor antigen 1 (CFA/I) that controls expression of CFA/I pilin. *Infection and Immunity*, **58**, 874–8.

Clark, C. A., Heuzenroeder, M. W. & Manning, P. A. (1992). Colonization factor antigen CFA/IV (PCF8775) of human enterotoxigenic *Escherichia coli*: nucleotide sequence of the CS5 determinant. *Infection and Immunity*, **60**, 1254–57.

Coote, J. G. (1991). Antigenic switching and pathogenicity: environmental effects on virulence gene expression in *Bordetella pertussis*. *Journal of General Microbiology*, **137**, 2493–503.

Coote, J. G. & Brownlie, R. M. (1988). Genetics of virulence of *Bordetella pertussis*. In *Pathogenesis and Immunity in Pertussis*, A. C. Wadlaw & R. Parton, eds, pp. 39–74. Chichester: John Wiley & Sons.

Cornelis, G., Sluiters, C., Lambert de Rouvroit, C. & Michiels, T. (1989). Homology between VirF, the transcriptional activator of the *Yersinia* virulence regulon, and AraC, the *Escherichia coli* arabinose operon regulator. *Journal of Bacteriology*, **171**, 254–62.

Dalrymple, B. & Mattick, J. S. (1987). An analysis of the organization and evolution of type 4 fimbrial (MePhe) subunit proteins. *Journal of Molecular Evolution*, **25**, 261–9.

de Graaf, F. K., van der Woude, M. J. & Klaasen, P. (1991). Regulation of the biosynthesis of K99 and 987P fimbriae. In *Microbial Surface Components and Toxins in Relation to Pathogenesis*, FEMS Symposium No. 51, pp. 55–60. New York: Plenum Press.

de Haan, L. A. M., Willshaw, G. A., van der Zeijst, B. A. M. & Gaastra, W. (1991). The nucleotide sequence of a regulatory gene present on a plasmid in an enterotoxigenic *Escherichia coli* strain of serotype O167:H5. *FEMS Microbiology Letters*, **83**, 341–6. .

Dempsey, J. A. F., Litaker, W., Madhure, A., Snodgrass, T. L. & Cannon, J. A. (1991). Physical map of the chromosome of *Neisseria gonorrhoeae* FA1090 with locations of genetic markers including *opa* and *pil* genes. *Journal of Bacteriology*, **173**, 5476–86.

Deretic, V., Dikshit, R. Konyecsni, W. M., Chakrabarty, A. M. & Misra, T. K.

(1989). The *algR* gene, which regulates mucoidy in *Pseudomonas aeruginosa*, belongs to a class of environmentally responsive genes. *Journal of Bacteriology*, **171**, 1278–83.

Deretic, V. & Konyecsni, W. M. (1989). Control of mucoidy in *Pseudomonas aeruginosa*: transcriptional regulation of *algR* and identification of the second regulatory gene, *algQ*. *Journal of Bacteriology*, **171**, 3680–8.

Donachie, W., Griffiths, E. & Stephen, J. (1988). *Bacterial Infections of Respiratory and Gastrointestinal Mucosae*, Special Publications of the Society for General Microbiology, vol. 24. Oxford: IRL Press.

Dorman, C. J., Ní Bhriain, N., & Higgins, C. F. (1990). DNA supercoiling and environmental regulation of virulence gene expression in *Shigella flexneri*. *Nature (London)*, **334**, 789–92.

Doyle, R. J. & Rosenberg, M. (1990). *Microbial Cell Surface Hydrophobicity*. Washington, DC: American Society for Microbiology.

Elleman, T. C., Hoyne, P. A., Stewart, D. J., McKern, N. M. & Peterson, J. E. (1986). Expression of pili from *Bacteroides nodosus* in *Pseudomonas aeruginosa*. *Journal of Bacteriology*, **168**, 574–80.

Forsman, K., Göransson, M. & Uhlin, B. E. (1989). Autoregulation and multiple DNA interactions by a transcriptional regulatory protein in *E. coli* pili biogenesis. *EMBO Journal*, **8**, 1271–7.

Fulks, K. A., Marrs, C. F., Stevens, S. P. & Green, M. R. (1990). Sequence analysis of the inversion region containing the pilin genes of *Moraxella bovis*. *Journal of Bacteriology*, **172**, 310–16.

Gaastra, W., Jordi, B. J. A. M., Mul, E. M. A., Hamers, A. M., McConnell, M. M., Willshaw, G. A., Smith, H. R. & van der Zeijst, B. A. M. (1990). A silent regulatory gene *cfaD'* on region 1 of the CFA/I plasmid NTP113 of enterotoxigenic *Escherichia coli*. *Microbial Pathogenesis*, **9**, 285–91.

Glasgow, A. C., Hughes, K. T. & Simon, M. I. (1989). Bacterial DNA inversion systems. In *Mobile DNA*, D. E. Berg & M. M. Howe, eds, pp. 637–59. Washington, DC: American Society for Microbiology.

Göransson, M., Berit, S., Sondén, B., P., Dagberg, B., Forsman, K., Emanuelson, K. & Uhlin, B. E. (1990). Transcriptional silencing and thermoregulation of gene expression in *Escherichia coli*. *Nature (London)*, **344**, 682–5.

Göransson, M., Forsman, K. & Uhlin, B. E. (1988). Functional and structural homology among regulatory cistrons of pili-adhesin determinants in *Escherichia coli*. *Molecular and General Genetics*, **212**, 412–17.

Göransson, M., Forsman, K. & Uhlin, B. E. (1989). Regulatory genes in the thermoregulation of *Escherichia coli* pili gene transcription. *Genes and Development*, **3**, 123–30.

Gross, R., Aricò, B. & Rappuoli, R. (1989). Families of bacterial signal-transducing proteins. *Molecular Microbiology*, **3**, 1661–7.

Haas, R., Facius, D., Gibbs, C. P., Rudel, T., van Putten, J. P. M. & Meyer, T. F. (1990). Gonococcal pilin variation and modulation of cellular adherence. In *Proceedings of the 7th International Pathogenic* Neisseria *Conference*, M. Achtman, ed., pp. 585–90. Wilhelm de Gruyter: Berlin.

Haas, R., Schwartz, H. & Meyer, T. F. (1987). Release of soluble pilin antigen coupled with gene conversion in *Neisseria gonorrhoeae*. *Proceedings of the National Academy of Sciences, USA*, **84**, 9079–83.

Haas, R., Veit, S. & Meyer, T. F. (1992). Silent pilin genes of *Neisseria gonorrhoeae* MS11 and the occurrence of related hypervariant sequences among other gonococcal isolates. *Molecular Microbiology*, **6**, 197–208.

Hacker, J. (1990). Genetic determinants coding for fimbriae and adhesins in extraintestinal *Escherichia coli. Current Topics in Microbiology and Immunology*, **15**, 1–27.

Hamers, A. M., Pel, H. J., Willshaw, G. A., Kusters, J. G., van der Zeijst, B. A. M. & Gaastra, W. (1989). The nucleotide sequence of the first two genes of the CFA/I fimbrial operon of human enterotoxigenic *Escherichia coli*. *Microbial Pathogenesis*, **6**, 297–309.

Heuzenroeder, M. W., Neal, B. L., Thomas, C. J., Halter, R. & Manning, P. A. (1989). Characterization and molecular cloning of the PCF8775 CS5 antigen from an enterotoxigenic *Escherichia coli* 0115:H40 isolated in Central Australia. *Molecular Microbiology*, **3**, 303–10.

Hibberd, M. L., McConnell, M. M., Willshaw, G. A., Smith, H. R. & Rowe, B. (1991). Positive regulation of colonization factor antigen I (CFA/I) production by enterotoxigenic *Escherichia coli* producing the colonization factors CS5, CS6, CS7, PCF09, PCF0159:H4 and PCF0166. *Journal of General Microbiology*, **137**, 1963–70.

Higgins, C. F., Dorman, C. J., Stirling, D. A., Waddell, L., Booth, I. R., May, G. & Bremer, E. (1988). A physiological role for DNA supercoiling in the osmotic regulation of gene expression in *S. typhimurium* and *E. coli. Cell*, **52**, 569–84.

Hübner, P. & Arber, W. (1989). Mutational analysis of a prokaryotic recombinational enhancer element with two functions. *EMBO Journal*, **8**, 577–85.

Hultgren, S. J., Normark, S. & Abraham, S. N. (1991). Chaperone-assisted assembly and molecular architecture of adhesive pili. *Annual Review of Microbiology*, **45**, 383–415.

Huo, L., Martin, K. J. & Schleif, R. (1988). Alternative DNA loops regulate the arabinose operon in *Escherichia coli. Proceedings of the National Academy of Sciences, USA*, **85**, 5444–8.

Iglewski, B. H. & Clark, V. L. (1990). *Molecular Basis of Bacterial Pathogenesis, The Bacteria*, vol. XI. New York: Academic Press.

Irani, M. H., Orosz, L. & Adhya, S. (1983). A control element within a structural gene: the *gal* operon of *Escherichia coli. Cell*, **32**, 783–8.

Jalajakumari, M. B., Thomas, C. J., Halter, R. & Manning, P. A. (1989). Genes for biosynthesis and assembly of CS3 pili of CFA/II enterotoxigenic *Escherichia coli*: novel regulation of pilus production by bypassing an amber codon. *Molecular Microbiology*, **3**, 1685–95.

Jann, K. & Jann, B. (1990). *Bacterial Adhesins, Current Topics in Microbiology and Immunology*, vol 151. Springer Verlag, Berlin.

Jonsson, A. -B., Nyberg, G. & Normark, S. (1991). Phase variation of gonococcal pili by frameshift mutation in *pilC*, a novel gene for pilus assembly. *The EMBO Journal*, **10**, 477–88.

Jordi, B. J. A. M., van Vliet, A. H. M., Willshaw, G. A., van der Zeijst, B. A. M. & Gaastra, W. (1991). Analysis of the first two genes of the CS1 fimbrial operon in human enterotoxigenic *Escherichia coli* of serotype 0139:H28. *FEMS Microbiology Letters*, **80**, 265–70.

Karjalainen, T. K., Evans, D. G. & Evans, D. J., Jr., Graham, D. Y. & Lee, C.-H. (1991). Iron represses the expression of CFA/I fimbriae of enterotoxigenic *Escherichia coli. Microbial Pathogenesis*, **11**, 317–23.

Karjalainen, T. K., Evans, D. G., So, S. & Lee, C.-H. (1989). Molecular cloning and nucleotide sequence of the colonization factor antigen I gene of *Escherichia coli. Infection and Immunity*, **57**, 1126–30.

Kawula, T. H. & Orndorff, P. E. (1991). Rapid site-specific DNA inversion in

Escherichia coli mutants lacking the histone-like protein H-NS. *Journal of Bacteriology*, **173**, 4116–23.

Klaasen, P. & de Graaf, F. K. (1990). Characterization of FapR, a positive regulator of expression of the 987P operon in enterotoxigenic *Escherichia coli*. *Molecular Microbiology*, **4**, 1179–83.

Krogfelt, K. A. (1991). Bacterial adhesion: Genetics, biogenesis and role in pathogenesis of fimbrial adhesins of *Escherichia coli*. *Reviews of Infectious Diseases*, **13**, 721–35.

Krogfelt, K. A. & Klemm, P. (1988). Investigation of minor components of *Escherichia coli* type 1 fimbriae: protein chemical and immunological aspects. *Microbial Pathogenesis*, **4**, 231–8.

Kuehn, M. J., Henser, J., Normark, S. & Hultgren, S. J. (1992). P pili in uropathogenic *E. coli* arc composite fibres with distinct fibrillar adhesive tips. *Nature (London)*, **356**, 252–5.

Kustu, S., Santero, E., Keener, J., Popham, D. & Weiss, D. (1989). Expression of σ^{54} (*ntrA*)-dependent genes is probably united by a common mechanism. *Microbiological Reviews*, **53**, 367–76.

Lambert de Rouvroit, C., Sluiters, C. & Cornelis, G. R. (1992). Role of trancriptional activator, VirF, and temperature in the expression of the pYV plasmid genes of *Yersinia enterocolitica*. *Molecular Microbiology*, **6**, 395–409.

Lark, D. L. (1986). *Protein–Carbohydrate Interactions in Biological Systems: The Molecular Biology of Microbial Pathogenicity*, FEMS Symposium No. 31. London: Academic Press.

Lazdunski, A., Guzzo, J., Filloux, A., Bally, M. & Murgier, M. (1990). Secretion of extracellular proteins by *Pseudomonas aeruginosa*. *Biochimie*, **72**, 147–56.

Lejeune, P. & Danchin, A. (1990). Mutations in the *bglY* gene increase the frequency of spontaneous deletions in *Escherichia coli* K-12. *Proceedings of the National Academy of Sciences, USA*, **87**, 360–3.

Lepper, A. W. D. & Barton, I. J. (1987). Infectious bovine keratoconjunctivitis: seasonal variation in cultural, biochemical and immunoreactive properties of *Moraxella bovis* isolated from the eyes of cattle. *Australian Veterinary Journal*, **64**, 33–9.

Low, D., Robinson, E. N., Jr., McGee, Z. A. & Falkow, S. (1987). The frequency of expression of pyelonephritis-associated pili is under regulatory control. *Molecular Microbiology*, **1**, 335–46.

McConnell, M. M., Thomas, L. V., Willshaw, G. A., Smith, H. R. & Rowe, B. (1988). Genetic control and properties of coli surface antigens of colonization factor antigen IV (PCF 8775) of enterotoxigenic *Escherichia coli*. *Infection and Immunity*, **56**, 1974–80.

McGillivray, D. M., Coote, J. G. & Parton, R. (1989). Cloning of the virulence regulatory (*vir*) locus of *Bordetella pertussis* and its expression in *Bordetella bronchiseptica*. *FEMS Microbiology Letters*, **65**, 333–8.

MacGregor, D., Coote, J. G., Duggleby, C. J. & Parton, R. (1991). Serotype variation in *Bordetella pertussis* is governed by cis-acting elements. *FEMS Microbiology Letters*, **78**, 333–8.

Manning, P. A., Timmis, K. N. & Stevenson, G. (1985). Colonization factor antigen II [CFA/II] of enterotoxigenic *Escherichia coli* : molecular cloning of the CS3 determinant. *Molecular and General Genetics*, **200**, 322–7.

Marrs, C. F., Rozsa, F. W., Hackel, M., Stevens, S. P. & Glasgow, A. C. (1990). Identification, cloning and sequencing of *piv*, a new gene involved in inverting the pilin genes of *Moraxella lacunata*. *Journal of Bacteriology*, **172**, 4370–7.

Marrs, C. F., Ruehl, W. W., Schoolnik, G. K. & Falkow, S. (1988). Pilin gene phase variation of *Moraxella bovis* is caused by an inversion of the pilin genes. *Journal of Bacteriology*, **170**, 3032–9.

Marrs, C. F., Schoolnik, G., Koomey, J. M., Hardy, J., Rothbard, J. & Falkow, S. (1985). Cloning and sequencing of a *Moraxella bovis* pilin gene. *Journal of Bacteriology*, **163**, 132–9.

Manning, P. A., Kaufmann, A., Roll, U., Pohlner, J., Meyer, T. F. & Haas, R. (1991). L-pilin variants of *Neisseria gonorrhoeae* MS11. *Molecular Microbiology*, **5**, 917–26.

Mattick, J. S., Bills, M. M., Anderson, B. J., Dalrymple, B., Mott, M. R. & Egerton, R. (1987). Morphogenetic expression of *Bacteroides nodosus* fimbriae in *Pseudomonas aeruginosa*. *Journal of Bacteriology*, **169**, 33–41.

Meyer, T. F. (1990). Variation in pilin and opacity-associated protein in pathogenic *Neisseria* species. In *Molecular Basis of Bacterial Pathogenesis, The Bacteria*, vol. XI, ed. B. H. Iglewski & V. L. Clark, pp. 137–53. New York: Academic Press.

Meyer, T. F., Gibbs, C. P. & Haas, R. (1990). Variation and control of protein expression in *Neisseria*. *Annual Review of Microbiology*, **44**, 451–77.

Meyer, T. F. & Haas, R. (1988). Phase and antigenic variation by DNA arrangements in procaryotes. In *Transposition*, Symposium of the Society for General Microbiology, vol. 43, A. J. Kingsman, K. F. Chater & S. M. Kingsman, eds, pp. 193–219. Cambridge: Cambridge University Press.

Meyer, T. F. & van Putten, J. P. M. (1989). Genetic mechanisms and biological implications of phase variation in pathogenic neisseriae. *Clinical Microbiology Reviews*, **2**, Supplement, S139–45.

Miller, V. L., DiRita, V. J. & Mekalanos, J. J. (1989). Identification of *toxS*, a regulatory gene whose product enhances ToxR-mediated activation of the cholera toxin promoter. *Journal of Bacteriology*, **171**, 1288–93.

Miller, V. L. & Mekalanos, J. J. (1985). Genetic analysis of the cholera toxin-positive regulatory gene *toxR*. *Journal of Bacteriology*, **163**, 580–5.

Miller, V. L. & Mekalanos, J. (1988). A novel suicide vector and its use in construction of insertion mutations: osmoregulation of outer membrane proteins and virulence determinants in *Vibrio cholerae* requires *toxR*. *Journal of Bacteriology*, **170**, 2575–83.

Miller, V. L., Taylor, R. K. & Mekalanos, J. J. (1987). Cholera toxin transcriptional activator ToxR is a transmembrane DNA binding protein. *Cell*, **48**, 271–9.

Mirelman, D. (1986). *Microbial Lectins and Agglutinins: Properties and Biological Activity*. New York: John Wiley & Sons.

Monack, D. M., Aricò, B., Rappuoli, R. & Falkow, S. (1989). Phase variants of *Bordetella bronchiseptica* arise by spontaneous deletions in the *vir* locus. *Molecular Microbiology*, **3**, 1719–28.

Moxon, E. R. & Weiser, J. N. (1991). Studies on the genetic basis of *Haemophilus influenzae* pathogenicity. In *Microbial Surface Components and Toxins in Relation to Pathogenesis*, E. Z. Ron & S. Rottem, eds, pp. 171–7. New York: Plenum Press.

Normark, S., Båga, M., Göransson, M., Lindberg, F. P., Lund, B., Norgren, M. & Uhlin, B.-E. (1986). Genetics and biogenesis of *Escherichia coli* adhesins. In *Microbial Lectins and Agglutinins: Properties and Biological Activity*, D. Mirelman, ed., pp. 113–43. New York: John Wiley & Sons.

Normark, S., Lindberg, F., Lund, B., Båga, M., Ekbäck, G., Göransson, M., Mörner, S., Norgren, M., Marklund, B.-I., & Uhlin, B. E. (1985). Minor pilus components acting as adhesins. In *Protein–Carbohydrate Interactions in Biological Systems: The Molecular Biology of Microbial Pathogenicity*, FEMS Symposium No. 31, D. L. Lark, ed., pp. 3–12. London: Academic Press.

Nunn, D., Bergman, S. & Lory, S. (1990). Products of three accessory genes, *pilB*, *pilC*, and *pilD*, are required for biogenesis of *Pseudomonas aeruginosa* pili. *Journal of Bacteriology*, **172**, 2911–19.

Ott, M., Schmoll, T., Goebel, W., van Die, I. & Hacker, J. (1987). Comparison of the genetic determinant coding for the S-fimbrial adhesin (*sfa*) of *Escherichia coli* to other chromosomally encoded fimbrial determinants. *Infection and Immunity*, **55**, 1940–3.

Oudega, B. & De Graaf, F. K. (1988). Genetic organization and biogenesis of adhesive fimbriae of *Escherichia coli*. *Antonie van Leeuwenhoek. Journal of Microbiology and Serology*, **54**, 285–99.

Paranchych, W. (1990). Molecular studies on *N*-methylphenylalanine pili. In *Molecular Basis of Bacterial Pathogenesis, The Bacteria*, vol. XI, B. H. Iglewski & V. L. Clark, eds, pp. 61–78. New York: Academic Press.

Paranchych, W. & Frost, L. S. (1988). The physiology and biochemistry of pili. *Advances in Microbial Physiology*, **29**, 53–114.

Parton, R. (1989). *Bordetella pertussis* toxins and the human host. *Current Opinion in Infectious Diseases*, **2**, 788–95.

Parton, R. (1991). Changing perspectives on pertussis and pertussis vaccination. *Reviews in Medical Microbiology*, **2**, 121–8.

Perez-Casal, J., Swartley, J. S. & Scott, J. R. (1990). Gene encoding the major subunit of CS1 pili of human enterotoxigenic *Escherichia coli*. *Infection and Immunity*, **58**, 3594–600.

Proctor, R. A. (1987). *Fibronectin and the Pathogenesis of Infections. Reviews of Infectious Diseases*, **9**, Supplement 4.

Rappuoli, R., Pizza, M., Podda, A., De Magistris, M. T. & Nencioni, L. (1991). Towards third-generation whooping cough vaccines. *Trends in Biotechnology*, **9**, 232–8.

Riegman, N., Kusters, R., van Veggel, H., Bergmans, H., Van Bergen en Henegouwen, P., Hacker, J. & van Die, I. (1990). F1C fimbriae of a uropathogenic *Escherichia coli* strain: genetic and functional organization of the *foc* gene cluster and identification of minor subunits. *Journal of Bacteriology*, **172**, 1114–20.

Robinson, A., Duggleby C. J., Gorringe, A. R. & Livey, I. (1986). Antigenic variation in *Bordetella pertussis*. In *Antigenic Variation in Infectious Diseases*, Special Publications of the Society for General Microbiology, Volume 19, T. H. Birkbeck & C. W. Penn, eds, pp. 147–61. Oxford: IRL Press.

Ron, E. Z. & Rottem, S. (1991). *Microbial Surface Components and Toxins in Relation to Pathogenesis*, FEMS Symposium No. 51. New York: Plenum Press.

Roth, J. A. (1988). *Virulence Mechanisms of Bacterial Pathogens*. Washington, DC: American Society for Microbiology.

Rozsa, F. W. & Marrs, C. F. (1991). Interesting sequence differences between the pilin gene inversion regions of *Moraxella lacunata* ATCC 17956 and *Moraxella bovis* Epp63. *Journal of Bacteriology*, **173**, 4000–6.

Ruehl, W. W., Marrs, C. F., Fernandez, R., Falkow, S. & Schoolnik, G. K. (1988). Purification, characterization and pathogenicity of *Moraxella bovis* pili. *Journal of Experimental Medicine*, **168**, 983–1002.

Sakai, T., Sasakawa, C., Makino, S., Kamata, K. & Yoshikawa, M. (1986*a*). Molecular cloning of a genetic determinant for congo red binding ability which is essential for the virulence of *Shigella flexneri*. *Infection and Immunity*, **51**, 476–82.

Sakai, T., Sasakawa, C., Makino, S. & Yoshikawa, M. (1986*b*). DNA sequence and product analysis of the *virF* locus responsible for congo red binding and cell invasion in *Shigella flexneri* 2a. *Infection and Immunity*, **54**, 395–402.

Sakai, T., Sasakawa, C. & Yoshikawa, M. (1988). Expression of four virulence antigens of *Shigella flexneri* is positively regulated at transcriptional level by the 30 kiloDalton *virF* protein. *Molecular Microbiology*, **2**, 589–97.

Savelkoul, P. H. M., Willshaw, G. A., McConnell, M. M., Smith, H. R., Hamers, A. M., van der Zeijst, B. A. M. & Gaastra, W. (1990). Expression of CFA/I fimbriae is positively regulated. *Microbial Pathogenesis*, **8**, 91–9.

Schmoll, T., Morschhäuser, J., Ott, M., Ludwig, B., van Die, I. & Hacker, J. (1990*a*). Complete genetic organization and functional aspects of the *Escherichia coli* S fimbrial adhesin determinant: nucleotide sequence of the genes *sfa B, C, D, E, F. Microbial Pathogenesis*, **9**, 331–43.

Schmoll, T., Ott, M., Oudega, B. & Hacker, J. (1990*b*). Use of a wild-type gene fusion to determine the influence of environmental conditions on the expression of the S fimbrial adhesin in an *Escherichia coli* pathogen. *Journal of Bacteriology*, **172**, 5103–11.

Scott, J. R., Wakefield, J. C., Russell, P. W., Orndorff, P. E. & Froehlich, B. J. (1992). CooB is required for assembly but not transport of CS1 pilin. *Molecular Microbiology*, **6**, 293–300.

Seifert, H. S. & So, M. (1988). Genetic mechanisms of bacterial antigenic variation. *Microbiological Reviews*, **52**, 327–36.

Shaw, C. E., Peterson, K. W., Sun, D., Mekalanos, J. J. & Taylor, R. K. (1988). TCP pilus expression and biogenesis by classical and El Tor biotypes of *Vibrio cholerae* O1. In *Molecular Mechanisms of Microbial Adhesion*, L. Switalski, M. Höök & E. Beachey, eds, pp. 23–35. New York: Springer Verlag.

Smith, H. R., Cravioto, A., Willshaw, G. A., McConnell, M. M., Scotland, S. M., Gross, R. J. & Rowe, R. (1979). A plasmid coding for the production of colonisation factor antigen I and heat-stable enterotoxin in strains of *Escherichia coli* of serogroup O78. *FEMS Microbiology Letters*, **6**, 255–60.

Smith, S. G. J., Coleman, D. C. & Smyth, C. J. (1991). Localization of the CS1 fimbrial operon, the *rns* gene and *rns* gene homologues in enterotoxigenic *Escherichia coli* of serotype O6:K15:H16 or H-. In *Molecular Recognition in Host–Parasite Interactions: Mechanisms of Viral, Bacterial and Parasite Infections*, FEMS Symposium, Porvoo, Finland, August, 1991, Abstract.

Smyth, C. J. (1982). Two mannose-resistant haemagglutinins on enterotoxigenic *Escherichia coli* of serotype O6:K15:H16 or H- isolated from travellers' and infantile diarrhoea. *Journal of General Microbiology*, **128**, 2081–96.

Smyth, C. J. (1986). Fimbrial variation in *Escherichia coli*. In *Antigenic Variation in Infectious Diseases*, Special Publications of the Society for General Microbiology, Volume 19, T. H. Birkbeck & C. W. Penn, eds, pp. 95–125. Oxford: IRL Press.

Smyth, C. J. (1988). Fimbriae: their role in virulence. In *Immunochemical and Molecular Genetic Analysis of Bacterial Pathogens*, FEMS Symposium No. 40, P. Owen & T. J. Foster, eds, pp. 13–25. Elsevier: Amsterdam.

Smyth, C. J., Boylan, M., Matthews, H. M. & Coleman, D. C. (1991). Fimbriae of human enterotoxigenic *Escherichia coli* and control of their expression. In *Microbial Surface Components and Toxins in Relation to Pathogenesis*, FEMS Symposium No. 51, E. Z. Ron & S. Rottem, eds, pp. 37–53. New York: Plenum Press.

Stanbridge, T. N. & Preston, N. W. (1974). Variation of serotype in strains of *Bordetella pertussis. Journal of Hygiene, Cambridge*, **73**, 305–10.

Stern, A. & Meyer, T. F. (1987). Common mechanism controlling phase and antigenic variation in pathogenic neisseriae. *Molecular Microbiology*, **1**, 5–12.

Stibitz, S., Aaronson, W., Monack, D. & Falkow, S. (1989). Phase variation in *Bordetella pertussis* by frameshift mutation in a gene for a novel two-component system. *Nature (London)*, **338**, 266–9.

Stibitz, S. & Yang, M.-S. (1991). Subcellular localization and immunological detection of proteins encoded by the *vir* locus of *Bordetella pertussis*. *Journal of Bacteriology*, **173**, 4288–96.

Strom, M. S., Nunn, D. & Lory, S. (1991). Multiple roles of the pilus biogenesis protein PilD: involvement of PilD in excretion of enzymes from *Pseudomonas aeruginosa*. *Journal of Bacteriology*, **173**, 1175–80.

Switalski, L., Höök, M. & Beachey, E. (1988). *Molecular Mechanisms of Microbial Adhesion*. New York: Springer Verlag.

Taha, M. K., So, M., Seifert, H. S., Billyard, E. & Marchal, C. (1988). Pilin expression in *Neisseria gonorrhoeae* is under both positive and negative transcriptional control. *EMBO Journal*, **7**, 4367–78.

Taylor, R. K., Miller, V. L., Furlong, D. B. & Mekalanos, J. J. (1987). Use of *pho* gene fusions to identify a pilus colonization factor coordinately regulated with cholera toxin. *Proceedings of the National Academy of Sciences USA*, **84**, 2833–7.

Tennent, J. M., Hultgren, S., Marklund, B.-I., Forsman, K., Göransson, M., Uhlin, B. E. & Normark, S. (1990). Genetics of adhesin expression in *Escherichia coli*. In *Molecular Basis of Bacterial Pathogenesis*, The Bacteria, Vol. XI, B. H. Iglewski & V. L. Clark, eds, pp. 79–110. London: Academic Press.

Twohig, J., Boylan, M. & Smyth, C. J. (1988). Expression of CS fimbriae in *Escherichia coli* strains of human and animal origin belonging to O-serovar 6, K-serovar 15 or H-serovar 16. *FEMS Microbiology Letters*, **56**, 327–30.

Uhlin, B.-E., Båga, M., Göransson, M., Lindberg, F. P., Lund, B., Norgren, M. & Normark, S. (1985). Genes determining adhesin production in uropathogenic *Escherichia coli*. *Current Topics in Microbiology and Immunology*, **118**, 163–78.

Walker, M. J., Rohde, M., Brownlie, R. M. & Timmis, K. N. (1990). Engineering upstream transcriptional and translational signals of *Bordetella pertussis* serotype 2 fimbrial subunit protein for efficient expression in *Escherichia coli: in vitro* autoassembly of the expressed product into filamentous structures. *Molecular Microbiology*, **4**, 39–47.

Walker, M. J., Guzmán, C. A., Rohde, M. & Timmis, K. N. (1991). Production of recombinant *Bordetella pertussis* serotype 2 fimbriae in *B. parapertussis* and *B. bronchiseptica*: utility of *Escherichia coli* gene expression signals. *Infection and Immunity*, **59**, 1739–46.

Wardlaw, A. C. & Parton, R. (1988). *Pathogenesis and Immunity in Pertussis*. Chichester: John Wiley & Sons.

Weiser, J. N., Love, J. M. & Moxon, E. R. (1989). The molecular mechanism of phase variation of *H. influenzae* lipopolysaccharide. *Cell*, **59**, 657–65.

Weiss, A. A. & Falkow, S. (1983). The use of molecular techniques to study microbial determinants of pathogenicity. *Philosophical Transactions of the Royal Society, London*, **B303**, 219–25.

Weiss, A. A. & Falkow, S. (1984). Genetic analysis of phase change in *Bordetella pertussis*. *Infection and Immunity*, **43**, 263–9.

Weiss, A. A. & Hewlett E. L. (1986). Virulence factors of *Bordetella pertussis*. *Annual Review of Microbiology*, **40**, 661–86.

Weiss, A. A., Hewlett, E. L. Myers, G. A. & Falkow, S. (1983). Tn5-induced mutations affecting virulence factors of *Bordetella pertussis*. *Infection and Immunity*, **42**, 33–41.

Weiss, A. A., Hewlett, E. L., Myers, G. A. & Falkow, S. (1984). Pertussis toxin and extracytoplasmic adenylate cyclase as virulence factors of *Bordetella pertussis*. *Journal of Infectious Diseases*, **150**, 219–22.

White-Ziegler, C. A., Blyn, L. B., Braaten, B. A. & Low, D. A. (1990). Identification of an *Escherichia coli* genetic locus involved in the thermoregulation of the *pap* operon. *Journal of Bacteriology*, **172**, 1775–82.

Willems, R., Paul, A., van der Heide, H. G. J., ter Avest, A. R. & Mooi, F. R. (1990). Fimbrial phase variation in *Bordetella pertussis*: a novel mechanism for transcriptional regulation. *EMBO Journal*, **9**, 2803–9.

Willshaw, G. A., McConnell, M. M., Smith, H. R. & Rowe, B. (1990*a*). Structural and regulatory genes for coli surface associated antigen 4 (CS4) are encoded by separate plasmids in enterotoxigenic *Escherichia coli* strains of serotype O25.H42. *FEMS Microbiology Letters*, **68**, 255–60.

Willshaw, G. A., Smith, H. R., McConnell, M. M., Gaastra, W., Thomas, A., Hibberd, M. & Rowe, B. (1990*b*). Plasmid-encoded production of coli surface-associated antigen 1 (CS1) in a strain of *Escherichia coli* serotype O139.H28. *Microbial Pathogenesis*, **9**, 1–11.

Willshaw, G. A., Smith, H. R., McConnell, M. M. & Rowe, B. (1988). Cloning of genes encoding coli-surface (CS) antigens in enterotoxigenic *Escherichia coli*. *FEMS Microbiology Letters*, **49**, 473–78.

Willshaw, G. A., Smith, H. R., McConnell, M. M. & Rowe, R. (1991). Cloning of regulator genes controlling fimbrial production by enterotoxigenic *Escherichia coli*. *FEMS Microbiology Letters*, **82**, 125–30.

Willshaw, G. A., Smith, H. R. & Rowe, B. (1983). Cloning regions encoding colonisation factor antigen I and heat-stable enterotoxin in *Escherichia coli*. *FEMS Microbiology Letters*, **16**, 101–6.

SALMONELLA GENETICS AND VACCINE DEVELOPMENT

S. CHATFIELD[1], J. L. LI[1], M. SYDENHAM[1], G. DOUCE[2] AND G. DOUGAN[2]

[1]*Vaccine Research Unit, Medeva Group Research, Department of Biochemistry, Imperial College of Science, Technology and Medicine, Wolfson Laboratories, London SW7 2AY*
[2]*Department of Biochemistry, Imperial College of Science, Technology and Medicine, Wolfson Laboratories, London SW7 2AY*

INTRODUCTION

With the exception of *Escherichia coli* K12, the genetics of *Salmonella typhimurium* has arguably been studied more intensively than any other bacterium. The *S. typhimurium* genome is now well characterised and, in general terms, is organised in a similar manner to *E. coli* with chromosomal DNA showing around 90% homology (Neidhardt, 1987). The genetic manipulation techniques available for use in salmonellae are well advanced and DNA can be transferred between strains using various approaches including transformation, conjugation, transduction and now electroporation. Despite an increasing understanding of the genetic organisation of the *S. typhimurium* chromosome, it has only recently been possible, with a few exceptions, to associate particular genes with virulence. Work on the genetics of virulence of other *Salmonella* serotypes is even less advanced than *S. typhimurium*. Salmonellae are sophisticated pathogens that are able to cause a variety of diseases and syndromes in man and domestic animals ranging from systemic infections such as typhoid, to limited infections of the gut such as gastroenteritis or food poisoning. Although disease symptoms are varied, it is generally assumed that tissue invasiveness is a common trait of these infections; this is self-evident for disseminating infections, but has not been formally proven for *S. typhimurium*-induced acute gastroenteritis. However, using a Hep-2 cell based assay, Douce *et al.* (1991) established a correlation between strains of *S. typhimurium* of known virulence (in the context of gastroenteritis) and invasiveness: virulent strains were more invasive than avirulent strains. Evidence from animal studies suggests that most pathogenic *Salmonella* strains have the ability to penetrate the mucosal barriers of the intestine, enter host tissues and invade and grow within both professional (Fields *et al.*, 1986; Buchmeier and Heffron, 1989) and non-professional phagocytic cells (Takeuchi, 1967). Many *Salmonella* do not penetrate significantly beyond the lamina propria and mesenteric lymph nodes of the gut wall whereas others become systemic and cause invasive, disseminated disease. The potential to cause systemic infection is a trait of most *Salmonella* strains but, in some cases, infection spread is limited by host immune defence functions. When this natural immunity breaks down in immunocompromised individuals such as AIDS sufferers, they become susceptible to systemic infections by strains such as *S. typhimurium* that normally cause gut-associated gastroenteritis (Cellum *et al.,* 1987). Species such as *S. typhi* can cause systemic infections even in fully immunocompetent individuals, and thus must harbour additional genetic information.

As well as advanced genetic manipulation systems being available, pathogenic salmonellae can also be studied in convenient *in vivo* and *in vitro* models. Many mice strains are highly susceptible to infections by certain isolates of many serotypes including *S. typhimurium*, *S. dublin* and *S. enteritidis* (Collins, 1974). These mouse virulent isolates normally cause invasive, systemic infections which, in many ways, resemble typhoid. Mice do not develop gastroenteritis or diarrhoeal disease and can not realistically be used as a model for food poisoning. The murine model is often referred to as mouse typhoid or murine salmonellosis. The murine model has been used with success to study both natural and acquired resistance to salmonellosis. Some murine genes such as *ity* and others within the MHC dramatically affect the susceptibility of mice to salmonellosis (Hormaeche, 1979; O'Brien *et al.,* 1979, 1980; Lissner, Swanson & O'Brien, 1983; Hormaeche, Harrington & Joysey, 1985; Hormaeche & Maskell, 1989).

The murine model is now being used by geneticists to identify *Salmonella* genes required for virulence. One approach is to look for genes which, when mutated, attenuate a virulent isolate. A more indirect approach is to carry out initial screening using *in vitro* methods such as tissue culture invasiveness prior to *in vivo* testing for attenuation (Galan & Curtiss, 1989). Using these approaches, a rapidly increasing number of genes are being identified which are apparently required to establish infection or cause disease. Although the murine model is readily accessible, and is yielding valuable information, caution must be taken when extrapolating data obtained in the mouse to other animal hosts. In many cases, mutations which lead to attenuation in the mouse model also show attenuation in other animal species. However, this correlation may not be absolute. Also, attenuating lesions will behave differently in different *Salmonella* strain backgrounds (O'Callaghan *et al.,* 1988; Benjamin *et al.,* 1991*b*).

In addition to the mouse, other mammals have been used to study *Salmonella* infection. Chickens are highly susceptible to infections by certain *Samonella* serotypes including *S. gallinarum* and *S. pullorum* (Barrow, 1990; Cooper *et al.,* 1992). Attempts have been made to use the rabbit to study gut-associated salmonellosis, but this system has caused problems mainly because of the great variability in the susceptibility of individual rabbits. Larger animals, such as cattle, are natural host species for many *Salmonella* isolates but their availability is limited.

Salmonella isolates vary greatly in their ability to establish infections in these different host species. To some degree, this is a reflection of the serotype of the strain. For example, *S. typhi* is almost exclusively a human pathogen. Thus, although salmonellae harbour many common virulence-associated genes, significant differences must exist either between individual alleles or in the additional virulence genes possessed. Even today, little is known about how host specificity is determined in the salmonellae. This is an area where much progress should be made over the next few years.

Despite the obvious limitations, pathogenic salmonellae are ideal organisms for studying the genetics of virulence. In the next sections, some of the progress made to date will be discussed.

EARLY STUDIES

Most of the early studies on the genetics of virulence in *Salmonella* relied on the use of spontaneous or chemically induced mutations. This obviously limited the type of approaches that could be used, and detail that could be obtained. Early work (Bacon, Burrows & Yates, 1951) showed that certain auxotrophic mutations led to the attenuation of *S. typhi* for mice. Strains dependent on purines and para-aminobenzoic acid (PABA) for growth were less virulent than the parental strain. This work was later followed up using more refined genetic approaches (Hoiseth & Stocker, 1981). Another class of mutants that attracted early interest were rough variants. Rough mutants, named after their characteristic colonial morphology were easy to identify, and are the result of mutations in the lipopolysaccharide (LPS) biosynthetic pathway. Different mutations can affect LPS synthesis at various stages, and it was found that mutants defective in *o*-side chain synthesis were attenuated. Germanier carried out a detailed study in the early 1970s on the effects of LPS mutations on virulence and subsequent immunogenicity. In this study he identified *galE* mutants, which are defective in the enzyme UDP glucose-4-epimerase, as potential vaccine candidates. He found that *S. typhimurium galE* mutants were attenuated and highly immunogenic (Germanier & Furer, 1971). Some *galE* mutants are highly sensitive to exogenous galactose, lysing if galactose is present in the growth medium. This may be due to the build-up of toxic biosynthetic intermediates in *galE* strains metabolising galactose. Germanier used the data obtained in the murine model as a guide to develop a live, oral typhoid vaccine based on *galE* mutants of *S. typhi*. He used chemical mutagenesis of *S. typhi* Ty2 to isolate *galE* mutants which were stable on subculturing. This strain, named Ty21a, was then tested as a live, oral typhoid vaccine (Gilman *et al.*, 1977), initially in volunteers and eventually in field trials around the world (Wahdan *et al.*, 1982). Ty21a is now sold under licence in many countries. Although efficacious, the vaccine requires multiple doses in order to induce immunity, and harbours multiple mutations, which were induced during mutagenesis, in addition to the *galE* lesion. Recently, doubts have been expressed about whether *galE* is actually an attenuating lesion in *S. typhi*. A defined *galE* mutant of *S. typhi* Ty2 has been constructed using a cloned *galE* gene. The cloned gene was mutated in *E. coli* and returned to the *Salmonella* chromosome using homologous recombination, thus replacing the wild-type *galE* gene. This defined mutant strain was fed to several volunteers, some of whom developed typhoid symptoms (Hone *et al.*, 1988). This clearly illustrates the potential dangers of extrapolating too much data from amongst different *Salmonella* serotypes and host animal species.

Other early work on *Salmonella* virulence involved the use of transductional and *hfr* crosses. This approach was used to construct hybrid strains harbouring regions of the *Salmonella* chromosome – replaced with heterologous DNA from other *Salmonella* serotypes and from other enterobacteriaceae. This approach has limited use because of the large regions of DNA normally exchanged, and has been largely replaced by the more defined techniques of transposon mutagenesis and site-directed gene replacements. However, this early work was used to map several virulence-associated genes including the *S. typhi* Vi-capsule synthesis loci *viaA* and *viaB* (Johnson & Baron, 1969).

AUXOTROPHIC MUTANTS

Auxotrophic mutants of salmonellae were first shown to be attenuated in the early 1950s (Bacon *et al.*, 1951). Their original observation was not followed up until the early 1980s with more defined techniques. Para-amino benzoic acid (PABA)-dependent mutants of *S. typhi* were known to be attenuated. PABA and other aromatic compounds are synthesised via the chorismate biosynthetic pathway. This pathway is not present in mammals who rely on exogenous aromatic metabolites in their diet. The levels of PABA and possibly other aromatic compounds are tightly controlled in mammalian tissues which may go some way towards explaining why aromatic dependent mutants grow poorly *in vivo*. Stocker transduced an *aroA*::Tn*10* lesion from the weakly pathogenic *S. typhimurium* LT2 into the fully mouse virulent SL1344. He isolated several stable *aroA* mutants of SL1344 and other mouse virulent salmonellae (Hoiseth & Stocker, 1981). All were highly attenuated in mice proving conclusively the requirement for a functional *aroA* gene *in vivo*. *aroA* mutant strains were also found to be excellent single-dose oral vaccines.

Later work in the authors' laboratory showed that other genes in the chorismate biosynthetic pathway, including *aroC* and *aroD*, were also essential for *in vivo* growth. Single *aro* or even double and triple *aro mutants* were attenuated to a similar level in mice and, like single *aroA* mutants, were excellent single dose oral vaccines (Dougan *et al.*, 1988; Miller I. A. *et al.*, 1989). This early work has led to the development of genetically defined candidate oral typhoid vaccines based on double *aro* mutants of *S. typhi* Ty2 (Hone *et al.*, 1991; Chatfield *et al.*, 1992).

Purine dependent mutants are also known to be attenuated. However, unlike *aro* mutants, different *pur* mutations give rise to different degrees of attenuation if the mutations are compared in the same *Salmonella* strain background and in the same inbred mice (McFarland & Stocker, 1987). For example, *purA* mutants are much more attenuated than *purE* mutants. This observation can be explained partly by the branched nature of the purine biosynthetic pathway. Combining *purA* and *aroA* mutations in the same strain leads to a higher level of attenuation than either of the two attenuating lesions alone (O'Callaghan *et al.*, 1988). This is different from the observation with double *aro* mutants where attenuation is similar to that with single *aro* mutants. A measure of the degree of difference in attenuation can be monitored after oral feeding of the strains to mice. *S. typhimurium aroA* mutants colonise the reticuloendothelial system of mice for several weeks whereas *aroA, purA* mutants do not penetrate significantly past the mesenteric lymph nodes. Thus, using simple genetic approaches, strains can now be fashioned to different levels of attenuation. This is obviously very important if live, oral vaccines are to be designed which are immunogenic but well tolerated.

MUTANTS OF *SALMONELLA* UNABLE TO ENTER OR GROW WITHIN EUCARYOTIC CELLS

There has been an ongoing debate as to whether salmonellae reside within eucaryotic cells during infection and whether this is an essential ecological niche for *in vivo* survival. Although this subject is still controversial, it is now clear that *Salmonella* can invade and grow within eucaryotic cells *in vitro* and *in vivo* (Leung & Finlay, 1991; Finlay & Falkow, 1988; Buchmeier & Heffron, 1989). One of the first attempts to analyse this using a genetic approach was by monitoring the survival of a bank of Tn*10* mutants of *S. typhimurium* in a macrophage cell-line (Fields *et al.*, 1986). Tn*10* mutants were generated using random mutagenesis of the genome. A large number of isolates were selected that had a decreased ability to survive in this cell-line. Upon initial characterisation, many had no detectable phenotypic alteration other than their inability to survive inside macrophages, while others were auxotrophs. When these isolates were examined in the murine model, they were found to have reduced virulence compared to the parental strain. Thus there was a good correlation between an inability to survive in macrophages and an ability to cause infection in mice. Two limitations of this study were that the macrophage line was in a differentiated state, and that the *S. typhimurium* strain used for mutagenesis was partially rough.

Nevertheless, the experiments were a breakthrough in this kind of global analysis of *Salmonella* virulence.

More groups began to study the interaction of *Salmonella* with cultured mammalian cells. *Salmonella* could clearly bind to these cells, and enter the cells inside vacuoles, where, unlike *Listeria*, they remained and replicated (Finlay *et al.*, 1988). Salmonellae were also found to have the ability to transcytose across differentiated, polarised epithelial cells growing on filters in tissue culture medium. Finlay and coworkers generated a mutant bank of a highly invasive *S. cholerae-suis* using the novel transposon TnphoA (Finlay *et al.*, 1988). TnphoA is a derivative of Tn5 which has the 5' terminal repeat partially replaced by the *E. coli* alkaline phosphatase open reading sequence lacking a promoter and signal sequence. If TnphoA integrates into a gene, encoding a protein which is normally secreted, an in-frame fusion can generate a hybrid protein combining the signal sequence and 5' end of the inactivated protein and the alkaline phosphatase. This hybrid protein may be transported to the cell periplasm where enzyme activity is generated following removal of the signal sequence. Using this approach, it is possible to enrich for transposon insertions in genes encoding secreted proteins which are more likely to be involved in cell penetration mechanisms than cytoplasmic proteins. The bank was screened in the MDCK monolayer model for mutants defective in transcytosis and/or cell attachment. The mutants that were isolated could be grouped into at least six different groups using physical chromosomal mapping techniques. Some of these mutants were defective in LPS biosynthesis implicating a role for this molecule in the transcytosic pathway of *Salmonella*, at least for *S. cholerae-suis*.

Again these mutants were assessed for their virulence for mice. Interestingly, many but not all of the mutants were attenuated to some degree. Some mutants which had clearly lost the ability to transcytose MDCK cells *in vitro* were still highly virulent for mice. These clearly illustrate the need to closely correlate *in vitro*- and *in vivo*-derived observations. Galan and Curtiss used cosmid cloning to identify a region of the *S. typhimurium* chromosome which was able to complement a natural variant of *S. typhimurium* that was defective in cell invasion (Galan & Curtiss, 1989). They constructed a mutant of a mouse-virulent, cell-invasive *S. typhimurium* using the cloned gene. They mutated the gene in *E. coli* and returned the mutation by allelic exchange to the *S. typhimurium* chromosome. The mutant had a slightly reduced virulence for the mouse when given orally but not following parenteral challenge. A number of different genes have now been identified in this locus, and work is currently under way to define their role in the cell invasion process (Galan & Curtiss, 1991). Several teams are now looking at the cell invasion mechanisms utilised by pathogenic salmonellae. A group at Walter Reed have cloned a region of the *S. typhi* chromosome which can confer on *E. coli* K12 the ability to penetrate eucaryotic cells (Elsinghorst, Baron & Kopecko, 1989). This region was isolated originally on a large cosmid and is also a complex locus. Some of the host specificity functions of salmonellae may be able to be identified by carefully characterising these regions.

THE *SALMONELLA* VIRULENCE PLASMID

Many, but not all, salmonellae harbour a large plasmid, whose absolute size varies between serotypic groups. This used to be referred to as the cryptic plasmid. Jones and colleagues were the first to demonstrate a role for this plasmid in *Salmonella* virulence (Jones *et al.*, 1982). They used transposon tagging to facilitate curing of the cryptic plasmid from virulent strains of *S. typhimurium*, and showed that plasmid-free variants were highly attenuated. Several groups then used transposon mutagenesis and hybridisation studies to show that the cryptic plasmids of different *Salmonella* serotypes encoded a common virulence associated region. This region has homologues in different serotypes. Further genetic analysis is now in progress to fully define this region. The exact role of the region in virulence is still in some doubt. Plasmid-free salmonellae are less invasive in terms of their ability to replicate in the reticuloendothelial system of mice. It is believed that the plasmid-free variants may be less able to survive the attentions of macrophages than the fully virulent plasmid-containing strains. Clearly, although the plasmid is important for *Salmonella* virulence, it is far from being the whole story. Indeed, *S. typhi* does not harbour a virulence plasmid or encode regions highly homologous to the common virulence region. However, further analysis may reveal a homologue of this region in *S. typhi*.

GLOBAL REGULATORS AND VIRULENCE

Pathogenesis is obviously an extremely complicated process. Geneticists can identify genes which play a role in virulence using mutagenesis. The gene can have a direct role perhaps encoding a protein involved in attachment and invasion or an enzyme involved in a key metabolic pathway. It may also have an indirect role controlling the expression of genes encoding the more 'conventional' virulence proteins. As pathogens enter the host, they will encounter dramatically different microenvironments which will exert different stresses and demands on the cells. In order to respond rapidly, the pathogen must be able to sense environmental conditions and co-ordinately control the expression of blocks of genes whose expression is required under certain conditions but not others (for a general review see Griffiths, 1991; Mekalanos, 1991). Work in other pathogens, particularly in *Bordetella pertussis* has shown that key genes could control the expression of several associated proteins. The *bvg* locus of *B. pertussis* controls the expression of several virulence proteins including pertussis toxin and adenylate cyclase in response to environmental stimulation (Scarlato *et al.*, 1990). The *bvg* locus serves as an example of a two-component global regulatory system (Arico *et al.*, 1991).

Perhaps the first people to study global regulators in *Salmonella* were Curtiss and colleagues in St Louis. They constructed derivatives of a mouse-virulent *S. typhimurium* which harboured Tn*10* insertions in the *cya* and *crp* genes encoding proteins involved in the regulation of cyclic AMP levels in bacteria. Adenylate cyclase and the cyclase-binding protein are involved in the regulation of expression of many operons in *E. coli* and other enterobacteriacae. They were able to show that *crp cya* mutants of *Salmonella* were highly attenuated in mice. They were able to invade mouse tissues but did not overwhelm the animals. These mutants were also found to be highly immunogenic and candidate vaccine strains (Curtiss & Kelly, 1987).

Part of the adaptive response of bacterial pathogens such as *Salmonella* to conditions of high osmolarity involves the preferential expression of one type of porin, OmpC, over another type, OmpF (Nikaido & Vaara, 1985). Expression of the *ompC* and *ompF* genes is regulated coordinately at the level of transcription (Hall & Silhavy, 1979) involving an environmental sensor, EnvZ, an inner membrane protein of the histidine kinase class, and a cytoplasmically located transcriptional activator, OmpR. An *ompR*::Tn*10* lesion was introduced into a mouse-virulent *S. typhimurium* strain SL1344 and the effects on the virulence of *ompR* were monitored. SL1344 *ompR* but not SL1344 *ompC* or SL1344 *ompF* mutants were highly attenuated and excellent single-dose oral vaccines (Dorman *et al.*, 1989). Further studies revealed that an *ompC ompF* double mutant does not mimic the behaviour of an *ompR* mutant. Although it is attenuated to a similar level orally, it showed little loss of virulence when given intravenously (Chatfield *et al.*, 1991). This suggests that other *ompR* regulated genes may play a role in the early stages of the infection process. Further work is ongoing to identify other *ompR* regulated genes.

Several groups have identified the *pho* regulon as a two-component system essential for *Salmonella* virulence (Miller, Kukral & Mekalanos, 1989*b*). The *pho* regulon responds to phosphate levels and other environmental areas such as low pH and the intracellular environment of the macrophage (Miller, 1991). Experiments in mice have shown *pho* mutants are attenuated. Some of the *pho* regulated genes have now been identified (Miller *et al.*, 1989*b*). One, the *pagC* gene is thought to encode a secreted or envelope protein essential for survival in macrophages (Pulkkinen & Miller, 1991). Pho mutants also have increased susceptibility to the low molecular weight defensins expressed by macrophages as a component of their antibacterial repertoire. However, attenuation of the *pagC*- strains does not result from an increased sensitivity to defensins, as this strain is as resistant to native defensins as the parent strain (Miller & Mekalanos, 1990). This implies that another *pag* gene is involved in resistance. Several *prg* (*phoP* repressed genes) are also important in virulence, as mutants which express *phoP* constitutively are unable to survive within macrophages and show reduced virulence in mice (Miller & Mekalanos, 1990). Other regulatory systems involved in *Salmonella* virulence include the chemotaxis genes. In cell culture-based assays, mutations in the *che* genes of *S. typhimurium* results in the production of a hyperinvasive phenotype (Khoramian-Falsafi *et al.*, 1990; Lee, Jones & Falkow, 1992).

OTHER *SALMONELLA* VIRULENCE GENES

An increasing number of loci have now been implicated as having some role in *Salmonella* virulence. Attempts to define a role for the flagella have in general met with frustration. Some *fla* mutants are fully virulent although some groups have, in the past, claimed that some non-flagellated strains are avirulent for mice (Carsiotis *et al.*, 1984). Part of the problem may be that flagella synthesis requires the co-operation of many genes, and mutations in some may lead to pleiotropic effects. However, it is now generally accepted that flagella are not essential for virulence (Lockman & Curtiss, 1990).

A heat-shock protein homologue has been identified in *Salmonella* which is essential for full virulence in the host (Johnson *et al.*, 1991). The *htrA* gene encodes a serine-protease which is localised in the periplasm of salmonellae. *htrA* mutant strains have an increased susceptibility to oxidative stress and are impaired in their ability to grow in host tissues. Defined *htrA* mutants of *S. typhimurium* are excellent oral vaccines in mice (Chatfield *et al.*, 1992). Other genes implicated as playing a role in *Salmonella* virulence include *hemA* (Benjamin *et al.*, 1991*a*) and *mviA* (Benjamin *et al.*, 1991*b*). *mviA* maps close to the trp operon of *S. typhimurium*, and the active gene may play a role in depressing the virulence of *Salmonella* in an *Ity*-related manner.

It is likely that other genes which are essential for virulence will be identified in the next few years. It may be fruitful to start screening in other host systems to make comparisons on the virulence of *Salmonella* between species. The availability of genetically modified immunocompromised strains of mice may eventually lead to a small animal *S. typhi* model.

SALMONELLA GENETICS AND VACCINE DEVELOPMENT

Attenuated *Salmonella* strains are attracting a great deal of attention as vaccine delivery systems for the oral delivery of antigens. Ty21a is currently marketed as a live, oral typhoid vaccine, licensed for human use. As has already been pointed out, Ty21a has the disadvantage of requiring multiple doses in order to induce an acceptable level of immunity. A search is under way for an attenuated *S. typhi* strain that can be used safely as a single-dose oral typhoid vaccine. Such a strain would be useful in its own right but could also be of potential value as a carrier for delivering heterologous antigens to the mammalian immune system.

Many of the attenuating lesions described above could be considered for incorporation into an oral typhoid vaccine strain. Stocker and colleagues constructed mutants of S. typhi that harboured aroA and purA mutations. Those strains were fed to volunteers at the vaccine testing centre in Baltimore but, although apparently well tolerated, they were considered to be too weakly immunogenic to warrant further studies (Levine et al., 1987). Together with Myron Levine at Baltimore, mutants of S. typhi harbouring deletions in aroC and aroD (Hone et al., 1991) and aroA and aroC (Chatfield et al., 1992) have been constructed. Two independent strains harbouring aroC and aroD lesions were tested for safety and immunogenicity in volunteers. One of the strains, based on a Ty2 background, was well tolerated and highly immunogenic (Tacket et al., 1992). Immunogenicity in terms of seroconversion to o-antigen, and numbers of antibody secreting cells from gut-derived lymphocytes, were very good. An S. typhi crp cya double mutant was also found to be immunogenic but caused some reactogenicity in the volunteers. However, further careful evaluation of these strains is required before too many conclusions can be drawn with respect to their potential as oral typhoid vaccines.

Many groups have been using attenuated Salmonella strains as experimental oral vaccines for the delivery of heterologous antigens to the immune system (Chatfield, Strugnell & Dougan, 1989). Salmonellae are attractive carriers because they can be used orally, and they are able to induce a variety of immune responses in the host, including secretory and humoral antibodies and various cell-mediated responses. It should be stressed that, at present, these are experimental vaccines. Earlier work using Ty21a as a carrier was disappointing because of genetic instability and poor immune responses to the heterologous antigens. A variety of antigens derived from bacteria, viruses and parasites have now been delivered using this system. The delivery of interleukin 1β by S. typhimurium aro mutants has also been reported (Carrier et al., 1992). One of the main drawbacks of this approach is the potential instability of gene expression of the heterologous antigen. Stability can be increased by placing the foreign gene on the Salmonella chromosome (Strugnell et al., 1990), although immunogenicity due to lower levels of expression may be compromised. More sophisticated expression systems will need to be developed if this system is eventually going to prove useful.

CONCLUSION

Quite clearly significant advances are being made in identifying Salmonella genes required for the in vivo survival of the organism. Several different categories of genes have been identified although, as yet, little is known about the actual proteins which interact with mammalian cells during infection. Some of this work is assisting the development of candidate oral *Salmonella* vaccines of defined genetic composition. Further significant advances should be expected in the future.

REFERENCES

Arico, B., Scarlato, V., Monack, D. M., Falkow, S. & Rappuoli, R. (1991). Structural and genetic analysis of the *bvg* locus in *Bordetella* species. *Molecular Microbiology*, **5**, 2481–91.

Bacon, G. A., Burrows, T. N. & Yates, M. (1951). The effect of biochemical mutation on the virulence of *Bacterium typhosum*: the loss of virulence of certain mutants. *British Journal of Experimental Pathology*, **32**, 85–96.

Barrow, P. A. (1990). Experimental fowl typhoid in chickens induced by a virulence plasmid-cured derivative of *S. gallinarum*. *Infection and Immunity*, **58**, 2283–8.

Benjamin Jr, W. H., Hall, P. & Briles, D. E. (1991a). A *hemA* mutation renders *Salmonella typhimurium* avirulent in mice, yet capable of, eliciting protection against intravenous infection with *S. typhimurium*. *Microbial Pathogenesis*, **11**, 289–96.

Benjamin Jr, W. H., Yother, J., Hall, P. & Briles, D. E. (1991b). The *Salmonella typhimurium* locus *mviA* regulates virulence in Ity^s but not Ity^r mice: functional *mviA* results in avirulent; mutant (non-functional) *mviA* results in virulence. *Journal of Experimental Medicine*, **174**, 1073–83.

Buchmeier, N. A. & Heffron, F. (1989). Intracellular survival of wild type *S. typhimurium* and macrophage sensitive mutants in diverse populations of macrophages. *Infection and Immunity*, **57**, 1–7.

Carrier, M. J., Chatfield, S. N., Dougan, G., Nowicka, V. T. A., O'Callaghan, D., Beesley, J. E., Milano, S., Cillais, E. & Liew, F. Y. (1992). Expression of human interleukin-1B in *Salmonella typhimurium*: a model system for the delivery of recombinant therapeutic proteins *in vivo*. *Journal of Immunology*, **148**, 1176–81.

Carsiotis, M., Weinstein, D. L., Karch, H., Holder, I. A. & O'Brien, A. D. (1984). Flagella of *Salmonella typhimurium* are a virulence factor in infected C57BL/6J mice. *Infection and Immunity*, **46**, 814–18.

Cellum, C. L., Chaisson, R. E., Rutherford, G. W., Lowell Barnhart, J. & Echenberg, D. F. (1987). Incidence of salmonellosis is patients with AIDS. *Journal of Infectious Diseases*, **156**, 998–1002.

Chatfield, S. N., Strugnell, R. A. & Dougan, G. (1989). Live salmonellae as vaccines and carriers of foreign antigenic determinants. *Vaccine*, **7**, 495–8.

Chatfield, S. N., Dorman, C. J., Hayward, C. & Dougan, G. (1991). Role of *ompR*-dependent genes in *Salmonella typhimurium* virulence: mutants deficient in both OmpC and OmpF are attenuated *in vivo*. *Infection and Immunity*, **59**, 449–52.

Chatfield, S. N., Fairweather, N., Charles, I., Pickard, D., Levine, M., Hone, D., Posada, M., Strugnell, R. A. & Dougan, G. (1992). Construction of a genetically defined *Salmonella typhi* Ty2 *aroA*, *aroC* mutant for the engineering of a candidate oral typhoid–tetanus vaccine. *Vaccine*, **10**, 53–60.

Chatfield, S. N., Strahan, K., Pickard, D., Charles, I. G., Hormaeche, C. E. & Dougan, G. (1992). Evaluation of *Salmonella typhimurium* strains harbouring defined mutations in *htrA* and *aroA* in the murine salmonellosis model. *Microbial Pathogenesis*, in press.

Collins, F. M. (1974). Vaccines and cell-mediated immunity. *Bacterial Reviews*, **38**, 371–402.

Cooper, G. L., Venables, L. M., Nicholas, R. A. J., Cullen, G. A. & Hormaeche, C. E. (1992). Vaccination of chickens with chicken-derived *Salmonella enteritidis* phage type 4 *aroA* live oral *Salmonella* vaccines. *Vaccine*, **10**, 247–54.

Curtiss III, R. & Kelly, S. M. (1987). *Salmonella typhimurium* deletion mutants lacking adenylate cyclase and cyclic AMP receptor protein are avirulent and immunogenic. *Infection and Immunity*, **55**, 3035–43.

Dorman, C. J., Chatfield, S., Higgins, C. F., Hayward, C. & Dougan, G. (1989). Characterization of porin and *ompR* mutants of a virulent strain of *Salmonella typhimurium*: *ompR* mutants are attenuated *in vivo*. *Infection and Immunity*, **57**, 2136–40.

Douce, G. R., Amin, I. I. & Stephen, J. (1991). Invasion of HEp-2 cells by strains of *Salmonella typhimurium* of different virulents in relation to gastroenteritis. *Journal of Medical Biology*, **35**, 349–57.

Dougan, G., Chatfield, S., Pickard, D., Bester, B., O'Callaghan, D. & Maskell, D. (1988). Construction and characterization of vaccine strains of *Salmonella* harbouring mutations in two different *aro* genes. *Journal of Infectious Diseases*, **158**, 1329–35.

Elsinghorst, E. A., Baron, L. S. & Kopecko, D. J. (1989). Penetration of human intestinal epithelial cells by *Salmonella*: molecular cloning and expression of *Salmonella typhi* invasion determinants in *Escherichia coli*. *Proceedings of the National Academy of Sciences, USA*, **86**, 5173–7.

Fields, P. I., Swanson, R. V., Haidaris, D. G. & Heffron, F. (1986). Mutants of *Salmonella typhimurium* that cannot survive within the macrophage are avirulent. *Proceedings of the National Academy of Sciences*, **83**, 5189–93.

Finlay, B. B. & Falkow, S. (1988). Comparison of the invasion strategies used by *Salmonella cholerae-suis*, *Shigella flexneri* and *Yersinia enterocolitica* to enter cultured animal cells: endosome acidification is not required for bacterial invasion or intracellular replication. *Biochemie*, **70**, 1089–99.

Finlay, B. B., Gumbiner, B. & Falkow, S. (1988). Penetration of *Salmonella* through a polarized MDCK epithelial cell monolayer. *Journal of Cell Biology*, **107**, 221–30.

Finlay, B., Stanbach, M. N., Francis, C. L., Stocker, B. A. D., Chatfield, S., Dougan, G. & Falkow, S. (1988). Identification and characterization of Tn*phoA* mutants of *Salmonella* which are unable to pass through a polarized MDCK epithelial cell monolayer. *Molecular Microbiology*, **2**, 757–66.

Galan, J. E. & Curtiss III, R. (1989). Cloning and molecular characterization of genes whose products allow *Salmonella typhimurium* to penetrate tissue culture cells. *Proceedings of the National Academy of Sciences, USA*, **86**, 6383–7.

Galan, J. E. & Curtiss III, R. (1991). Distribution of the *invA*, *-B*, *-C*, and *-D* genes of *Salmonella typhimurium* among other *Salmonella* serovars: *invA* mutants of *Salmonella typhi* are deficient for entry into mammalian cells. *Infection and Immunity*, **59**, 2901–8.

Germanier, R. & Furer, E. (1971). Immunity in experimental salmonellosis. II. Basis of the avirulence and protective capacity of *galE* mutants of *Salmonella typhimurium*. *Infection and Immunity*, **4**, 663–73.

Gilman, R. H., Hornick, R. B., Woodward, W. E., DuPont, H. L., Snyder, M. J., Levine, M. M. & Libonati, J. P. (1977). Evaluation of a UDP galactose-4-epimeraseless mutant of *Salmonella typhi* as a live oral vaccine. *Journal of Infectious Diseases*, **136**, 717–23.

Griffiths, E. (1991). Environmental regulation of bacterial virulence – implications for vaccine design and production. *Tibtech*, **9**, 309–15.

Hall, M. N. & Silhavy, T. J. (1979). Transcriptional regulation of *Escherichia coli* K-12 major outer membrane protein 1b. *Journal of Bacteriology*, **140**, 342–50.

Hoiseth, S. K. & Stocker, B. A. D. (1981). Aromatic-dependent *Salmonella typhimurium* are non-virulent and effective as live vaccines. *Nature*, **291**, 238–9.

Hone, D. M., Attridge, S. R., Forrest, B., Morrona, R., Daniels, D., LaBrooy, J. T., Bartholomeusz, R. C. A., Shearman, D. J. C. & Hackett, J. (1988). A *galE* (Vi antigen negative) mutant of *Salmonella typhimurium* Ty2 retains virulence in humans. *Infection and Immunity*, **56**, 1326–33.

Hone, D. M., Harris, A. M., Chatfield, S., Dougan, G. & Levine, M. M. (1991). Construction of genetically-defined double *aro* mutants of *Salmonella typhi*. *Vaccine*, **9**, 810–16.

Hormaeche, C. E. (1979). Natural resistance to *Salmonella typhimurium* in different inbred mouse strains. *Immunology*, **37**, 311–18.

Hormaeche, C. E., Harrington, K. A. & Joysey, H. S. (1985). Natural resistance to salmonellae in mice: control by genes within the major histocompatibility complex. *Journal of Infectious Diseases*, **152**, 1050–6.

Hormaeche, C. E. & Maskell, D. J. (1989). Influence of the *ity* gene on *Salmonella* infection. *Research in Immunology*, **140**, 791–3.

Jones, G. W., Rabert, D. K., Svinarich, D. M. & Whitfield, H. J. (1982). Association of adhesive, invasive, and virulent phenotypes of *Salmonella typhimurium* with autonomous 60-megadalton plasmids. *Infection and Immunity*, **38**, 476–86.

Johnson, E. M. & Baron, L. S. (1969). Genetic transfer of the Vi antigen from *Salmonella typhosa* to *Escherichia coli*. *Journal of Bacteriology*, **99**, 358–9.

Johnson, K., Charles, I., Dougan, G., Pickard, D., O'Gaora, P., Costa, G., Ali, T., Miller, I. & Hormaeche, C. (1991). The role of a stress-response protein in *Salmonella typhimurium* virulence. *Molecular Microbiology*, **5**, 401–7.

Khoramian-Falsafi, T., Haryooma, S., Kutsukake, K. & Pechere, J. C. (1990). Effect of motility and chemotaxis on the invasion of *Salmonella typhimurium* into HeLa cells. *Microbial Pathogenesis*, **9**, 47–53.

Lee, C. A., Jones, B. D. & Falkow, S. (1992). Identification of *Salmonella typhimurium* invasion locus by selection for hyperinvasive mutants. *Proceedings of the National Academy of Sciences, USA*, **89**, 1847–51.

Leung, K. Y. & Finlay, B. B. (1991). Intracellular replication is essential for the virulence of *Salmonella typhimurium*. *Proceedings of the National Academy of Sciences, USA*, **88**, 11470–4.

Levine, M. M., Herrington, D., Murphy, J. R., Morris, J. G., Losonsky, G., Tall, B., Lindberg, A. A., Svenson, S., Baqar, S., Edwards, M. F. & Stocker, B. A. D. (1987). Safety, infectivity, immunogenicity and *in vivo* stability of two attenuated auxotrophic mutant strains of *Salmonella typhi*, 541Ty and 543Ty, as live oral vaccines in humans. *Journal of Clinical Investigation*, **79**, 888–902.

Lissner, C. R., Swanson, R. N. & O'Brien, A. D. (1983). Genetic control of the innate resistance of mice to *Salmonella typhimurium*: expression of the *ity* gene in peritoneal and splenic macrophages *in vitro*. *Journal of Immunology*, **131**, 3006–13.

Lockman, H. A. & Curtiss, R. (1990). *Salmonella typhimurium* mutants lacking flagella or molility remain virulent in BALB/c mice. *Infection and Immunity*, **58**, 137–43.

McFarland, W. C. & Stocker, B. A. D. (1987). Effect of different purine auxotrophic mutations on mouse virulent of a Vi-positive strain of *Salmonella dublin* and of two strains of *Salmonella typhimurium*. *Microbial Pathogenesis*, **3**, 129–41.

Mekalanos, J. J. (1991). Environmental signals controlling expression of virulence determinants in bacteria. *Journal of Bacteriology*, **174**, 1–7.

Miller, I. A., Chatfield, S., Dougan, G., DeSilva, L., Joysey, H. S. & Hormaeche,

C. E. (1989a). Bacteriophage P22 as a vehicle for transducing cosmid gene banks between smooth strains of *Salmonella typhimurium*: use in identifying a role for *aroD* in attenuating virulent *Salmonella* strains. *Molecular and General Genetics*, **215**, 312–16.

Miller, S. I., Kukral, A. M. & Mekalanos, J. J. (1989b). A two-component regulatory system (*phoP phoQ*) controls *Salmonella typhimurium* virulence. *Proceedings of the National Academy of Sciences, USA*, **86**, 5054–8.

Miller, S. I. & Mekalanos, J. J. (1990). Constitutive expression of the PhoP regulon attenuates *Salmonella* virulence and survival within macrophages. *Journal of Bacteriology*, **172**, 2485–90.

Miller, S. I. (1991). PhoP/PhoQ: macrophage-specific modulators of *Salmonella* virulence? *Molecular Microbiology*, **5**, 2073–8.

Neidhardt, F. C. (ed.), (1987). *Escherichia coli* and *Salmonella typhimurium*; Cellular and Molecular Biology. American Society for Microbiology, Washington DC.

Nikaido, H. & Vaara, M. (1985). Molecular basis of bacterial outer membrane permeability. *Microbiology Reviews*, **49**, 1–32.

O'Brien, A. D., Scher, I., Campbell, G. H., MacDermott, R. P. & Formal, S. B. (1979). Susceptibility of CBA/N mice to infection with *Salmonella typhimurium*: influence of the X-linked gene controlling B-lymphocyte function. *Journal of Immunology*, **123**, 720–4.

O'Brien, A. D., Rosenstreich, O. L., Scher, I., Campbell, G. H., MacDermott, R. P. & Formal, S. B. (1980). Genetic susceptibility to *Salmonella typhimurium* in mice. Role of the *lps* gene. *Journal of Immunology*, **124**, 20–4.

O'Callaghan, D., Maskell, D., Liew, F. Y., Easmon, C. S. F. & Dougan, G. (1988). Characterisation of aromatic- and purine-dependent *Salmonella typhimurium*; attenuation, persistence and ability to induce protective immunity in BALB/c mice. *Infection and Immunity*, **56**, 419–23.

Pulkkinen, W. S. & Miller, S. I. (1991). A *Salmonella typhimurium* virulence protein is similar to a *Yersinia enterocolitica* invasion protein and a bacteriophage lambda outer membrane protein. *Journal of Bacteriology*, **173**, 86–93.

Scarlato, V., Prugnola, A., Arico, B. & Rappuoli, R. (1990). Positive transcriptional feedback at the *bvg* locus controls expression of virulence factors in *Bordetella pertussis*. *Proceedings of the National Academy of Sciences, USA*, **87**, 6753–7.

Strugnell, R. A., Maskell, D., Fairweather, N. F., Pickard, D., Cockayne, A., Penn, C. & Dougan, G. (1990). Stable expression of foreign antigens from the chromosome of *Salmonella typhimurium* vaccine strains. *Gene*, **88**, 57–63.

Tacket, C. O., Hone, D. M., Losonsky, G. A., Guers, L., Edelman, R. & Levine, M. M. (1992). Clinical acceptability and immunogenicity of CVD908 *Salmonella typhi* vaccine strain. *Vaccine*, in press.

Takeuchi, A. (1967). Electron microscopic studies of experimental *Salmonella* infection. I. Penetration into the intestinal epithelium by *Salmonella typhimurium*. *American Journal of Pathology*, **50**, 109–36.

Wahdan, M. H., Serie, C., Cerisier, Y., Sallam, S. & Germanier, R. (1982). A controlled field trial of live *Salmonella typhi* strain Ty21a oral vaccine against typhoid: three year results. *Journal of Infectious Diseases*, **145**, 292–7.

SPECIES INDEX

SUBJECT INDEX